Advances in Natural and Technological Hazards Research

Volume 45

More information about this series at http://www.springer.com/series/6362

Jeanette L. Drake • Yekaterina Y. Kontar
John C. Eichelberger • T. Scott Rupp
Karen M. Taylor
Editors

Communicating Climate-Change and Natural Hazard Risk and Cultivating Resilience

Case Studies for a Multi-disciplinary Approach

Springer

Editors
Jeanette L. Drake
College of Business
University of Michigan-Dearborn
Dearborn, Michigan, USA

John C. Eichelberger
University of Alaska Fairbanks
Fairbanks, Alaska, USA

Karen M. Taylor
Department of Communication
University of Alaska Fairbanks
Fairbanks, Alaska, USA

Yekaterina Y. Kontar
International Arctic Research Center
University of Alaska Fairbanks
Fairbanks, Alaska, USA

T. Scott Rupp
International Arctic Research Center
University of Alaska Fairbanks
Fairbanks, Alaska, USA

ISSN 1878-9897 ISSN 2213-6959 (electronic)
Advances in Natural and Technological Hazards Research
ISBN 978-3-319-20160-3 ISBN 978-3-319-20161-0 (eBook)
DOI 10.1007/978-3-319-20161-0

Library of Congress Control Number: 2015953863

Springer Cham Heidelberg New York Dordrecht London

Printed on acid-free paper

Springer International Publishing AG Switzerland is part of Springer Science+Business Media (www.springer.com)

To teachers everywhere who taught us to ask before we ever pick up pen and paper: "Who is our audience? What are our objectives? What is the medium?" and "Is this recycled paper?" And to students of climate change sciences—who will become our biggest teachers.

Foreword

Scientific communication is a process of knowledge transfer when scientists interact with each other, the public, policymakers, science writers, or others. The communication of scientific knowledge on climate change and disaster risks is known to be a challenging task. How can scientists, science writers, and others properly communicate science on climatic and environmental changes and/or integrated disaster risk science to convince individuals or national governments that climate is changing under natural and also anthropogenic influences and that disaster risks are associated mainly with vulnerability and exposure? The famous mathematician Israel Gelfand stated that proper communication of mathematical knowledge is important and provided an apt analogy:

> Mathematics is a way of thinking in everyday life . . . It is important not to separate mathematics from life. You can explain fractions even to heavy drinkers. If you ask them, 'Which is larger, 2/3 or 3/5?' it is likely they will not know. But if you ask, 'Which is better, two bottles of vodka for three people, or three bottles of vodka for five people?' they will answer you immediately. They will say two for three, of course. (Chang 2009, para. 25)

A similar point may be made about the communication of scientific knowledge. It is important not to separate science from life.

This book *Communicating Climate-Change and Natural Hazard Risk and Cultivating Resilience: Case Studies for a Multi-disciplinary Approach* responds to the need for better communication by thoroughly discussing many aspects of communication and resilience. This volume emphasizes disaster risk and crisis communication principles and practices. The book begins with several chapters discussing the role of communication in fostering resilient communities and emphasizing collaboration of natural and social scientists and other stakeholders to reframe the conversation on climate change and disaster risks. The succeeding chapters discuss communication practices before, during, and after disasters happen, particularly, new trends and best practices in communication of probabilistic forecasts and predictions, damage mitigation, and recovery. The concluding chapters deal with various approaches to the communication of climate change and disaster risks among a variety of different audiences.

The book includes contributions from experts in natural or social sciences dealing with risk communication and/or disaster management related to natural or technological phenomena. The book editors are prominent and early career natural and social scientists and experts in climate change, disaster risks, and scientific knowledge communication. They have edited a comprehensive book that everyone in the relevant fields would be well advised to read.

Institute of Earthquake Prediction Theory and Mathematical Geophysics
Russian Academy of Sciences, Moscow, Russia
Institute of Applied Geosciences
Karlsruhe Institute of Technology
Karlsruhe, Germany

Alik Ismail-Zadeh, Ph.D., D.Sc.

Reference

Chang K (2009, Oct 7) Israel Gelfand, math giant, dies at 96, The New York Times. http://www.nytimes.com/2009/10/08/science/08gelfand.html?_r=0. Accessed 24 Mar 2015

Preface

Probably no aspect of reducing disaster risk and promoting resilience has changed as much in recent years as communication. This is because of multiple forces, including increased evidence and urgency of the risks and ramifications; heightened politicization of the issues; new and revolutionary pathways for communication; stealth rhetoric; and the realization by governments, earth and physical scientists, and communities what communication experts and social and behavioral scientists have long known about effective communication—it ain't easy.

Communicating Disaster Risk Is as Challenging as the Mitigation Itself

As natural and technological disasters increase in scope, intensity, and frequency, this truism necessitates even more entrepreneurial, transdisciplinary teams such as the ones behind this book. Although we do it everyday and the layperson may be lulled into taking it for granted, *communication* (aka human interaction) has never been an easy art, an exact science, or a mastered process—least not when it involves a complex subject such as climate change. This convergence of phenomena has inspired this volume on the topic—one that encompasses case studies, best practices, and myriad perspectives. One that puts social scientists side-by-side with geophysicists to examine new trends in communicating science, policy, and practices in predicting, monitoring, mitigating, and recovering from natural and technological hazards.

Clearly Communicating Climate Matters

With the focus shifting from resistance to resilience, we realize natural disasters and much of the current and anticipated damage from climate change occur because people and property are in the wrong place at the wrong time. Engineered risk reduction, for example by constructing seawalls, levees, and sabo works, is sometimes necessary but often prohibitively expensive and/or disadvantageous. The easiest solution is to not be in the wrong place to begin with. This means planning of community development to minimize vulnerability—something requiring continuous dialog among planners, the community, natural hazards and climate change experts, and risk communication professionals. Failing that, because often there are compelling reasons for living in high-risk areas, the important thing is to get people and movable assets out of the way in an extreme event. Such facilitation during a crisis is a special topic within hazard communication. Simplicity and considerable exchange of knowledge in advance are required and are at the heart of crisis management and communication. In fact, with loss of power and telecommunication infrastructure, traditional modes of communication may become impossible, demonstrating why the members of the impacted community must already know what to do.

Communication, Like any Science, Is Fallible

Ineffectual risk and crisis communication shows time and again that no component of the deceptively simple communication model may be taken for granted—not the media, not the message, not the milieu, and least not the members who are sending, receiving, encoding, decoding, and interacting. The 2011 Tohoku earthquake and tsunami event in Japan tragically illustrates how much more progress in communication is needed, even in arguably the most hazard-aware and prepared country in the world. Only about half the at-risk population received the tsunami warning, and even some who did get the message made decisions that proved fatal. Some disregarded the risk messages and went to the coast to watch the tsunami or searched for family members instead of going immediately to safety. This case calls for a reexamination of the distribution channels as well as the framing strategies.

Emergency Messages Demand Persuasion and Precision

The same transformation from analog to digital technology and vast increase in computing power that make two-way communication among everyone possible also permits sophisticated, near-real-time forecasts that may prove to be inadequate. The model for the Tohoku event grossly underpredicted the size of the tsunami.

Modeling of the 2010 Eyjafjallajökull eruption, not a human tragedy but a huge financial loss, grossly overpredicted the risk to aviation. In both cases, it can be argued that communicating the results of the forecasting models made things worse because the data were flawed. Of course, the communication process relies at all times on the accuracy of the message and credibility of the sender, which are prone to human error and miscalculation, particularly with regard to forecasting.

Precision Notwithstanding, Not All Communication Is Created Equal

Communication is political, and, many would argue, all communication is persuasive in its intent. Both claims surface regularly in communicating climate change, including recently in Florida, where Korten (2015) reported that Department of Environmental Protection officials "have been ordered not to use the term *climate change* or *global warming* in any official communications, emails, or reports, according to former DEP employees, consultants, volunteers and records" (para. 3). Filling the supersized news hole and 24/7 news appetite has become impossible for gutted newsrooms, which rely more and more on public relations. This scenario provides both an opportunity and a threat in terms of communicating risk and resilience. Although simply providing audiences with more knowledge is, in itself, not likely to overcome the political divide on climate change, a new study has found that conservatives and liberals become less polarized when the perception of harm increases (Hart et al. 2015).

Noise Complicates the Equation

Despite the loud debate about the proper role of government now occurring in the United States, few would assert that assessing natural hazards and managing the crises they cause are not government responsibilities. Individual politicians can opine about what they believe or don't believe, but operational agencies do not have that luxury. Somehow, the operational government must have credibility in science and policy modulated by the delicate concept of uncertainty, which is to say avoiding the tempting but absurd claim to infallibility. Credibility of government cannot necessarily be based as it once was on the government knowing more, because increasingly hazard-monitoring data are available to all in real time. Additionally, government—entities at various scales and sometimes including multiple countries—cannot reach the majority of a population on its own and relies, as it has done for centuries, on the systems and expertise of the professional media and now, increasingly, on the capacity and capabilities of social media.

Twitter Exemplifies the Fluidity of the Transactional Model

Over the last decade, social media platforms have sometimes served as a lifeline during crises around the world. But in the digital public sphere, too, noise can muddle the message. It appears that neither government nor the media have fully adapted to this recent and evolving phenomenon of everyone talking to everyone via social networks of the Internet, where *reason* for deliberation is disappearing particularly when the subject is science. In fact, Labarre (2013) said comments are bad for science, explaining why the 141-year-old science and technology magazine *Popular Science* cut off discussion on new articles on its website stating that even a cacophonous minority has the power to skew public perception. A hallmark of excellent communication has been its two-way, inclusive nature, but as this magazine surmised even that notion is turned on its head by the times and trend of online discussions to eschew logos and ethos (JP Drake and Drake 2014).

Rhetorical Mountains Remain

Scientists and science communicators must get creative rhetorically in overcoming science deniers and the burden of proof in science. In the wake of natural and technological disasters, standards for scientific argumentation become increasingly important, and an awareness of rhetorical strategies is helpful. Paroske (2009) identifies the *epistemological filibuster*, which "occurs when one side in a science policy controversy exploits uncertainty over how thoroughly to deliberate as a means to preclude the resolution of that issue in government action" (p. 151). Paroske (2012) also identifies the *presumptive breach*, a state of tension "in the face of an emergent crisis, where we are legitimately convinced that regulation is warranted and yet are unable to craft arguments to justify policy adoption that meet an imposed burden of proof" (p. 474). Compelling storytelling is de rigueur in person, in print, and online.

Collective Impact Relies on *New Trends in Communicating Risk and Cultivating Resilience*

The nascent digital media landscape means that emergency managers have the potential to access far more comprehensive information about what is happening to real people in real time, while at the same time receiving rapid feedback on their actions or inaction. And it means exponentially greater possibilities for multidisciplinary collaboration around the world. That is the premise of this volume and the promise of collective impact (Kania and Kramer 2011) such as the efforts being undertaken by those behind this book.

This volume is dedicated to advancing hazard and crisis communication strategies in the brave new world of everyone talking to everyone. After laying a foundation by overviewing the principles of crisis communication and the role of framing in communicating risk, the authors in Part I examine the role of communication in helping or hindering resiliency. In Part II, experts address new ways to predict, prepare, and prevent the worst. In Part III, contributing authors consider how to communicate through change, uncertainty, disaster, and recovery. We conclude in Part IV with best practices for communicating climate change among diverse audiences.

By the numbers, this book represents 55 authors from academe, science, government, NGOs, and business. It spans seven countries, three continents, and dozens of disciplines. Knowing that numbers never tell the whole story, we invite you to explore *Communicating Climate-Change and Natural Hazard Risk and Cultivating Resilience: Case Studies for a Multi-disciplinary Approach.*

University of Michigan-Dearborn, Dearborn, MI, USA Jeanette L. Drake
University of Alaska Fairbanks, Fairbanks, AK, USA Yekaterina Y. Kontar
University of Alaska Fairbanks, Fairbanks, AK, USA John C. Eichelberger
University of Alaska Fairbanks, Fairbanks, AK, USA T. Scott Rupp
University of Alaska Fairbanks, Fairbanks, AK, USA Karen M. Taylor

References

Drake JP, Drake JL (2014) One blog at a time, students slog through complex issues. Ohio Social Stud Rev 51(1):52–63. http://edhd.bgsu.edu/ossr/journal/index.php/ossr/index. Accessed 11 Apr 2015

Hart PS, Nisbet EC, Myers TA (2015) Public attention to science and political news and support for climate change mitigation. Nat Clim Change. doi: 10.1038/nclimate2577

LaBarre S (2013) Why we're shutting off our comments. Popular Sci. http://www.popsci.com/science/article/2013-09/why-were-shutting-our-comments. Accessed 8 Apr 2015

Kania J, Kramer M (2011) Collective impact: Large-scale social change requires broad cross-sector coordination, yet the social sector remains focused on the isolated intervention of individual organizations. Stanford Social Innov Rev. http://www.ssireview.org/articles/entry/collective_impact. Accessed 8 Apr 2015

Korten T (2015) In Florida, officials ban term 'climate change'. http://www.miamiherald.com/news/state/florida/article12983720.html. Accessed 8 Apr 2015

Paroske M (2009) Deliberating international science policy controversies: uncertainty and AIDS in South Africa. Quart J Speech 95(2):148–170

Paroske M (2012) Overcoming burdens of proof in science regulation: ephedra and the FDA. Rhetoric Public Affairs 15(3):467–498

Acknowledgments

This earthy endeavor would not have been possible without the talents of a great team. We are grateful to the 55 individuals from agricultural and environmental chemistry, astronomy, atmospheric sciences, biology, business, chemistry, climate engineering, climate governance, communication studies, computer science, crisis communication, economics, education, engineering, environmental communication, environmental engineering, environmental science, forest ecology, geography, geohazards, geology, geophysics, geosciences, journalism, law, marine chemistry, media studies, natural resources, physics, political science, public health, public policy, and public relations who contributed their expertise to this transdisciplinary text working virtually around our common denominator—planet Earth.

Special appreciation to our publisher who made this work possible and to the motivation behind the monograph Yekaterina Y. Kontar, who brings to the table a unique and essential blend of expertise in social, behavioral, and earth sciences.

Contents

About the Editors and Authors

Ian Kaye Alejandrino (B.S., Civil Engineering, University of the Philippines-Diliman, 2013) is an M.S. candidate in civil engineering-geotechnical at the University of the Philippines-Diliman and is a senior science research specialist at DOST Project NOAH. His research focuses on landslides and geotechnical properties of soil to enhance awareness and improve engineering solutions.

Minda R. Berbeco (Ph.D., Tufts University, 2011) is a programs and policy director at the National Center for Science Education, specializing in climate change. She is also the California Science Teacher Association Region 2 director and a visiting scholar at the University of California Museum of Paleontology. During her doctoral work in biology, she studied the impact of climate change on forest systems.

Christina Carlson (A.B., Duke University, 2011) is a policy research assistant in the Center for Science and Democracy at the Union of Concerned Scientists. Her research focuses primarily on how science is used and misused in public policy. Prior to joining UCS, she worked on national and regional ocean policy issues as a program associate and fellow at Meridian Institute in Washington, D.C. She has been a contributing author to several reports on topics including business response to climate change and sugar and health policy. Her research at Duke focused on ocean acidification.

Ilan Chabay (Prof., Ph.D., Chemical Physics, University of Chicago, 1972) is a senior fellow at IASS, focusing on transdisciplinary research and coleading the Sustainable Modes of Arctic Resource-driven Transformations (SMART) project. He chairs an international research alliance on Knowledge, Learning, and Societal Change (klscproject.org) and examines models and narratives for decision-making on complex system issues.

Danny Cullenward (Ph.D., Environment and Resources, Stanford University, 2013; J.D., Stanford Law School, 2013) is the Philomathia Research Fellow at the University of California, Berkeley. A lawyer and economist by training, his research focuses on the design and implementation of science-based climate policy.

Melanie Davis (B.S., University of Southampton, 2000) is the climate services manager for the Climate Forecasting Unit at the Catalan Institute of Climate Sciences (IC3). She has 11 years' experience in renewable energy business and policy and more recently in climate research. Her core role has been the translation of technical science to end users and decision-makers. This diverse background enables her to bridge the gap between different communities to develop actionable services from end user-driven research. She has contributed to the development of the first European-funded climate services projects, where she leads various work packages, and is also coordinating a national climate and energy project, funded by the Spanish Ministerio de Economía y Competitividad. She is actively collaborating with climate research institutions and private companies worldwide and is a member of the Climate Services Partnerships (CSP) and European Energy Research Association (EERA).

Kristina Dahl (Ph.D., Massachusetts Institute of Technology/Woods Hole Oceanographic Institute, 2005) is a climate scientist whose research, analysis, and communication services help make the impacts of climate change on our environment and society more tangible to the public and policymakers. With a background in paleoclimate research, she formerly coordinated the Rutgers Climate and Environmental Change Initiative and served as an online educator for the American Museum of Natural History.

Francisco Doblas-Reyes (Ph.D., Universidad Complutense of Madrid, 1996) is a researcher in the fields of climate dynamics and climate prediction. Besides a long list of more than 100 scientific peer-reviewed papers with high citation impact, he is currently leading numerous research and computational projects, mostly funded by the European Commission, the Spanish Ministerio de Economía y Competitividad, and private partners. He is actively collaborating with climate research institutions across several continents, including some Global Producing Centres. He serves on numerous international steering committees that design the research programs of institutions like the World Meteorological Organisation or the Intergovernmental Panel for Climate Change.

Jeanette L. Drake (Ph.D., Communication Studies, Bowling Green State University, 2004) is a writer, educator, and consultant specializing in environmental communication. She teaches at the University of Michigan-Dearborn and investigates the influence of framing on public dialogue and public policy. Her work has been widely published including her theory on the Subterfusion of Innovation, which first appeared in the *Ohio Communication Journal*, and her action scholarship that has been featured in several textbooks. Accredited in public relations, she received a bachelor's degree in social and behavioral sciences and a master's degree in journalism—both from the Ohio State University. She has taught and developed undergraduate, graduate, and certificate programs in public relations, respectively, at the University of Findlay and Kent State University in Ohio, USA, and the University of Social Sciences and Humanities in Ho Chi Minh City, Vietnam. In 2008, she was elected to the College of Fellows of the Public Relations Society of America for her impact on public relations.

Rodrigo Narod Eco (B.S., Computer Science, University of the Philippines-Los Baños, 2006) is an M.S. candidate in geology at the University of the Philippines-Diliman and is a chief science research specialist at DOST Project NOAH. His research focuses on remote sensing, landslides, and volcanic hazards.

John C. Eichelberger (Ph.D., Stanford University, 1974) is one of the genuine visionaries in the field of geosciences, and his contributions to the understanding of volcanic and magmatic processes have been spectacular and long-lasting. Several of his papers are among a small number of the most influential works of the past half century. These include his Nature papers on magma mixing and the role of permeability in eruption of silicic magmas. These are among classic works that offered elegant and inspirational answers to fundamental questions concerning eruption triggers and eruptive behavior. In addition, he is internationally recognized for his work in two other domains: scientific drilling and Arctic geosciences.

Nancy Fresco (Ph.D., University of Alaska Fairbanks, 2006) is a research professor and network coordinator at the Scenarios Network for Alaska and Arctic Planning (SNAP) at the International Arctic Research Center at the University of Alaska Fairbanks (UAF). Her work focuses on linking SNAP data to the needs of stakeholders and interpreting the results of complex modeling efforts. Her background is in biology, forest ecology, and environmental education. She completed her undergraduate work at Harvard in 1994 and was granted a master's degree from the Yale School of Forestry and Environmental Studies in 1999, before earning her doctorate in biology from UAF in 2006, as part of an interdisciplinary program in Regional Resilience and Adaptation. She is particularly interested in bridging the gaps between academia, practical applied science, policy, and public understanding. Prior to working at UAF, she spent 6 years doing research and outreach for an environmental nonprofit and 2 years as a Peace Corps volunteer.

Gretchen Goldman (Ph.D., Georgia Institute of Technology, 2011) is a lead analyst in the Center for Science and Democracy at the Union of Concerned Scientists. Her research interests focus on how science informs policymaking. She came to UCS from the Georgia Institute of Technology where she was a postdoctoral research fellow working on statistical modeling of urban air pollution for use in epidemiologic studies of acute human health effects. Her recent work has focused on the business response to climate change, scientific integrity at government agencies, and a public right-to-know in science-based policies. She also holds a master's degree in environmental engineering from Georgia Tech and a bachelor's in atmospheric science from Cornell University.

Sarah A. Green (Ph.D., Marine Chemistry, Massachusetts Institute of Technology and Woods Hole Oceanographic Institution Joint Program, 1992) is professor of chemistry at Michigan Technological University, where she served as department chair from 2004 through 2013. As a Jefferson Science Fellow in 2013–2014, she served in the US State Department as a senior science advisor for the Bureau of East Asia and Pacific Affairs, covering environmental issues in the region. Dr. Green has served on the Michigan Green Chemistry Roundtable since 2008. She has conducted research on sensors, photochemistry, tobacco smoke, aquatic and forest organic carbon, and the Great Lakes. Her interest in climate communication stems from public outreach on the issue and led, in 2013, to her participation in the Skeptical Science Consensus Project.

Emily Grubert (M.S., Environmental and Water Resources Engineering, the University of Texas at Austin, 2011) is a Ph.D. student in environment and resources at Stanford University who studies the integration of social and environmental needs for energy policy in the USA, particularly in the resource extraction industry.

Barbara Haya (Ph.D., Energy and Resources, University of California, Berkeley, 2010) is a research fellow at the Stanford Law School. Her research focuses on the outcomes of carbon offsets and credit trading programs and how program incentives affect decisions within the particular financial, political, legal, and cultural contexts in which they are implemented.

Richard Hum (M.A., Rural Development, University of Alaska Fairbanks, 2013) is a Ph.D. student in the Department of Communication at the University of Alaska Fairbanks. His research focuses on communication networks that form around issues of community resilience during times of rapid social-ecological change. The main focus of his work is on cross-scale social-ecological relationships and the potential benefits of social media to increase communicative connectivity across scales. He uses data mining, network, and content analysis methodologies to empirically explore these topics.

Alik T. Ismail-Zadeh (D.Sc., Geophysics, Russian Academy of Sciences, Russia, 1997) is a chief scientist of the Russian Academy of Sciences, Moscow, Russia, and a senior research fellow at the Karlsruhe Institute of Technology, Karlsruhe, Germany. His research interests cover mathematical geophysics and geodynamics, particularly seismic hazards and disaster risks.

Peter H. Jacobs (B.A., English Language and Literature, the College of William and Mary, 2003) is a graduate student at George Mason University in the Department of Environmental Science and Policy. His research focus is using paleoclimatic data to better inform policymaking. In addition to his primary research, he is also interested in the science of how people respond to the issue of climate change from the perspectives of communication, psychology, and sociology. This has led to his participation in various projects relating to the scientific consensus on man-made global warming, principally with volunteers from the Skeptical Science team.

Ari Jokimäki (B.Sc., Computer Engineering, Espoo-Vantaa Institute of Technology, 2000) is a volunteer with the Skeptical Science online community and climate blogger. His main interests are in cataloging the peer-reviewed literature on different aspects of climate change and in investigating new publications in the field. He has coauthored several peer-reviewed refutations of studies that misunderstand or misrepresent the scientific literature on climate.

Kelly Klima (Ph.D., Engineering and Public Policy, Carnegie Mellon University, 2011) is a research scientist at Carnegie Mellon University. Her research focuses on community resilience, adaptation, hazard mitigation, climate, extreme weather, and risk communication. In addition to working with the City of Pittsburgh, she is currently serving as an adaptation subject matter expert in New Jersey. Previously, she worked at the Center for Clean Air Policy (CCAP), where she helped New York and Washington, D.C., advance their adaptation planning. An active member of nine professional societies including the American Society of Adaptation Professionals (ASAP), she serves on the Natural Hazard Mitigation Association (NHMA) Board of Directors and the American Geophysical Union (AGU) Executive Council.

Yekaterina Y. Kontar (M.B.A., the University of Findlay, 2009) is currently a Ph.D. student and research assistant at the University of Alaska Fairbanks (UAF). Her primary research encompasses physical and climatic processes behind river ice breakups as well as their social impacts in Alaska and in Russia. The main goal of Kontar's research is to define effective interdisciplinary approaches to hydrological hazard mitigation and response and climate change adaptation that could promote robust Arctic communities that are able to withstand the shock of disaster. Prior to starting her graduate studies at UAF, Kontar worked as an education program associate at the ANDRILL (Antarctic geological Drilling) science management office. She assisted ANDRILL scientists in the creation of two sets of educational modules entitled *Environmental Literacy Framework with a Focus on Climate Change* and *Antarctica Climate Secrets*. These modules were aimed at a middle school audience of teachers and students. In addition, she organized and facilitated local and national education and outreach events with the goal to communicate geo- and climate science to nontechnical audiences of different ages.

Sumiko Kubo (D.Sc.., Tokyo Metropolitan University, 1995) is a professor in geography at the School of Education, Waseda University. She teaches physical geography at WU as a faculty member since 2001. Her major interests are on geomorphology and quaternary environmental changes of fluvial plains. Her research field includes Kanto Plain in Japan, the Brahmaputra-Jamuna River Plain in Bangladesh, the Thu Bon River Plain in Vietnam, the Mekong River Basin in Laos and Cambodia, the Godavari and Krishna river plains in India, etc. in relation to human history and natural hazards.

Lindsay E. Kuntzman (M.A., Journalism and Mass Communication, Kent State University, 2009) is assistant director of marketing at Kent State University, where she develops and executes marketing communication strategies for the Department of

Alumni Relations and manages mobile initiatives. As a graduate research assistant, she conducted in-depth studies of apologia, best practices in crisis communication, social media marketing, and communication theories.

Alfredo Mahar Lagmay (Ph.D., Geology, University of Cambridge, 2000) is a professor at the University of the Philippines-Diliman and is the executive director of DOST Project NOAH. His research focuses on volcano tectonics, remote sensing, geohazards, and the formulation of national policies to reduce the ill effects of geohazards.

Mark G. Lawrence (Prof., Ph.D., Georgia Institute of Technology in Atlanta, USA, 1996) is scientific director at the Institute for Advanced Sustainability Studies (IASS). His research focuses on the interface of air pollution and climate change and on the uncertainties and risks of proposals for climate engineering. Until 2011 he worked at the Max Planck Institute for Chemistry in Mainz, Germany.

Rachel Lowe (Ph.D., University of Exeter, 2011) is lead researcher for Climate Services for Health in the Climate Impacts and Dynamics Unit at the Catalan Institute of Climate Sciences (IC3). Her research involves modeling climate-sensitive disease risk and finding novel ways to communicate probabilistic forecasts to public health decision-makers. Also a coordinator of the Climate Services Partnership (CSP), she graduated with a B.Sc. in meteorology and oceanography degree from the University of East Anglia in 2004 with first class honors. In 2007, she completed an M.Sc. with distinction in geophysical hazards at University College London where she received a Graduate Masters Award before going on to obtain her doctorate in mathematics (Ph.D. thesis: "Spatio-temporal modeling of climate-sensitive disease risk: towards an early warning system for dengue in Brazil"). She was also a network facilitator for the Leverhulme Trust-funded project EUROBRISA and a visiting scientist at the International Centre for Theoretical Physics in Italy.

Alexandra Masaitis (M.S., St. Petersburg State University, 2009) is a research assistant in the Department of Natural Resources and Environmental Science, University of Nevada, Reno (UNR). She has a master of science degree in geography and geo-ecology from St. Petersburg State University, St. Petersburg, Russia, and a master of science degree in international relations in the field of environmental protection with the St. Petersburg State University, in the Department of International Relations. Currently she is working on a research project started in August 2012 at UNR to develop a global interactive education network about GNA standards as well as to investigate the details of mine development.

Alicia M. Mason (Ph.D., University of Oklahoma, 2010) is an associate professor in the Department of Communication at Pittsburg State University. She is a former broadcaster, who worked with local media outlets KOAM and KKOW in SE Kansas, prior to receiving a Ph.D. from the University of Oklahoma. Her academic interests in risk and crisis communication span organizational/business, public health, and environmental contexts. She was the 2013 recipient of the Outstanding Undergraduate Research Mentor Award and is the faculty representative for the Bachelors of Integrated Studies: Sustainability, Society and Resource Management [SSRM] degree program.

Achim Maas (M.A., University of Bradford, UK, 2006) works as coordinator of the research cluster *Sustainable Interactions with the Atmosphere* at the Institute for Advanced Sustainability Studies (IASS) in Potsdam, Germany. His main research areas are environmental and particular climate governance, as well as the challenges of the Anthropocene.

Mark McCaffrey (M.A., Education Leadership and Policy Studies, University of Northern Colorado, 1994) is a programs and policy director at the National Center for Science Education, specializing in climate change education. He is the author of a newly published book from Corwin Press: *Climate Smart & Energy Wise*. Mark played a leadership role in the development of *Climate Literacy: Essential Principles of Climate Science* and *Energy Literacy: Essential Principles and Fundamental Concepts for Energy Education* and is a founding member of the Climate Literacy and Energy Awareness Network (CLEAN).

Glenn C. Miller (Ph.D., University of California Davis, 1977) is a professor of environmental and resource science at the University of Nevada, Reno (UNR), where he also serves as director of the Graduate Program in Environmental Sciences and Health. Along with his doctorate in agricultural and environmental chemistry, he has a B.S. in chemistry from the University of California, Santa Barbara (1972). His current research focuses on Nevada mining issues, including precious metals, water quality, acid mine remediation using anaerobic sulfate reducing system, and the reduction of conflict in mine permitting.

Holly M. Miller (B.S., Colorado State University, 1999) is a social scientist with the US Geological Survey's Fort Collins Science Center. She researches the human dimensions of natural resources, primarily through large-scale surveys and quantitative data analysis. Her recent work has focused on the users, uses, and values of Landsat imagery, including three surveys of thousands of Landsat users and multiple case studies of specific applications of the imagery. Her other work has focused on the values, attitudes, and behaviors of the American public regarding wildlife and wildlife management, most recently in regards to waterfowl and wetlands management.

Naoko Nagumo (Ph.D., the University of Tokyo, 2011) is a research specialist in the International Centre for Water Hazard and Risk Management (ICHARM), Public Works Research Institute (PWRI), Japan. She came to ICHARM, PWRI, from the University of Tokyo where she was a project research associate working on educational activities for natural environmental studies and geomorphological and geographical research of Cambodian rivers and Khmer archaeological sites. Now she has started risk management work for water-related hazards in Southeast Asian countries. She also holds a master's degree in education and a bachelor's degree in geography from Waseda University.

Timothy R. Newman (M.S., Northrop University, 1988) is the US Geological Survey's Land Remote Sensing Program coordinator. The program operates the Landsat satellites and provides the nation's portal to the largest archive of remotely sensed land data in the world. He is responsible for providing program policy, oversight, and guidance; formulating and executing budgets; interfacing with senior administration and congressional staff; building international earth observation partnerships; and interacting with the aerospace and remote sensing industries. He also oversees the development of remote sensing science and applications.

Aditi Rao (M.S. Communications, Ithaca College, 2015) is currently working towards a career as a consultant. She also holds a Master's and Bachelor's degree in Biochemistry from the University of Mumbai, India. Previously, she worked as an academic editor for 4 years. She served as a publication consultant for ESL researchers, scientists, and doctors, and edited their papers before submission to their desired journals.

Ken Rice (Ph.D., Physics, University of KwaZulu-Natal, 1998) is a theoretical and computational astrophysicist who works at the Institute for Astronomy, University of Edinburgh. He uses numerical simulations to study disks around young stars and how their evolution influences the subsequent formation of stars and planets. He is particularly interested in the earliest stages of star formation, when disks are massive and when much of the planet building process probably occurs. He is also involved in a project that aims to characterize extrasolar planets that have been detected by NASA's Kepler satellite. His background in physics and astronomy has led—recently—to an interest in climate science and how it is communicated.

Leslie A. Richardson (Ph.D., Colorado State University, 2011) is an economist with the US Geological Survey's Fort Collins Science Center. She received her doctorate in agricultural and resource economics from CSU, where she was named an international presidential fellow. Her dissertation, conducted for the US Forest Service, focused on applying nonmarket valuation techniques to quantify the economic cost of human health effects from wildfire smoke exposure. A journal article resulting from this research was awarded the 2014 Sören Wibe Prize for considerable development in empirical knowledge or methodology in the field of forest economics. Her current research involves assisting federal agencies with the economic valuation of nonmarketed resources, including ecosystem services and information sources such as satellite imagery. Her research methods include primary data collection through socioeconomic surveys, as well as applying benefit transfer methods based on secondary data.

Xavier Rodó (Ph.D., University of Barcelona, 1997) leads the Unit for Climate Dynamics and Impacts (UDIC) at the Catalan Institute of Climate Sciences (IC3), of which he was founding director. He formerly created the Climate Research Laboratory at the Barcelona Science Park (Univ. Barcelona). He began engineering studies to complete his doctorate on the simulation of responses in extreme ecosystems under climate stress. He was a visiting fellow at Princeton University and Scripps/UCSD and an affiliated scientist at COLA, Maryland. He taught ecology, advanced statistics, and sustainability sciences. He has published over 70 papers in top-ranked journals and participates in diverse international committees and panels. His research tries to use innovative approaches to address problems at the interfaces of climate, ecology, statistics, and human health with a focus on enhancing predictability.

Alexis Rosamilia (M.S, Communications, Ithaca College, 2013) is a working professional within the greater New York City area. She currently resides on a Wellness Committee and focuses her research on healthy practices for desk-job employees.

T. Scott Rupp (Ph.D., Forest Ecology, University of Alaska Fairbanks, 1998) is the director of the Scenarios Network for Alaska+Arctic Planning in the International Arctic Research Center at the University of Alaska Fairbanks. He also serves as the university director for the Department of Interior's Alaska Climate Science Center and as coprincipal investigator for the NOAA-funded Alaska Center for Climate Assessment and Policy. He is a well-established forest ecologist with specialized experience in ecosystem modeling. His recent work has focused on modeling the response of subarctic ecosystems to changes in global climate and disturbance regimes. He has authored or coauthored more than 65 peer-reviewed journal articles and book chapters.

Sarah J. Ryker (Ph.D., Carnegie Mellon University, 2008) is the US Geological Survey's deputy associate director for climate and land use change. She is responsible for science and policy leadership, and management oversight, of the USGS Climate Research and Development Program, Land Change Science Program, Land Remote Sensing Program, National Climate Change and Wildlife Science Center, and LandCarbon assessment of biological carbon sequestration. She returned to USGS in 2011 from the White House Office of Science and Technology Policy's research center, where she led a team focused on energy and environmental research and policy.

Julia Schmale (Ph.D., Johannes-Gutenberg-University Mainz, Germany, 2011) was the ClimPol project leader at the Institute for Advanced Sustainability Studies (IASS) in Potsdam, Germany, and now is with Paul Scherrer Institute in Switzerland. She applies a transdisciplinary research approach at the science, policy, and society interface focusing on the linkages between air pollution and climate change. Her regional foci are Europe and remote regions such as the Arctic or Central Asia. Her background is in atmospheric sciences.

Simon Schneider (Dipl. Geophysics, Johann Wolfgang Goethe University, Frankfurt on Main, 2002) has been public outreach manager of the research and development program GEOTECHNOLOGIEN in Potsdam since 2005. He has served as coeditor for the online magazine *The Euroscientist*, has studied public relations at the University for Management and Communications UMC Potsdam, and is working on a Ph.D. project at the Freie Universität Berlin. He earned a German diploma in geoelectrical measurements in 1995 from the Johann Wolfgang Goethe University Frankfurt, where he worked in science communication and education with the Geophysics and Earth Science departments. He participated in international surveys as scientist and media relations specialist, traveling to Hawaii Volcanoes National Park, Lake Bosumtwi, and Ghana and aboard the research vessel MS Sonne in the southern waters of the Atlantic and Indian Ocean.

Erika von Schneidemesser (Ph.D., University of Wisconsin-Madison, USA, 2009) is a project scientist at the Institute for Advanced Sustainability Studies (IASS) focusing on urban air quality and air pollutant emissions inventories, as well as the linkages between air quality and climate change. She works at the interface of science and policy.

Larisa O. Serbina (M.S., Colorado State University, 2011) is an economist with the US Geological Survey's Fort Collins Science Center. She conducts applied research on the value of public goods such as government-generated data, rivers, land, and other natural resources. In studying the value of Landsat imagery, she focuses on water resources, agriculture, and the private tech sector. Her recent work contributes a regional economic assessment to the Cumulative Effect Assessment on the Yellowstone River.

Sophie Steffen (M.A., International Studies, University of Barcelona, 2014) completed her undergraduate degree in media studies and political science at the University of Regensburg in 2012. Her master's thesis, "Challenges in Communicating Climate Forecast Information to Decision-Makers in 2014," was supervised by Dr. Lowe and Miss Melanie Davis at IC3.

Aaron Strong (M.A., Environmental Policy, Tufts University, 2011) is a Ph.D. candidate in environment and resources at Stanford University, where he studies the ecology of ecosystem responses to anthropogenic global change and nascent governance approaches to managing the impacts of such changes on the carbon and nitrogen cycles.

Toshihiko Sugai (D.Sc., the University of Tokyo, 1994) is a professor in physical geography in the Department of Natural Environmental Studies within the Graduate School of Frontier Sciences at UT. Previously, he was a senior researcher at the Geological Survey of Japan (GSA) until 1999. His study interest is long-term natural environmental changes and their relations with human activities based on fluvial and tectonic geomorphology and quaternary geology.

A. Ester Sztein (Ph.D., University of Maryland, 1999) is the assistant director of the Board on International Scientific Organizations of the National Academy of Sciences. In this capacity, she manages the US National Committees for geological sciences, geodesy, and geophysics; quaternary research; and soil sciences and has managed the committee on psychological science. An active participant in the geoscience community, she works with US and international professional societies on projects ranging from geoscience and climate change outreach to medical geology. She specializes in multi- and interdisciplinary science. As a research scientist at the University of Maryland, she pioneered the study of hormone metabolism in land plants within an evolutionary context. With the Amazon Conservation Team, she participated on bio-cultural conservation efforts in Latin America. She also earned a B.Sc./M.Sc. in biology from the University of Buenos Aires, Argentina.

Karen M. Taylor (Ph.D., Communication with certification in History and Philosophy of Science, University of Pittsburgh, 2002) is an assistant professor at University of Alaska Fairbanks. Her research uses organizational communication research methods to study the institutional contexts for scientific knowledge production, utilization, and dissemination.

Kristin Timm (M.S., Science Communication, University of Alaska Fairbanks, 2014) is the Science Communications Lead at the Scenarios Network for Alaska and Arctic Planning (SNAP) at the International Arctic Research Center at the University of Alaska Fairbanks. In her position she oversees strategic internal and external communications for the institute, engages in communication project management, and is trying to develop some small practice-inspired communication research projects. She has a master's degree in science communication from the University of Alaska Fairbanks, where she studied the role of communication in climate change adaptation. Before working in an academic setting, she spent over a decade doing environmental and science communication with federal agencies and a nonprofit. Loving both the practical applications of communication and research, she is interested in informing science communication theory from practice and is particularly interested in how climate change communication impacts decision-making in the resource management and private sectors.

James R. Triplett (Ph.D., the University of Kansas, 1976) is emeritus professor and chair in the Biology Department at Pittsburg State University. After earning his doctorate, he joined the faculty in the Department of Fisheries and Wildlife at the Ohio State University, leaving in 1981 for a position in the Biology Department at PSU, where he served as chair from 1985 to 2008. He was elected chair of the Grand/Neosho River Committee in 1990 and appointed chair of the Special Commission on Surface Water Quality in 1978 by Governor Bill Graves. In 2000, he was presented the FOX Award by the Kansas Water Office for work with Kansas Water Resources issues and appointed to the Vision 21st Century Task Force. He currently works in the area of natural resource policy and management with an emphasis on sustainability. He serves as president of the Grand Lake O' the Cherokees Watershed Alliance Foundation and chair of the Neosho Basin Advisory Committee.

Bärbel Winkler (Abitur, 1983, Friedrich-Schiller-Gymnasium Marbach) lives and works in Germany. She has always had an interest in environmental issues and has been active as a volunteer at her local zoo and a conservation group on climate issues for many years. She joined the Skeptical Science team in 2010, serving as a translator for German-language content. Since 2013 she has been coordinating the translation efforts for all languages and is a blogger for the site. She is a coauthor of "Quantifying the consensus on anthropogenic global warming in the scientific literature" (Cook et al. 2013).

Cory Young (Ph.D., Bowling Green State University, 2001) is an associate professor and chair of the Graduate Program in Communications in the Department of Strategic Communication at Ithaca College in upstate New York. She is primarily responsible for teaching corporate and organizational communications, specializing in crisis communication and social media. She has been published in various communication journals (*Case Studies in Strategic Communication*, *Journal of Visual Literacy*, *Prism*, *Journal of Communication Studies*), books, and trade publications (*PR News Media*) and has presented her research at regional, national, and international academic conferences. She also holds a master's degree of social science in culture studies from the University of Birmingham, England.

Chapter 1
Introduction: An Overview of Crisis Communication

Lindsay E. Kuntzman and Jeanette L. Drake

Abstract Crisis communication, a subspecialty of public relations, proffers a vast body of knowledge initially fueled by two events—the Tylenol crisis of 1982 and the Exxon Valdez crisis of 1989. In their wake, these examples underscored that communication becomes the lifeblood when managing emergency situations. Opposite sides of the same coin, these two classic case studies show companies, organizations, and governments what to do and what *not* to do in times of crisis. In responding to cyanide poisoning found in its product, Johnson & Johnson provided an early exemplar of prompt action and open communication. On the other hand, Exxon's slow and clumsy response to a devastating oil spill off the coast of Alaska made corporations and communicators sit up, take notice, and make crisis communication a priority. Although the cases throughout this book deal specifically with crises associated with climate change and natural and technological hazards, a general overview of the crisis communication literature provides a useful underpinning. Thus, the purpose of this introduction is not unraveling the skeins of crisis communication research that exist, but rather to summarize key principles beginning with a brief backdrop explaining why communication matters during crisis.

Keywords Crisis communication • Framing

L.E. Kuntzman, M.A. (✉)
Kent State University, Kent, OH, USA
e-mail: lekuntzman@gmail.com

J.L. Drake, Ph.D.
College of Business, University of Michigan-Dearborn, Dearborn, MI, USA
e-mail: dr.jldrake@gmail.com

J.L. Drake et al. (eds.), *Communicating Climate-Change and Natural Hazard Risk and Cultivating Resilience*, Advances in Natural and Technological Hazards Research 45, DOI 10.1007/978-3-319-20161-0_1

1.1 Framing Matters in Communicating Crisis

Few issues surpass climate change in heating up a conversation. Despite near consensus in the scientific community, the issue remains politically charged. As a result, earth scientists face not only physical but also rhetorical challenges in their mission to mitigate effects of global warming. Thus, as important as the issue itself is how the issue is presented or *framed*. Entman (1993) clarifies:

> To frame is to select some aspects of a perceived reality and make them salient in a communicating text in such a way so as to promote a particular problem definition, causal interpretation, moral evaluation, and/or treatment recommendation. (p. 52)

Framing is central to communicating about risks and crises associated with climate change and natural hazards. Framing occurs intentionally and unintentionally via a culture, a communicator, a listener, and the media. Framing is neither inherently good nor malevolent so, for an audience, it is paramount to discern the source and motivation behind a message. For an actor intent on influencing social good, it is vital to understand how framing works (see, e.g., Reese et al. 2003).

Because framing is not neutral, the actors with the most resources—including money, power, communication savvy, and access to media—are better able to present and maintain the dominant frame. Thus, knowledge of crisis communication principles and practices can and should inform how we communicate climate change and natural hazard risk and cultivate resilience. Historically, media reported on natural disasters primarily as major events occurred, which is called *episodic* framing and limits information to a superficial discussion. More and more science and environmental journalists have adopted *thematic* framing, in which they follow a particular issue, such as climate change, more deeply over time so as to connect the dots of a complex phenomenon. Framing occurs in thought; individual words, sentences, or paragraphs; in whole texts; visual images; and throughout a culture. In the midst of a natural disaster, for instance, media reports are replete with framing techniques—naming (e.g., global warming, climate change); presenting victims, villains, and vindicators; showing images, making metaphors—presenting the situation in a certain way or not presenting a situation at all. Through it all, it is easy to forget what is not shown, what is not said. The most powerful framing of all silences an issue or keeps it out of the news and out of the public eye, which, for some, is at the heart of crisis communications. An ethical public relations perspective and a more critical lens place public well-being even above company reputation, knowing in the long run that responsibly and responsively addressing the former will take care of the latter.

This century has seen extreme weather conditions producing more frequent and more intense natural disasters around the world. Communication plays a major role for every crisis at every juncture—prediction, prevention, preparation, mitigation, recovery, and, more and more important, resilience. There is a mandate for twenty-first-century earth and social scientists to work together in helping individuals and communities develop the capacity to recover quickly from natural disasters, thus the need for a multidisciplinary perspective. We begin with a better understanding of crisis communication.

1.2 Crises Defined and Explained

Individuals experience personal crises every day—a car that stops running, a past-due bill, or a fast-approaching deadline for work—but these emergencies differ from the type of situations a crisis communication plan addresses. While definitions of a crisis vary from one source to another, a common theme incorporates a disruption of activities that, potentially, lead to devastating consequences.

A crisis may be defined as follows:

- "...an untimely but predictable event that has actual or potential consequences for stakeholders' interests and the organization's reputation" (Millar and Heath 2004, p. 2).
- "...a major occurrence with a potentially negative outcome affecting the organization, company, or industry as well as its publics, products, services, or good name" (Fearn-Banks 2002, p. 2).
- "...perception of an unpredictable event that threatens important expectancies of stakeholders and can seriously impact an organization's performance and generate negative outcomes" (Coombs 2007, p. 2).
- "...an organizational crisis is a specific, unexpected, and nonroutine event or series of events that create high levels of uncertainty and threaten or are perceived to threaten an organization's high-priority goals" (Ulmer et al. 2007, p. 7).

Millar and Heath (2004) noted that the following characteristics appeared frequently in crisis communication literature (p. 19):

- A crisis occurs quickly.
- A crisis requires quick reaction.
- A crisis impacts business operations.
- A crisis causes uncertainty and scrutiny.

The literature differs in considering a crisis a predictable or unpredictable event. Coombs (2007) clarified this issue by writing "...a crisis is unpredictable but not unexpected" (p. 3). For example, airlines may not predict when an airplane crash will occur, but airlines could reasonably expect a crash to happen at some time.

Warning signs often precede a crisis making it a more predictable event. These warning signs (also called *prodromes*) appear in many forms: disgruntled employees, an increase in calls from the media, similar complaints from customers about a product, or another company within the industry experiencing a crisis (Fearn-Banks 2002, p. 9).

Although crises differ from one another, "...they tend to cluster into identifiable types" (Coombs 2007, p. 65). Two types emerge: *intentional* and *unintentional.*

Intentional crises result from an initiated, purposeful act (Ulmer et al. 2007, p. 9).

Examples from Ulmer et al. (2007) include the following (p. 9):

- Terrorism
- Sabotage
- Workplace violence

- Poor employee and management relationship
- Hostile takeovers
- Unethical leadership

Unintentional crises do not originate from an intention to harm an organization (Ulmer et al. 2007, p. 11). This book focuses on unintentional crises; examples include the following:

- Natural disasters
- Disease outbreaks
- Unforeseeable technical interactions
- Product failure
- Downturns in the economy

Regardless of type, all crises follow similar life cycles—a beginning, middle, and end. Different research models explain the different stages and life spans of a crisis. Coomb's (2007) three-stage approach identifies the *precrisis*, *crisis*, and *post-crisis* stages (p. 17).

1.3 Stage I: Precrisis

Detection, prevention, and preparation made up the first stage in Coombs' model (p. 17). Management aspires to detect and prevent a crisis before it actually transpires. However, when detection or prevention fails and a crisis strikes, preparation becomes crucial.

Before summarizing the literature about detection, prevention, and preparation, it must be acknowledged that many organizations try to practice two-way communication. Public relations researchers described two-way communication as a "… dialogue involving communicator and receiver" (Pfau and Wan 2006, p. 102). That dialogue occurs through face-to-face communication, websites, social media, phone conversations, and so on.

On the other hand, an organization engaging in one-way communication simply disseminates a message and does not attend to feedback from key publics. Some experts argued that crisis communication starts before a crisis hits an organization. Communicating with key publics before an emergency promotes a better sense of partnership and benefits the company (Ulmer et al. 2007, p. 35). The scope of this overview does not include that precrisis aspect.

1.3.1 Detection

Detection entails *scanning* the environment (Coombs 2007, p. 22). The environment means the setting in which an organization functions. For a company, the environment may include the office, the city where the office is located, the industry in which the organization operates, etc. When scanning the environment, individuals identify "…as many warning signs as possible" (Coombs 2007, p. 22). The sources individuals use to scan the environment include a variety of materials: newspapers, trade publications, newsletters, public opinion polls, websites, blogs, discussion groups, financial audits, accident records, etc. With so many potential sources to scan, the public relations practitioner or crisis manager selects those most relevant to crises, such as accident reports or an online message board for customers.

A theoretical approach lends credence to the idea that scanning the environment prevents crises. Systems theory focuses on "…interacting units that endure through time within an established boundary by responding and adjusting to change pressures from the environment to achieve and maintain goal states" (Cutlip et al. 2006, p. 176). Part of a company's system may include employees, consumers, and investors. The system contains interdependent parts, such as employees' dependency on consumers buying the company's products.

Systems theory describes the organization's response to its environment on a continuum that ranges from *relatively closed systems* to *relatively open systems.* Relatively closed systems "…do not adapt to external change and eventually disintegrate" (p. 181). In contrast, relatively open systems monitor the environment to predict and anticipate potential changes. By acting as a relatively open system, a company notices changes in its environment that indicate potential conflicts.

The following example of a relatively open system illustrates the theory and warning signs. Rumors circulated online that Febreze, a Procter & Gamble product, killed animals. The company started receiving e-mails and phone calls from customers. Procter & Gamble addressed consumer concerns by devoting a portion of the Febreze website to disproving the rumors. The online rumors, e-mails, and phone calls supplied warning signs of a potential crisis brewing. By scanning its environment (which included online sources, e-mail, and phone traffic), Procter & Gamble avoided a disaster (Coombs 2007, p. 36). The company received and exchanged information with consumers demonstrating a relatively open system.

1.3.2 Prevention

After detecting warning signs, prevention becomes imperative. Prevention consists of "…two components: (1) change and (2) monitoring" (Coombs 2007, p. 50). For example, if management learns employees no longer follow proper procedures when handling dangerous chemicals, then a refresher course on procedures intends to change the employees' actions (p. 50). Once employees take their refresher course, management monitors their behavior to determine if the course effectively changed their actions, thereby preventing a crisis. Monitoring acts as "…a form of focused tracking; it keeps a close watch on the warning signs that have the greatest potential to become crises" (p. 22).

Not every warning sign results in crisis. The crisis manager assesses the potential of a warning sign developing into a crisis by analyzing two factors: *likelihood* and *impact* (also called damage). Likelihood refers to the probability of a warning sign developing into a crisis, while impact refers to the extent the crisis disrupts the work process or damages property, people, reputation, etc. (pp. 39–40). A scale ranks likelihood and impact. Some suggest using a scale of one to five, while others suggest a scale of one to ten (Coombs 2007, p. 39; Fearn-Banks 2002, p. 24). Regardless, a higher number means greater likelihood and impact. Monitoring should still continue even if the warning signs seem unlikely to develop into a crisis. If the warning signs do lead to a crisis situation, preparation is critical for survival.

1.3.3 Determining Potential Crises

In preparing for a crisis, the crisis manager should consider possible scenarios. No harm exists in assuming certain crises might strike: an airline assumes a crash, a restaurant assumes food poisoning, a school assumes violence, etc. Asking all departments within an organization to provide a list of potential crises covers all aspects. After all, the custodial staff probably knows the heating system best, the workers on the line know the equipment best, and so forth. From that list, use a scale to rank crises on the list for likelihood or damage (Fearn-Banks 2002, p. 24). Fearn-Banks (2002) recommended drawing a bar graph to show and compare the likelihood and impact rankings for each crisis (p. 25). The graph helps show the most threatening crises.

1.3.4 Crisis Communication Plan

Crisis management attempts to minimize the impact. It uses factors such as prevention, preparation, response, and revision (Coombs 2007, p. 5). Communication is just a part of crisis management. Crisis communication provides "…the dialog between the organization and its publics prior to, during, and after the negative occurrence" (Fearn-Banks 2002, p. 2).

Ideally, management should create a crisis communication plan as soon as the business starts operations, before any crisis warning signs appear. The plan plays a large role in preparing an organization for a crisis.

A crisis communication plan "…states purposes, policies, and goals, then assigns employees to various duties" (Fearn-Banks 2002, p. 28). The plan gives guidelines for creating messages, communicating the message, and knowing when to communicate the message (Leighton and Shelton 2008, p. 28). Preparing the plan before a crisis alleviates potential mistakes and stress. Individuals typically draft a crisis plan when "…an organization has done any crisis preparation" (Coombs 2007, p. 89).

However, a plan does not create "…a magic insurance policy that protects an organization from a crisis" (p. 89). An organization must practice and revise the plan to make it useful.

The plan collects important data (such as background information on the company or employees' contact information) and provides a framework for a crisis team to use during a stressful situation (p. 90). Rules for creating a plan include keeping it short, but detailed (p. 89).

Experts agree on using the following elements to create a crisis communication plan:

Cover Page First, a cover page identifies the document and includes creation and revision dates (Fearn-Banks 2002, p. 29). The dates indicate the currency of the plan (Coombs 2007, p. 91).

Introduction Next, an introduction stresses the importance of the plan and is written by the CEO or a ghostwriter.

Acknowledgment The plan includes an acknowledgment page that takes "...the form of an affidavit" (Fearn-Banks 2002, p. 29). After employees sign this page, place it in personnel files.

Rehearsal Dates The dates indicate scheduled rehearsals (Coombs 2007, p. 91). The plan needs to be practiced at least annually, if not every six months (Fearn-Banks 2002, p. 29).

Purpose and Objectives The purpose statement and objectives explain what the company wants to achieve. This may include a timeline for resuming normal business operations and communicating accurate information (Fearn-Banks 2002, p. 30).

Key Publics A crisis affects different people involved in the organization. Listing all potential publics in the plan saves time during a stressful situation (Marra 2004, p. 312). Within the key public section, the plan includes methods for reaching different publics—such as a telephone tree for employees. Crisis team members document the information communicated to publics within this section (Coombs 2007, p. 93).

Crisis Communication Team This section provides a directory for the crisis team. It includes team members' names, titles, role on the crisis team (e.g., *spokesperson*), office phone number, home number, mobile phone number, e-mail address, home address, and any additional notes deemed important (e.g., travel plans out of the country might be included) (Fearn-Banks 2002, p. 35). This section also identifies the flow of communication (Leighton and Shelton 2008, p. 35). Often, the flow begins with someone identifying a crisis and notifying the crisis team leader or point person. The team leader then notifies the entire crisis team. Together, the team determines the course of action (p. 36).

Emergency Personnel Assistance from emergency personnel may be required; contact information for appropriate authorities should be included here (Fearn-Banks 2002, p. 35).

Media Similar to the list of key publics, this section documents all relevant media channels and the organization's usual press contacts. Placing this information in the document makes it more accessible. Within this section, the plan contains a media log. The sheet contains places to record basic information about media inquiries such as journalist name, publication, date or time called, telephone number, and who handled the inquiry (Leighton and Shelton 2008, p. 38). This log also aids in the evaluation of media relations during crisis.

Internet and Mobile Media A website provides a valuable channel for communicating with key publics. This section of the crisis communication plan details who posts information to the company's site and documents comments, number of hits, links, etc. (Fearn-Banks 2002, p. 39).

Crisis Communication Center This section specifies where to go during a crisis. Ideally, the plan includes several potential sites as certain crises (such as fires, tornadoes, etc.) may impact the normal location and includes information such as addresses, directions, and who gave permission to use the alternate locations (Fearn-Banks 2002, p. 36).

Background Materials Gathering materials before a crisis saves valuable time. Within this section, executive biographies, company profiles, quality assurance procedures, *fill-in-the-blank* press releases, and more are included (Fearn-Banks 2002, p. 37).

Key Messages In addition to preparing press releases in advance, prepare key messages within in the crisis communication plan. These key messages include the main points to make to key publics (Fearn-Banks 2002, p. 37). For example, one might include the company's mission statement and values.

Evaluation Form An evaluation form helps the team assess the response to the crisis. Completing the evaluation form soon after the crisis allows for better recall of events and planning for the next crisis situation (Fearn-Banks 2002, p. 40).

1.3.5 The Crisis Team

A crisis team enacts the crisis communication plan. An ideal team represents a variety of departments from the organization—public relations, human resources, quality assurance, marketing, etc. (Coombs 2007, p. 67). While some experts stated that the CEO belongs on the crisis team, others disagreed. Teams without the CEO include a high-ranking individual who does not have to ask permission to make decisions (Leighton and Shelton 2008, p. 29).

Members of the crisis team are assigned individual roles and responsibilities. The human body illustrates the team members' roles. Just as the brain, mouth, and hands help the human body to survive, a crisis team needs a brain, mouth, and hands to function properly (Leighton and Shelton 2008, p. 30).

The CEO (if included) and two to four other individuals serve as the brain for the crisis team. One of these individuals serves as the leader or point person. Before an emergency situation, the team must determine how the brain operates. Clear procedures (such as how to break a deadlock when voting for actions to take) for the brain's role and responsibilities help during the crisis (Leighton and Shelton 2008, p. 30).

The mouth of the crisis team consists of at least two spokespeople. Two individuals ensure a *backup* if one is unavailable and the ability to cover two locations if necessary (Leighton and Shelton 2008, p. 30). If a company operates in several states or countries, experts recommend selecting more than two spokespeople.

Additionally, Leighton and Shelton (2008) recommended appointing *specialist spokespeople* (p. 31). A specialist spokesperson adds credibility and provides more of an in-depth explanation regarding a crisis situation (p. 31). For example, a construction company that just experienced a building collapse might rely on its in-house architect to explain the structure of a building. The architect acts as a credible source and supports the company's normal spokesperson in delivering messages to audiences. When appointing specialist spokespeople, a crisis team should consider the types of crises it might experience (p. 31). For example, medical emergencies might require a physician as a spokesperson or a toy company might want to use someone in the company who has young children.

When selecting spokespeople, one also considers the company's key publics. A recent study examined how the ethnicity of a spokesperson affects key publics. A sample of Caucasian, Hispanic, African-American, and Asian American undergraduate students watched a news story in small groups. Each group received a news story with a spokesperson talking about a company. The companies were based in America, Japan, or Mexico. The stories differed in regard to the similarity between the company and the spokesperson. For example, some groups saw an American-based company with an American speaker, while others saw an American-based company with a Japanese speaker (Arpan 2002, p. 324).

From the study, spokespeople similar to participants were perceived as more credible (p. 332). Overall, Arpan (2002) cautioned that global organizations may want to consider using spokespeople with ethnic backgrounds similar to the majority of key publics (p. 333). Based on the limited, convenience sample used for this study, the results do not apply to a larger population. Nevertheless, it appears reasonable for a company to use a spokesperson relatable to its audiences.

Examples of effective and ineffective spokespeople appear throughout the media. The Indiana Methodist Hospital case study illustrates good spokesperson qualities. At the hospital, six premature infants received a drug overdose; three of the infants died. The hospital's CEO issued apologies and accepted responsibility for the error. He appeared empathetic, sincere, and regretful (Martin and Montgomery 2006, para. 2–3). The Exxon Valdez case demonstrates poor spokesperson qualities. After the Valdez ship ran aground, it started leaking oil and became the worst spill ever in American waters. The company spokesperson, however, said the company did not expect any environmental damage and downplayed the severity of the situation, while the company's CEO waited a week before he commented on the situation. When the CEO finally commented on the situation, he never apologized to the individuals affected (Ulmer et al. 2007, p. 68–69). Individuals were outraged and even canceled their company credit cards (Fearn-Banks 2002, p. 100). The Valdez case shows the importance of a spokesperson (particularly a CEO) responding quickly, empathetically, and accurately to a crisis.

The spokesperson becomes "...the voice of the organization during a crisis" (Coombs 2007, p. 78). While the brain makes decisions, the mouth manages "...the accuracy and consistency of the messages coming from the organization" (p. 79). The main function for the hands involves communicating. While the brain makes decisions, the hands develop and initiate communication to support the decisions. Typically, the "...lead in-house communication professional" serves as the communication leader. The key communication person relies on a support team (Leighton and Shelton 2008, p. 32). For large businesses, a chief communication officer acts as the point person with the public relations (PR) department providing support. For small companies, though, only one person may work as the sole communicator. In this situation, outside sources (such as PR agencies) could provide additional support (Fearn-Banks 2002, p. 34).

If no one at the company possesses the ability or experience to manage the situation, the crisis team should hire outside counsel immediately. Waiting to get assistance may damage a company or result in miscommunication. For example, the Sago Mine incident received continuous media coverage. After the explosion occurred on January 2, 2006, company officials waited for three days before bringing in Dix & Eaton, an Ohio-based PR firm. Dix & Eaton established a media strategy to ensure consistent messages (Dix and Eaton n.d., para. 2). However, before the agency arrived, miscommunication had already occurred with reports that all the miners were alive.

The hands act as a link between the brain and the mouth, approve messages, provide updates to the brain, develop media tools (e.g., press releases or media kits), update the website, and access media lists and contacts (Leighton and Shelton 2008, p. 32). Selecting members of the crisis team and assigning their roles eliminates confusion during a stressful situation. Individuals prepared to make decisions, act as communicators, and serve as the voice of the organization guide the company out of the crisis.

1.3.6 Education and Practice

During a crisis, the media want answers. Refusal to answer questions may lead the media to find other sources with potentially incorrect information. Before a crisis, individuals within the organization (particularly those fulfilling the spokesperson role) ought to practice handling interviews.

Experts exist who teach individuals "...how to be in front of a camera and deal with the press" (Leighton and Shelton 2008, p. 30). Some public relations practitioners conduct media training and some agencies offer training for crisis situations. Industry groups may provide sources on media training, too, such as seminars or videos (Anthonissen 2008, p. 13).

Media training teaches the following (Leighton et al. 2008, p. 135):

- How to build relationships with the media
- How to interact with the media
- How to deliver a message

- How to handle questions from the media
- How to handle anxiety or maintain control of emotions

From media training, spokespeople learn to avoid the phrase *no comment* since it fuels "…speculation that you're hiding something" (Leighton et al. 2008, p. 132). Use of that phrase "…is a form of silence, which is a very passive response" and "…65 % of stakeholders who hear or see 'no comment' equate it with…guilt" (Coombs 2007, p. 83). Telling the media one does not know the answer but will find out makes a better alternative to *no comment* (Coombs 2007, p. 83). Fearn-Banks (2002) deemed three responses acceptable to media inquires (p. 68):

- Here's all the information.
- We don't know everything. Here's what we know. We'll find out more for you.
- We'll find out and let you know.

The spokesperson puts a face on a company for the public. If he or she appears nervous and upset, the public may perceive the crisis as the business' fault. The public often views an inability to *deliver* a message as a sign of guilt. For example, the individual who lacks eye contact with the interviewer comes across as deceptive (Coombs 2007, p. 82). Practicing *delivery factors* allows the spokesperson to look attentive, compassionate, and in control while disseminating the company's message (Coombs 2007, p. 82). These delivery factors include making eye contact with the audience and camera, using appropriate hand gestures, varying tone of voice and facial expressions, and eliminating *ums* and *erhs* from his or her speech (Coombs 2007, p. 82). During an interview, act "…positive, concerned, empathetic, and apologetic, if necessary" (Fearn-Banks 2002, p. 72).

A crisis manager or PR practitioner can teach body language and nonverbal factors through demonstrations, mock-interview rehearsals, and video recordings of the spokesperson. Video recording makes a powerful tool in teaching body language to individuals. The recording shows how the individual appears in giving an interview or a speech (Wilcox 2005, p. 489). Often, individuals make nervous gestures unconsciously—the recording helps them see their body language so they can adjust accordingly.

To better prepare spokespeople, a crisis manager or PR practitioner might also conduct and videotape a mock news conference. The manager running the simulated news conference describes the crisis situation to the spokesperson and allows time for him or her to prepare talking points. Then employees within the organization pretend to be reporters and ask questions (Fearn-Banks 2002, p. 69).

Spokespeople must learn to avoid *trick questions*. Some of the trick questions described by Fearn-Banks (2002) included the following (p. 71):

- Leading questions ("You do agree that…?")
- Loaded questions ("Isn't it true that you knew…and failed to….?")
- Silence (The reporter waits for you to start spilling the details.)
- Chummy questions ("Hey, friend, off the record….?")

To counter negative questions from the media, say three positive remarks to "neutralize the negative statement" (Hyde 2007, p. 35). Do not repeat the negative question as that reinforces it. Additionally, do not use words such as *never*, *nothing*, or *none* as those words are absolutes and cause reporters to question the speaker's certainty (p. 35). Practice teaches spokespeople to avoid those trick questions.

Besides practicing how to handle the media, a crisis team must practice the plan. Training occurs in several steps. First, during an *orientation seminar*, the team reads the crisis communication plan. Second, during a *drill*, the team practices one component of the plan, such as the employee notification system. Third, the team holds a *tabletop exercise*. During this exercise, crisis team members discuss the plan and work through a mock crisis without time restrictions (Coombs 2007, p. 75).

Finch and Welker (2004) outlined a five-step process to use during the tabletop exercise (p. 198). First, management identifies a likely crisis. Second, the team assembles. In the third step, a *facilitator* tells the team the crisis and reminds it of the company's objectives (p. 198). Fourth, the team *solves* the crisis, by relying on the crisis communication plan. Finally, the team reflects on the outcome and any problems with the plan that need to be corrected (p. 199). Another component of training involves a *functional exercise*. With this method, the crisis team pretends the crisis is occurring in real time (creating time pressures) and interacts with others in the organization. For example, the team might interact with the human resources department.

Finally, training occurs through a *full-scale exercise*. This method mimics a real crisis as closely as possible. The team uses actual equipment, places people on-site and in the field, and has simulated injuries. This method of training involves time and financial resources. Coombs (2007) wrote that it should be used every few years (p. 75). Practicing an emergency drill results in better execution of the plan. For example, a practice run at an airport in Texas led employees to discover the crisis communication plan listed the wrong number to contact emergency personnel. Without the drill, airport employees might have wasted time during an actual crisis finding the correct number (p 96). Taking the time to practice the crisis communication plan provides an opportunity to feel how it works and make corrections.

1.4 Stage II: During the Crisis

Despite best efforts to prevent a crisis, one may still transpire. During the crisis, the plan guides the crisis team to fight and contain the situation. Fighting and containing the crisis includes three steps: instructing information, adjusting information, and reputation management (Coombs 2007, p. 133).

1.4.1 Instructing Information and the Initial Statement

Individuals' safety becomes the priority during a crisis situation. When a disaster strikes, uncertainty looms and people want facts. An organization instructs key publics on the situation and tells them how to protect themselves (Ulmer et al. 2007,

p. 21). For example, a business may tell community members that, due to a chemical spill, drinking water needs boiled before use. In addition to instructing people how to stay safe, the first statement shares the known facts (Augustine 1995, p. 22).

Typically, this initial statement relies on the mass media to relay it to key publics. The crisis team does not tailor this first statement to specific publics, but makes a statement with mass appeal. Communication after the first statement relies on direct channels with messages specific for key publics (Coombs 2007, p. 147). For example, the accounting or investor relations department sends letters or e-mails to investors about the company's financial situation (p. 148). For all communication during the crisis, experts advised "tell the truth and tell it fast" (Leighton and Shelton 2008, p. 31).

An experiment highlights the importance of quickly telling the truth. *Stealing thunder* is a term used to describe an organization disclosing a personal, negative situation before another source discloses it. To examine the concept of stealing thunder and "…message timing strategies in crisis management…" a study used journalists and college journalism students reading fictional crisis scenarios (Arpan and Pompper 2003, p. 296).

In the study, participants were assigned to one of two experimental conditions. Both conditions' stories used a company responsible for a chemical spill into the Chicago River. The researchers focused on three variables: (1) the credibility of the company's officials, (2) the crisis severity, and (3) the participants' interest in pursuing the story (Arpan and Pompper 2003, p. 298). The researchers found participants rated the officials' credibility higher when an organization's executives informed the public of a crisis (rather than the media informing the public). Participants considered the crisis equally severe whether company executives told the public about the situation or the news media broke the story. Additionally, participants wanted to pursue a story when company executives informed them of a crisis (as opposed to hearing about it from another media source) (p. 299).

Arpan and Pompper (2003) suggested the nature of the fictional crisis (the chemical spill) might have contributed to some of the findings (p. 301). However, the findings from this study revealed the importance in quickly providing information to key publics before other sources.

1.4.2 Adjusting Information and Creating Messages

The initial statement gives key publics information to protect themselves. After that statement, the crisis team adjusts information to reassure key publics about the crisis—the goal changes from ensuring people's safety to tending to their psychological well-being (Coombs 2007, p. 135). Informing key publics of an organization's actions to control and correct the situation provides a means of adjusting information. Psychological stress lessens when individuals believe the organization controls or knows how to fix the situation. Coombs (2007) cautioned managers against speculating about *corrective action* if they do not know how or why the crisis started

(p. 136). Coombs (2007) suggested a company use *renewal strategies* instead (p. 136).

For example, if an office burns down and management does not know what started the fire, the spokesperson can still say the company plans to rebuild the office (a renewal strategy). If faulty wiring caused the fire (and management knows), then the spokesperson shares the corrective action—perhaps the business plans for an electrician to redo the wiring.

Research shows that in highly stressful events, people "…can miss up to 80 % of message content" (Hyde 2007, p. 33). Difficulty in processing a message can occur due to a diminished ability to hear, understand, and remember in an emergency. To ensure that individuals hear, understand, and remember a message during a crisis, Hyde (2007) recommended using an approach called *message mapping* developed by the Center for Risk Communication (p. 33). Message mapping protects the psychological well-being of key publics.

Hyde (2007) identified seven steps in message mapping (p. 33):

- Identify stakeholders (key publics).
- Determine specific stakeholder concerns.
- Analyze concerns to find underlying concerns.
- Brainstorm with a team.
- Assemble supporting facts for messages.
- Plan dissemination of messages.

First, with message mapping the team identifies key publics to communicate with—different key publics hold different concerns in a crisis situation—and considers their concerns. During the brainstorming session, the team leader seeks everyone's input regarding the situation and then organizes and prioritizes the information. From brainstorming, the crisis team creates three key messages consisting of 7 to 12 words each with three facts supporting each message (Hyde 2007, p. 34). Keeping the key message to 7 to 12 words is important as Americans typically speak three words a second—a shorter message helps provide succinct sound bites (Hyde 2007, p. 34). Additionally, developing three supporting facts for each message helps elaborate and answer any questions key publics might pose. While the crisis communication plan may include skeletal key messages, the concept of message mapping tailors messages for the specific crisis to ease different key publics' minds.

1.4.3 Reputation Management

How an organization responds to the crisis impacts its reputation (Coombs 2007, p. 138).

In studying reputation management, attribution theory becomes relevant. Attribution theory focuses "…on the belief that people assign responsibility for negative, unexpected events" (Coombs 2007, p. 138). People feel a need to attribute the cause or source of a negative situation to something or someone. Typically, individuals consider a crisis to be a negative, unexpected event. Based on attribution theory, people will want to assign responsibility or blame for a crisis. The assignment of blame impacts how people view the organization that experienced the crisis (Coombs 2007, p. 138).

From attribution theory, a growing area of research called Situational Crisis Communication Theory (SCCT) developed (Coombs and Holladay 2004, p. 96). Situational Crisis Communication Theory offers four strategies to protect an organization's reputation, including denial, diminishment, repairing, and bolstering.

The *denial strategies* argue against a link between the organization and the crisis. Individuals cannot assign blame without connection between the organization and the crisis (Coombs 2007, p. 139). The *diminishment strategies* show the organization possessed little control over circumstances resulting in the crisis. The company offers excuses for the crisis when it uses diminishment strategies (Coombs 2007, p. 139). The *repair strategy* improves the organization's reputation by offsetting the negative event (perhaps through words or even monetary means) (Coombs 2007, p. 139). Finally, the *bolstering strategies* supplement the other three strategies. The bolstering strategies share the organization's good deeds or explain how it became a victim of the crisis (Coombs 2007, p. 141). Denying the existence of a crisis situation may lead to irreparable damage to an organization's reputation. Similarly, an intentional, preventable crisis (such as executives *cooking the books*) "…produces very strong attributions of organizational crisis responsibility," which damages reputations (Coombs 2007, p. 142). Combating the damage through rebuilding and bolstering strategies may prove most effective.

1.4.4 Apologia

Another crisis response strategy (a rebuilding posture) includes apology. Issuing apologies usually makes attorneys nervous as they often equate apology with an admission of guilt, which may open the company up for lawsuits. But, expressing empathy differs from accepting blame, and an expression of empathy may help regain trust from key publics. The phrase *I'm sorry* expresses empathy, but—depending on what one says after *I'm sorry*—not blame (Wojciesak et al. 2008, p. 11). Blame or admission of fault occurs if one says something to the effect of "I'm sorry, it was my fault," "I'm sorry I made a mistake," or "I'm sorry…if I just hadn't…" (p. 11).

While investigating the crisis situation, a suitable method for expressing empathy without accepting blame involves saying *I'm sorry* and then explaining the company launched an investigation and will provide updates (p. 49). It then becomes critical to communicate with key publics as the investigation proceeds.

1.4.5 Sources of Communication

An organization requires channels of communication to instruct, adjust, and manage one's reputation. When speed is critical, the Internet becomes a valuable tool; the team uses the company's website to keep key publics updated. In today's Web 2.0 world, the Internet encourages conversation; simply force-feeding or disseminating information no longer works. Rather than providing just the facts about the crisis on the company's blog or message board, the crisis team needs to prepare for a discussion online.

Occasionally, the discussion may become very negative. Individuals might feel angry and hurt by the crisis and post messages to that effect online. Lukaszewski (2007) encouraged responding in the following ways (p. 13):

- Keep responses brief and to the point (about 75–150 words).
- Don't speak for others—just your company.
- Be calm and positive.
- Establish credibility.
- Correct errors, but only on your site.

During a crisis, key publics connected to the Internet may "…help a company get back on track" (Bridgeman 2008, p. 177).

During two different crises, two companies (Texas Dow Employees Credit Union and Verity Credit Union) relied on blogs as a means of communication. The Texas Dow Employees Credit Union used its blog after Hurricane Ike in September 2008. The blog shared information about the company and posted information regarding power outages, blood donation sites, and local schools (Polaniecki 2008, p. 38).

For Verity Credit Union, after hackers got into the online banking sign-in screen, the company blog offered a quick and inexpensive way to warn members. Laurel McJannet, e-commerce manager for Verity Credit Union, described the blog as an effective crisis tool that is quickly updated (Polaniecki 2008, p. 40).

The crisis team does not need to wait until a crisis strikes to post information to the company's website. The team may create a *dark site* in preparing for a crisis. A dark site contains content about the company such as company and product background, executive biographies, or expert sources. During a crisis, the team activates the dark page and starts posting new information as the situation progresses. A section of the company's website may serve as the dark site or it may be a separate site (Coombs 2007, p. 101). Interestingly, only "…60 % of organizations in crisis use the Internet" (Coombs 2007, p. 101). The Internet presents an alternative channel with unlimited space to instruct key publics.

Mentions of an organization's website should also include details regarding online newsrooms and mobile communications. Wilcox (2005) described online newsrooms as pages dedicated to journalists (p. 382). These pages provide journalists with facts about a company, press releases, graphics, and contact information

(p. 382). In developing a crisis communication plan, it becomes necessary to consider if information related to the crisis will only appear within the newsroom or if it will appear in other sections of the website, social media platforms, and via mobile technologies. Additionally, the organization that creates a separate crisis website should consider linking to the online newsroom as it could provide journalists with needed facts about the company.

Of the myriad online platforms, Twitter, in particular, can help an organization effectively communicate with key audiences during times of crisis. Twitter is a free social networking and microblogging service. Individuals can send updates (called *tweets*) to family and friends over the web or by phone. It allows users to send messages of 140 characters. The tool was used effectively during a crisis in 2007 and many times since.

In October of 2007, wildfires began burning across Southern California, destroying over 1,000 homes and 500,000 acres of land. With the quick-spreading fires, it was difficult for those affected to access continuous news coverage. To provide residents with information, the LA Fire Department and Red Cross used Twitter. People also used it to connect with family and friends during the situation (Huyse 2007, para. 1–2). Twitter proved to be an effective tool to quickly give people information during a crisis.

In addition to relying on internal means, such as a website or an organization's Twitter feed, the media also play a key role in a crisis. The crisis communication plan includes the media contacts that the organization regularly interacts with, as well as contact information for other media outlets. In order to *steal thunder*, the crisis team may directly contact a media outlet. In situations resulting in injuries or death, the team should immediately call appropriate media outlets. In less urgent situations, the team might wait to gather facts and then contact the media (Fearn-Banks 2002, p. 67). In working with the media, crisis team members need to cooperate as "…the overall goal is to keep or get the public's trust through the media…" and the crisis team "…needs the media to communicate" (p. 68). Only spokespeople and others who received media training should speak to the media.

News conferences help an organization reach multiple media outlets at the same time. Two major reasons a company holds a news conference include: (1) to give all reporters the chance to hear something at the same time and (2) to allow for follow-up questions after an announcement. Management only organizes a news conference during a crisis when absolutely necessary—during the crisis, the goal is to stop making headlines (p. 69).

When holding a news conference, the team considers location. Ideally, they should hold the conference in a large facility. The room should include electric outlets for the media equipment and phone jacks (Wilcox 2005, p. 318). During the news conference, the spokesperson makes a few statements and then takes questions from journalists (p. 320).

1.4.6 Staying Calm

During stressful events, staying calm proves more challenging. To remain effective throughout the crisis situation, the team needs to use tools to de-stress. Loomis (2007) explained the concept of emotional intelligence as "…the ability to recognize and manage one's emotions in a productive manner, to use emotional information to enhance reasoning and problem solving and to positively influence the emotions of other people" (para. 4). Loomis (2007) identified three steps to controlling and understanding emotions during a crisis (para. 4).

First, one should identify personal thoughts, for example, "I can't do this" or "I have the ability to handle this." Second, the individual tries to determine if the thoughts are being helpful or hurtful (para. 12–13). Then, the individual focuses on physiological responses, noting an increase in respiration or perspiration and then relying on tricks such as deep breathing to lessen and control these physical reactions (para. 14).

Finally, individuals must pay attention to behavior. For example, anger might result from other team members not viewing the situation similarly (para. 15). No one possesses the ability to work continuously for days. Including basic amenities such as food and drinks in the crisis communication center helps the team's physical and mental wellness (Coombs 2007, p. 98).

1.4.7 Rumors as Crisis

In a review of crisis communication, it seems only relevant to discuss a few crises. Rumors, as a type of crisis, appear frequently within the literature. Doorley and Garcia (2007) defined rumors as a belief "…passed along from person to person, usually by word of mouth, without secure standards of evidence being present" (p. 29). A rumor typically spreads when individuals consider it news, relevant to their lives, and frightening (Fearn-Banks 2002, p. 44). Locating the source of a rumor proves difficult due to an ever-changing nature (Doorley and Garcia 2007, p. 28). According to Fearn-Banks (2002), "…rumors can cause the longest and most damaging of crises" (p. 41).

Different types of rumors exist. An *intentional rumor* should achieve something—perhaps an individual starts a rumor about a company's stock to make prices increase. A *premature-fact rumor* contains truth, but spreads before an organization makes an *official* statement—for example, employees may hear rumors of layoffs before the company officially announces layoffs (Fearn-Banks 2002, p. 44). A *malicious rumor* intends to hurt an organization. The *outrageous rumor* sounds so far-fetched people assume it must be true. The *nearly true rumor* contains partial truth and the *birthday rumor* appears annually (Fearn-Banks 2002, p. 45).

In the 1940s, two Harvard psychologists studied how rumors change. From an experiment they learned that as individuals pass on a rumor "…it tends to grow shorter, more concise and more easily told" (Doorley and Garcia 2007, p. 28). Doorley and Garcia (2007) identified two factors that influence rumors: importance and ambiguity (p. 29). To fight the rumor mill, management must "…diminish the importance assigned to the rumor…or eliminate the ambiguity around the factual basis of the rumor, or both" (p. 29). Fearn-Banks (2002) offered the following suggestions to fight rumors (p. 46–47):

- Establish a rumor center for employees to report rumors.
- Train employees on how to handle rumors.
- Strengthen relationships so key publics may doubt the rumor.
- Communicate with employees.
- Deny the rumor and use supporting information to make the denial.
- Let an expert source discredit the rumor.
- Do nothing (some rumors fade with time).

While rumors may appear challenging, companies do survive them. For example, when a rumor started that Bubble Yum (a brand of gum that broke sales records in the 1990s) contained spider eggs, ads in newspapers successfully disputed the rumor (Fearn-Banks 2002, p. 48).

1.4.8 Recalls as Crisis

For a business that sells goods, "…recalls can result in irreparable damage to a… brand or reputation" (Grabowski and Herzberg 2007, p. 13). The case of Audi of America provides an example. In 1986, Audi recalled the Audi 5000 as it went out of control when shifting gears. The company waited to initiate a recall and then, after recalling more than 100,000 vehicles, failed to change the positioning of the pedals as it promised to do. Instead, Audi installed a lock that drivers pressed before shifting gears. Accidents continued and a class-action lawsuit destroyed the company's sales and reputation (Smith et al. 1996).

To avoid devastating consequences, management must "think like a consumer" (Grabowski and Herzberg 2007, p. 14). When developing a product, management should research if consumers easily understand the product warnings or instructions. Asking these questions might eliminate possible problems due to human error (Grabowski and Herzberg 2007, p. 14). In the Audi case, data later revealed the out-of-control car actually resulted from driver error—not a mechanical problem (Smith et al. 1996).

Additionally, management should determine possible ways a product might fail. Grabowski and Herzberg (2007) stated that rigorous testing during product development stages and on the finished product are crucial (p. 14). Furthermore, conducting "…forced-failure tests" where researchers break parts on a product to see how it continues to operate may prove useful (p. 14).

Accurate records and documentation of product testing also become important. Having documentation such as the manufacturer responsible for the product or the batch the product came from can determine where a flaw occurred (p. 14). Management may want to train customer service associates to give special attention to complaints of defective merchandise (Smith et al. 1996, p. 83). In addition, creating a *feedback system* helps customer concerns reach the correct person (Grabowski and Herzberg 2007, p. 15).

In discussing recalls, the Merck Vioxx crisis becomes relevant. Merck's profitable drug posed cardiovascular risks to individuals taking it. While various studies documented these risks, Merck paid little attention to the concerns (Vlad et al. 2006, p. 365). After more studies, media coverage, and a class-action complaint, Merck issued a worldwide recall of Vioxx (p. 365). A study examined the communication strategies used by Merck following the recall. The study scrutinized company statements, news clips, and press conferences and found Merck typically relied on repairing strategies. Specifically, for the repairing strategies, the company expressed mortification, but not responsibility (p. 370). Researchers called this new strategy (where one expresses mortification over the crisis, but no responsibility), "rectification without assuming responsibility" (p. 370). Merck also attempted to ingratiate itself with consumers (a bolstering strategy) by stating it was concerned for their health (p. 370).

The research study also examined which communication strategies appeared most often in news coverage. Results from scanning print and broadcast stories revealed the media often reported the "rectification without assuming responsibility" strategy (p. 372). For example, stories centered on Merck's disbelief that one of its drugs resulted in health risks. The first week of the recall accounted for almost 20 % of the coverage within a 4-month period, indicating an overwhelming media interest (p. 373). Intense coverage continued for the four months studied. Researchers said a company must continue crisis management until media coverage lessens and the crisis resolves (p. 374). After an organization resolves its crisis, it enters the last stage of the crisis cycle: recovery.

1.5 Stage III: Postcrisis

The end of a crisis signifies a return to normal business procedures. Although the crisis appears over, the crisis team still fulfills four responsibilities. The responsibilities include communicating to key publics, cooperating with investigators, monitoring the crisis, and evaluating the response (Coombs 2007, p. 162).

1.5.1 Follow-Up Communication

For follow-up communication, the crisis team maintains relationships with key publics by answering lingering questions, explaining new information about the crisis, and explaining how the business plans to avoid a similar crisis in the future. If necessary, the team arranges assistance for employees (Coombs 2007, p. 162). For example, after witnessing workplace accidents or violence, employees may require psychological help (p. 136). To continue communication with key publics, the team uses the same channels employed during the crisis. Constant communication leads an organization back into its precrisis stage—where it used communication to establish positive relationships with the key publics.

1.5.2 *Investigation*

Cooperating with investigators benefits an organization in three ways. First, cooperation demonstrates to key publics that the organization practices openness and honesty. Second, cooperation may create a positive relationship between a company and the authorities conducting the investigation (Coombs 2007, p. 162). Finally, an investigation may reveal the true source or cause of the crisis.

Legal ramifications from a crisis may arise. The most common reasons for litigation include disgruntled employees, product recalls or faulty products, mergers and acquisitions, fraud, and white-collar crimes. An organization might institute a policy discouraging employees and board members from speaking with the media, but offer a prepared statement to ensure similar, appropriate messaging and to avoid the dreaded *no comment*. Heiby (2007) suggested a statement such as the following:

> We are aware of the situation and are currently investigating the circumstances, but we do not comment on ongoing litigation. Right now our thoughts are with the injured employee and his family. For further inquiries, please contact our PR director at… .(p. 43)

1.5.3 *Scanning and Monitoring*

In addition to cooperating with authorities, the crisis team continues to monitor the sources or factors that produced the crisis. For example, the team might scan posts on the company blog to assess consumer attitudes. The team should consider scanning both internal and external sources such as the company suggestion box, accident reports, trade publications, industry blogs, etc. Just like follow-up communication with key publics, scanning and monitoring the environment moves the company back into the precrisis stage (Coombs 2007, p. 163).

1.5.4 *Evaluation*

Finally, the team evaluates the response and the damage caused by the crisis. Ideally, the team evaluated its response to the crisis throughout the situation. Continuous evaluation allows the team to adjust messages and adapt as the situation changes.

Once the crisis enters the last stage of its life cycle, the team evaluates its actions through documents and the key publics' assessments (Coombs 2007, p. 152).

As mentioned in the precrisis section, the plan contains various documents such as a media log. Reviewing these items determines if the team responded promptly to journalists, if team members communicated the same messages, etc. Additionally, the crisis communication plan contains an evaluation form. The form includes questions about the notification system, strengths or weaknesses of the crisis response, strengths or weaknesses of the crisis team members, or other items considered relevant (p. 153). The form (or survey) can be used with internal or external key publics. Besides using a survey, the team could consider conducting focus groups or interviews with members of key publics. Analyzing the documents and forms shows what worked and where improvements need to be made for the future.

In addition to evaluating attempts to contain the crisis, crisis team members assess the actual damage. For example, they determine financial damage by comparing precrisis stock prices to the stock price after a crisis (p. 156). Determining the impact of a crisis on the organization's reputation proves more challenging to evaluate. As part of the reputation evaluation, the team measures *outputs* (paying special attention to the volume of media clips) and evaluates whether the media disseminated the business' message and the type of coverage received (Paine 2002, p. 7). Content analyzing media clips determine if the media framed the organization in positive or negative terms. If the media characterized the organization poorly, its reputation may suffer (Coombs 2007, p. 156).

Also, the team must consider how key publics viewed the media coverage. Feedback through polls or focus groups ascertains whether the key publics obtained messages through the media, believed the messages, or found them to be positive or negative (Paine 2002, para. 16). Sample questions for evaluating a company's reputation might include "how honest do you find organization X?" (Coombs 2007, p. 156). Evaluating the crisis response prepares an organization to survive a future crisis. The postcrisis stage detects and prepares for future crises, which segues back to the precrisis stage.

References

Anthonissen PF (ed) (2008) Crisis communication: practical PR strategies for reputation management and company survival. Kogan Page Limited, Philadelphia

Arpan L (2002) When in Rome? The effects of spokesperson ethnicity on audience evaluation of crisis communication. J Business Comm 39(3):314–339

Arpan LM, Pompper D (2003) Stormy-weather: testing "stealing thunder" as a crisis communication strategy to improve communication flow between organizations and journalists. Public Relat Rev 29:291–308

Augustine NR (1995) Managing the crisis you tried to prevent. Harvard business review. https://hbr.org/1995/11/managing-the-crisis-you-tried-to-prevent. Accessed 17 Jul 2015.

Bridgeman R (2008) Crisis communication and the net. In: Anthonissen PF (ed) Crisis communication: practical PR strategies for reputation management and company survival. Kogan Page, Philadelphia, pp 169–177

Coombs WT (2007) Ongoing crisis communication: planning, managing, and responding, 2nd edn. Sage Publications, Thousand Oaks

Coombs WT, Holladay SJ (2004) Reasoned action in crisis communication: an attribution theory-based approach to crisis management. In: Millar DP, Heath RL (eds) Responding to crisis: a rhetorical approach to crisis communication. Lawrence Erlbaum, Mahwah, pp 95–116

Cutlip SM, Center AH, Broom GM (2006) Effective public relations, 9th edn. Lawrence Erlbaum, Mahwah

Dix & Eaton (n.d.) Sago mine accident communications & counsel. http://www.dix-eaton.com/case_study_sago_feature_page.html. Accessed 18 Feb 2009

Doorley J, Garcia HF (2007) Rumor has it: understanding and managing rumors. Strategist 13(3):27–31

Entman RM (1993) Framing: toward clarification of a fractured paradigm. J Comm 43(4):51–58

Fearn-Banks K (2002) Crisis communications: a casebook approach, 2nd edn. Lawrence Erlbaum, Mahwah

Finch MR, Welker LS (2004) Informed organizational improvisation: a metaphor and method for understanding, anticipating, and performatively constructing the organization's precrisis environment. In: Millar DP, Heath RL (eds) Responding to crisis: a rhetorical approach to crisis communication. Lawrence Erlbaum, Mahwah, pp 189–200

Grabowski G, Herzberg JL (2007) Avoiding and managing product recalls. Risk Manage Mag 54(12):13–17

Heiby T (2007) Litigation and your board: assigning order to chaos. Strategist 13(3):42–43

Huyse K (2007) Red Cross uses social media tools to spread the word, deliver aid and raise money for California fires. Communication overtones. http://overtonecomm.blogspot.com/2007/10/red-cross-uses-social-media-tools-to.html. Accessed 23 Mar 2009

Hyde RC (2007) In crisis management, getting the message right is critical: considerations for effectively handling a crisis today. Strategist 13(3):32–35

Leighton N, Shelton T (2008) Proactive crisis communication planning. In: Anthonissen PF (ed) Crisis communication: practical PR strategies for reputation management and company survival. Kogan Page, Philadelphia, pp 24–43

Leighton N, Pellegrino S, Shelton T (2008) There is no substitute for media training. In: Anthonissen PF (ed) Crisis communication: practical PR strategies for reputation management and company survival. Kogan Page, Philadelphia, pp 132–141

Loomis L (2007) Crisis communications planning: the missing steps. PR tactics and strategist online. http://www.prsa.org/supportfiles/news/viewNews.cfm?pNewsID=972. Accessed 4 Jan 2009

Lukaszewski J (2007) Countering online aggression: surviving in a new media world where everybody is a journalist. PR Tactics 14(12):13

Marra FJ (2004) Excellent crisis communication: beyond crisis plans. In: Millar DP, Heath RL (eds) Responding to crisis: a rhetorical approach to crisis communication. Lawrence Erlbaum, Mahwah, pp 311–326

Martin D, Montgomery J (2006) Hospital apology could go a long way. Sorry works online. http://www.sorryworks.net/media51.phtml. Accessed 19 Feb 2009

Millar DP, Heath RL (eds) (2004) Responding to crisis: a rhetorical approach to crisis communication. Lawrence Erlbaum, Mahwah

Paine K (2002) How to measure your results in a crisis. The Institute for Public Relations, Gainesville

Pfau M, Wan H (2006) Persuasion: an intrinsic function of public relations. In: Botan CH, Hazelton V (eds) Public relations theory II. Lawrence Erlbaum, Mahwah, pp 101–136

Polaniecki R (2008) Crisis communication. Credit Un Mgmt 31(12):38–40

Reese SD, Gandy OH Jr, Grant AE (eds) (2003) Framing public life: perspectives on media and our understanding of the social world. Erlbaum, Mahwah

Smith NC, Thomas RJ, Quelch J (1996) A strategic approach to managing product recalls. Harv Bus Rev 74(4):102–113.

Ulmer R, Sellnow T, Seeger M (2007) Effective crisis communication: moving from crisis to opportunity. Sage Publications, Thousand Oaks

Vlad I, Sallot LM, Reber BH (2006) Rectification without assuming responsibility: testing the transgression flowchart with the Vioxx recall. J Public Relat Res 18(4):357–379

Wilcox DL (2005) Public relations writing and media techniques, 5th edn. Pearson Education, Boston

Wojciesak D, Saxton JW, Finkelstein M (2008) Sorry works! Disclosure, apology, and relationships prevent medical malpractice claims. AuthorHouse, Bloomington

Part I
Facing Crisis: The Role of Communication in Fostering Resilience or Fomenting Resistance

Chapter 2
Crisis and Risk Communications: Best Practices Revisited in an Age of Social Media

Cory Young, Aditi Rao, and Alexis Rosamilia

Abstract Over the last decade, scholars and practitioners have collaborated to codify a set of guiding principles, or best practices, within crisis communication "in an effort to improve quality and efficiency, inform practices, and … improve performance" (Seeger, J Appl Commun Res 34(3):232–244, 2006). Several factors, however, complicate the process by which these practices are determined, namely, the unpredictable nature of crises, organizational structures and cultures, divergent crisis communication goals (reduction of risk, image repair, informing multiple publics, etc.), and advances in technology. Social media and their vast array of networks have created multiple primary, secondary, and even tertiary stakeholder groups with competing and conflicting information needs, requiring organizations to create new practices of responsiveness. Coleman (J Brand Strategy 2(2):129–133, 2013) advocates for organizations to revise and update their best practices "in quieter times," based on "what is known to work" now in the digital age (p. 233). Therefore, the purpose of this chapter is to review best practices in crisis communication and to expand and extend the conversation to include social media.

Keywords Crisis • Risk • Social media • Best practices • Disasters

2.1 Introduction

The concept of *best practices* is a professional term associated with benchmarking and improving the effectiveness of organizational processes and practices. Within the field of communication, many scholars and practitioners have created varied

C. Young (✉)
Ithaca College, Ithaca, NY, USA
e-mail: young@ithaca.edu

A. Rao
MS in Communications, Ithaca College, Ithaca, New York, USA

A. Rosamilia
LandDesign, Charlotte, North Carolina

J.L. Drake et al. (eds.), *Communicating Climate-Change and Natural Hazard Risk and Cultivating Resilience*, Advances in Natural and Technological Hazards Research 45, DOI 10.1007/978-3-319-20161-0_2

lists of best practices in crisis and risk communication (Heath 2006; Korbas and Reich 2010; Reynolds 2006; Ropeik 2006; Sandman 2006; Seeger 2006; Veil and Husted 2012; Venette 2006). Best practice lists related to social media have also been published (FedEx and Ketchum 2011; Hobbs 2009; Hornick 2011; Mckintosh 2012; Moran and Gossieaux 2010).

These lists, however, have been primarily created in a vacuum, focusing on one variable of crisis or risk communication or social media; for example, health. Covello's (2003) article in the *Journal of Health Communication* presented a bulleted checklist of seven guidelines that "should be included in any public health risk and crisis communication plan" (p. 5). Gibbons et al. (2011) explored the best practices of social media (Web 2.0) to address health disparities. Other variables in which best practices related to crisis and risk communication or social media have been explored include journalism/mass media, environmental communication, and corporate and organizational communication (internal/external), among others. Yet, very few academic articles link the best practices in crisis and risk communication with social media. The purpose of this chapter then is to review articles on best practices in crisis and risk communication in general and on best practices in crisis and risk communication that integrate social media. Since many of the articles spring from a special issue of the *Journal of Applied Communication Research*, we'll begin here.

2.2 Best Practices in Crisis and Risk Communication

In 2006, the *Journal of Applied Communication Research* commissioned a special section on best practices in risk and crisis communication. This special issue, introduced by Dr. Steven Venette, was prompted by a collaborative effort between experts in the sciences, communication scholars, and the National Center for Food Protection and Defense (NCFPD), "to study risk and crisis communication with the goals of increasing the effectiveness of messages associated with foodborne hazards and training potential spokespeople to use developed communication strategies effectively" (p. 229). This organization was established by the Department of Homeland Security, in response to potential terrorist threats to the nation's systems of food production and distribution. The research revealed that well-intentioned competent spokespeople, when constructing and delivering risk and crisis messages, made "consistent, predictable errors when messages [were] not planned and when communication networks [were] not pre-established," especially when "pressure to present accurate, timely information [was] high" (p. 230). In response to this, the NCFPD created this list to codify best practices for risk and crisis communication:

1. Process approaches and policy approaches
2. Pre-event planning
3. Partnerships with the public
4. Listen to the public's concerns and understand the audience
5. Honesty and openness
6. Collaborate and coordinate with credible sources

7. Meet the needs of the media and remain accessible
8. Communicate with compassion, concern, and empathy
9. Accept uncertainty and ambiguity
10. Messages of self-efficacy

Venette invited seasoned crisis and risk communication scholars to respond to the list, thus sparking an important conversation within which scholars and practitioners could add their own contributions, beginning with Matthew Seeger, a communication scholar involved in the research. Seeger (2006) described each of the 10 best practices in more depth, detailing the interconnectedness "cautiously with a firm understanding of conceptual factors and situational variables," (p. 233) and the inherent ambiguity and inevitable uncertainty of crisis events (#9). Regarding the process approaches and policy development (#1), these aspects are tightly coupled and therefore should be integrated in all phases of a crisis and in all decision-making processes, especially pre-event planning (#2). Preventing crises is easier when organizations partner with their publics dialogically (#3), treating them "as a resource, rather than a burden" (p. 238) in creating messages of self-efficacy (#10). Dialoging with audiences necessitates listening for and understanding their concerns (#4) by managing issues and monitoring public opinions while "adapting messages to the public's dynamic needs and … concerns" (p. 239), in an honest, candid, open, compassionate, and empathic manner (#5 and #8).

In addition to partnering with the public, organizations should collaborate and coordinate messages with credible subject matter experts, agencies, networks, and stakeholders (#6) and engage the media proactively (#7) to maintain message consistency and reduce confusion and uncertainty when crises occur. For Seeger, these best practices represent anchors that individuals use to practice and improve efficiency and "are the principles or processes that underlie an effective crisis communication plan and effective crisis response" (p. 242).

Heath (2006) continues the conversation, agreeing with much of what Seeger advocates but adding a rhetorical twist—that of a narrative approach to crisis communication. Stories are vital for an organization's reputation and must be told with openness, honesty, and empathy, before media professionals frame their own narrative about a crisis event. Organizations should "work with others, including reporters and interested publics, to tell a story that exhibits factual accuracy and narrative coherence" (p. 248).

Sandman (2006) follows Heath with "some quibbles and additions" related to risk communication (p. 257). He points out that what is missing in the original list of ten best practices is advance communication and planning. Audience members are fearful and need the organization to "help them harness these fears in the form of appropriate precautionary action" and to help individuals not only act but actively participate in deciding the precautionary measures (p. 259). This augments Sandman's observations of a second gap in the list, that of trusting the public:

> To trust that most people are resilient and can bear dire warnings, awful events, and unpleasant truths; to trust that they will want to do the right thing … to trust that their ideas about how to best cope with crisis are likely to be worth hearing, worth implementing, and worth letting them implement. (p. 260)

Lastly, Sandman points out contradictions between some of the practices on the list, namely, that trying to have one coordinated voice for the organization's story is "overrated" and implies that "people can't handle expert disagreements" (p. 260–61). A common thread across the perspectives presented in this forum is the nature of communication with traditional media and with primary stakeholders during crises. A crucial fiber of this thread that is missing, however, is social media. This is not surprising given the fact that many social media platforms had not yet been created:

- Facebook—launched in 2004.
- YouTube—born in 2005.
- Twitter (2006)—the year that the special issue devoted to best practices in crisis and risk communication was published
- Instagram (2010)

The landscape of crisis communication has been transformed over the last two decades with the introduction of the Internet and social media platforms, generating 24/7 access to news and mobile/geo-located communication technologies, creating citizen journalists, and altering the speed at which practitioners respond and enact best practices in crisis and risk communication (Leavesley et al. n.d.).

González-Herrero and Smith's (2008) article on how the Internet changed the nature of public relations during crisis revealed the Internet as a double-edged sword: either it is a potential trigger of crises or the medium by which public relations practitioners can facilitate communication with stakeholders during a crisis. Either way, the integration of the Internet has impacted the way organizations manage crises: "on the one hand, existing plans need to be adapted to reflect how online communications could be used to prevent a conflicting scenario … on the other, organizations need to consider new scenarios" (p. 152) that emerge as companies integrate social media into their plans.

Veil and Husted (2012) argue that organizations should use existing best practices as an assessment tool, postcrisis, "to question whether their planning was sufficient and their strategies and responses met the needs of stakeholders" (p. 131). To this end, the next section discusses the integration of social media in crisis and risk communication and identifies complementary best practices.

2.3 Best Practices Integrating Social Media in Crisis and Risk Communication

Veil et al. (2011) acknowledge in their article "A work-in-process literature review: Incorporating social media in risk and crisis communication" published in the *Journal of Contingencies and Crisis Management* that the purpose is to "marry literature and examples of social media use with best practices in risk and crisis communication to demonstrate how communicators can embrace social media tools to better manage a risk or a crisis" (p. 111). To this end, the authors review and critique

Table 2.1 Summary of Veil et al's review and critique. Best practices for communicating crisis via social media

#	Social media strategy identified by Seeger (2006)
1	Integrate social media in risk and crisis communications
2	*Establish risk and crisis management policies* regarding the use of social media to encourage engagement with stakeholders, mitigate damages, and educate stakeholders about risks to reduce errors and confusion
3	*Preplan daily activities* and logistics such as uploading crisis documents (contacts, manuals, and statements), identifying technical requirements for multimedia informatics and graphics, monitoring the blogosphere, and evaluating the effectiveness of the usage of social media
4	Involve stakeholders and the media in decision making, by *listening* and understanding where people are getting their information. This requires using data gathering and analysis tools that integrate multimodal data sources "including on-site eyewitness video, interactivity on multiple social media sites, conventional and alternative news updates, and official notifications—into a coherent, time-based, replayable" environment (Palen et al. 2009, p. 12)
5	Use social media to *be honest, candid, and open* with information while acknowledging risk. Additionally, humanize communications with compassion, concern, empathy, and sensitivity
6	*Collaborate and coordinate* with credible sources. This requires an assessment of each medium's credibility and sharing messages through cross posting and re-tweeting
7	Use social media to meet and *sustain the informational needs of mainstream media*, as the "media [sic] is already using social media" (p. 119)
8	Accept uncertainty and ambiguity in the process of gathering and updating information
9	Provide messages of *self-efficacy—allowing active participation in social sites* to reduce the emotional uncertainty experienced by stakeholders
10	Conduct *demographic metrics* (e.g., age, ethnicity, socioeconomics)

the best practices identified in the special issue of the *Journal of Applied Communication Research* and summarized in Table 2.1, which are to integrate social media, establish policies, preplan activities, listen, be open, collaborate, sustain mainstream media, accept ambiguity, and invite participation. The authors conclude their article with an additional best practice—*conduct demographic metrics* on social media users (age, ethnicity, and literacy/socioeconomics) to account for cultural differences and adjust messages accordingly. The authors further acknowledge some potential negative outcomes of integrating social media while reminding us that reaching key publics during a crisis requires practitioners to use their audiences' media, whether tradition or social. In essence, practitioners need to become *bilingual* in their media and technological approaches to risk/crisis communication to improve effectiveness of messages.

Following Veil et al. (2011), Janoske et al. (2013), and Wendling et al. (2013) provide additional reports of best practices, drawing on large-scale disaster literature. Janoske et al. present the results from interviews and documents from a two-day workshop experts and practitioners attended to discuss past and future best practices. Their report confirmed many of the original best practices explained by Seeger (2006). The key tasks determined were identifying organizational constraints such as management mindset, staffing, time, and budgets and extending listening to postcrisis/recovery activities. In addition, conducting more research on "message templates, message reception and social media's ability to effectively communicate in a crisis" (p. 234); creating more opportunities for research and practitioners to "translate" each other's practices, by discussing case studies, "worst" and "realistic" practices; and challenging the assumption that best practices offer "a static, one-size-fits all approach" (p. 235) were noted as ways to improve the practices during disaster communication.

Wendling et al. (2013) also seek to challenge the one-size-fits-all approach by compiling strategies and practices governments and other emergency service agencies have generated using new media platforms within large-scale disasters. The reach, immediacy, impact, and success of social media in gathering and dissipating information during high-end disaster scenarios have led to emergency managers, researchers, and crisis relief organizations integrating social media within their traditional disaster crisis communication plans. New technologies increase the ability for disaster relief and response organizations to develop best practices around warning and alert systems, establishing backup modes of communication; coordinating feedback efforts with multiple stakeholders; and reducing costs, improving transparency of decision-making processes, managing perceptions, and correcting inaccurate information. Useful to those in the field are additional readings related to crisis communication and social media, as well as additional resources for best practices in crisis and disaster communication and social media (Additional readings[1,2,3,4,5,6,7,8,9,10,11,12,13,14,15,16,17,18,19,20,21,22,23,24,25,26,27,28,29,30,31,32,33,34,35,36,37,38,39,40,41,42] and Resources on social media and disaster communication[43,44,45,46,47,48,49,50,51,52]).

---

[1] 8 steps to planning for a social-media crisis (infographic). Retrieved from http://www.entreprenuer.com/article/227996. Accessed 22 Aug 2013.

[2] Baron G, Huyse KW (2009) 10 steps of crisis planning in a new media environment. [PowerPoint presentation]. Retrieved online from the Public Relations Society of America http://www.slideshare.net/kamichat/10-steps-integrating-social-media-into-crisis-plans

[3] Bell LM (2010, April) Crisis communication: the praxis of response. The Review of Communication. 10(2): 142–155. doi:10.1080/15358591003653577

[4] Bethard-Caplick D (2012, February 10) Beyond integrated MarComms: the next frontier for digital PR. Retrieved from the Public Relations Society of America.

[5] Borremans P (2010, July/August) Ready for anything: support or enhance your crisis communication plan with social media. Communication World, 31–33.

[6] Bottles K, Sherlock T (2011, March-April) Who should manage your social media strategy? PEJ, 68–72.

[7] Coombs T (2008) Crisis communication and social media. Retrieved from http://www.instituteforpr.org/crisis-communication-and-social-media/

[8] Crandall WR, Parnell JA and Spillan JE (2014) Crisis management: leading in the new strategy landscape, 2nd edn. Sage Publications, Thousand Oaks, CA.

[9] Daniels J (2010, November 11) Social media in crisis communication: a lesson learned by Charlotte Mecklenburg Library. Retrieved from the Nonprofit Technology Network website http://www.nten.org/blog/2010/11/11/social-media-crisis-communication-lessonlearned-charlotte-mecklenburg-library

2.4 Conclusion

Social media have already begun to replace traditional forms of media, making a journalist or reporter out of any individual with a smartphone or tablet. Incorporation of social media into crisis and disaster communications is inevitable. Social media

---

[10] Dozescu A (n.d.) Are you ready for a social media crisis? Retrieved from http://www.advanced-webranking.com/blog/social-media-crisis-management/

[11] Fearn-Banks K (2011) Crisis communications: a casebook approach, 4th edn. Lawrence Erlbaum, Mahwah, NJ.

[12] Ferguson SD (1999) Communication planning: an integrated approach. Sage Series in Public Relations. Sage Publications, Thousand Oaks, CA.

[13] Freberg K (2011) Assess, identify, integrate: how to add social media to your crisis communication plan. Vimeo presentation retrieved from http://imc.wvu.edu/community/imc_knowledge_base/social_media_and_crisis_communication

[14] Freberg K (2011, July 1) Geolocation & crisis communication: integration of crowdsourcing with social media. Retrieved from http://karenfreberg.com/blog/geolocation-crisis-communication-integration-of-crowdsourcing-with-social-media/

[15] Freberg K (2011, August 8) Is crisis + the future?:implementing Google + in crisis communications (benefits & challenges). Retrieved from the ComPRehension website at http://comprehension.prsa.org/?p=3280

[16] González-Herrero A, Pratt CB (1996) An integrated symmetrical model for crisis- communications management. Journal of Public Relations Research 8(2):79–105.

[17] Hennes Paynter Communications (2013) Creating plans. Retrieved from http://www.crisiscommunications.com/services/crisis-preparation/creating-plans/

[18] Hennes Paynter Communications (2013) Crisis response, crisis plans. Retrieved from http://www.crisiscommunications.com/services/crisis-preparation/creating-plans/

[19] Holmes W (n.d.) Crisis communications and social media: advantages, disadvantages, and best practices. [.pdf]. Unpublished paper retrieved from http://trace.tennessee.edu/cgi/viewcontent.cgi?article=1003&context=ccisymposium

[20] Ipsen M, Jönsson M (2013, May 27) Social media communication during a corporate brand crisis: the case of Fundus. [Master's Thesis].

[21] Jaques T (2007) Issue management and crisis management: an integrated, non-linear, relational construct. Public Relations Review 33: 147–157. doi: 10.1016/j.pubrev.2007.02.001.

[22] Jaques T (2009) Issue and crisis management: quicksand in the definitional landscape. Public Relations Review 35: 280–286. doi:10.1016/j.pubrev.2009.03.003.

[23] Johansen W, Aggerholm HK, Frandsen F (2012). Entering new territory: a study of internal crisis management and crisis communication in organizations. Public Relations Review 38: 270–279. doi:10.1016/j.pubrev.2011.11.008.

[24] Jordan-Meier J (2011) The four stages of highly effective crisis management: how to manage the media in the digital age. CRC Press, Boca Raton, FL.

[25] Keating M (2013) Integrated crisis management defined. Retrieved from PRWeb at http://www.prweb.com/releases/2004/05/prweb127847.htm.

[26] Kleinberg S (2013, July 25) How to handle a social media crisis: it may see like it, but it's not the end of the world—tips to clean up the mess and get back to normal. Tribune Newspapers. Retrieved from http://articles.chicagotribune.com/2013-07-25/features/ct-tribu-social-media-crisis-tips-20130725_1_auto-posts-six-tips-deep-breath

have the potential to make crisis communications more efficient. The advantages of using social media during natural disasters far supersedes its shortcomings, such as causing unnecessary panic and giving rise to rumors, which often seem to be the reasons behind organizations refraining from social media use. No other platform can match social media in terms of reaching large audiences globally within seconds, and for this reason alone, social media should be incorporated into disaster

[27] Korbas-Magal D, Reich Z (2010, June 22). The 'best practices' in risk and crisis communication: past, present, and future. Paper presented at the annual meeting of the International Communication Association, Suntec Singapore International Convention & Exhibition Centre, Suntec City, Singapore Online<PDF>. Retrieved from http://citation.allacademic.com/meta/p402804_index.html

[28] O'Neill S (n.d.) A social media crisis plan—Be prepared (to save your ass). Retrieved from http://firebrand.net/blog/social-media/social-media-crisis-plan/.

[29] Pang A, Jin Y, Cameron GT (n.d.) Final stage development of the integrated crisis mapping (ICM) model in crisis communication: the myth of low engagement in crisis. Retrieved from http://www.instituteforpr.org/integrated-crisis-mapping-in-crisis/

[30] Pang A, Nasrath B, Chong A (2014). Negotiating crisis in the social media environment: evolution of crises online, gaining credibility offline. Corporate Communications: An International Journal 19(1):96–118. doi: http://www.emeraldinsight.com/doi/abs/10.1108/CCIJ-09-2012-0064

[31] Perry DC, Taylor M, Doerfel ML (2003). Internet-based communication in crisis management. Management Communication Quarterly 17(2):206–232. doi: 10.1177/0893318903256227.

[32] Pratt CB, Bloom E (1997, October) Integrated communication campaigns for organizational crisis management in South Africa: implications for challenges today—and in the dawn of a new millennium. International Communication Gazette 59 311–329. doi: 10.1177/001654 9297059004005.

[33] Semple E (2009, August/September) Update your crisis comm plan with social media. SCM 13(5):7.

[34] Slabbert Y, Barker R (2011, December) An integrated crisis communication framework for strategic crisis communication with the media: a case study on a financial services provider. Communicatio: South African Journal for Communication Theory and Research 37(3):443–465. doi: 10.1080/02500167.2011.609996.

[35] Sturges DL (1994, February) Communicating through crisis: a strategy for organizational survival. Management Communication Quarterly 7(3):297–316. doi:10.1177/0893318994007003004.

[36] Taylor M, Perry DC (2005, February) Diffusion of traditional and new media tactics in crisis communication. Public Relations Review 31 209–217. doi: 10.1016/j.pubrev.2005.02.018.

[37] Thiessen A, Ingenhoff D (2011) Safeguarding reputation through strategic, integrated and situational crisis communication management: development of the integrative model of crisis communication. Corporate Communications: An International Journal 16(1):8–26. doi: 10.1108/13563281111100944.

[38] University of Wisconsin Oshkosh (2013) Integrated crisis communications. Retrieved from http://www.uwosh.edu/imc/media-relations/crisis-communications/

[39] Walaski P (2013, April) Social media: powerful tools for SH&E professionals. Professional Safety, 40–49.

crisis communications. That being said, technology will continue to evolve and organizations will have to adapt accordingly for maximum efficiency of their crisis communications plan as the inventory of crisis events is ever expanding.

Acknowledgments A version of this paper was presented at the 2013 National Communication Association annual convention, Washington, DC, November 21–24.

[40] Warner B (2011, October 27). Infographic: social media crisis communications decision tree. [Weblog]. Posted on SMI—Intelligence in Social Media. Retrieved from http://socialmediainfluence.com/2011/10/27/infographic-the-social-media-crisis-%20communications-decision-tree

[41] Weber Shandwick (2009) Inline communications report. Retrieved from http://webershandwick.co.uk/filelibrary/Inline-Report.pdf.

[42] Young C, Flowers A, Ren NZ (2011) Technology and crisis communication: emerging themes from a pilot study of U.S. public relations practitioners. Prism 8(1). Available online at http://www.prismjournal.org/fileadmin/8_1/Young_Flowers_Ren.pdf

[43] Alexander D (2014) Social media in disaster risk reduction and crisis management. Scientific and Engineering Ethics 20(3):717–33. doi: 10.1007/s11948-013-9502-z

[44] Bruns A, Liang YE (2012, April) Tools and methods for capturing Twitter data during natural disasters. First Monday, 17(4). doi: 10.5210%2Ffm.v17i4.3937.

[45] Dabner N (2012) 'Breaking ground' in the use of social media: a case study of a university earthquake response to inform educational design with Facebook. Internet and Higher Education 15 69–78. doi: 10.1016/j.iheduc.2011.06.001.

[46] Donahue A, Tuohy R (2006, July) Lessons we didn't learn: a study of the lessons of disasters, why we repeat them, and how we can learn from them. Homeland Security Affairs: The Journal of the Naval Postgraduate School Center for Homeland Defense and Security 2(2). Retrieved from http://www.hsaj.org/?fullarticle=2.2.4

[47] Federal Emergency Management Agency (2012). Social media for natural disaster response and recovery. Participant Handouts. Retrieved from https://www.firstrespondertraining.gov/catalog.do?a=nted

[48] Goldfine E (2011) Best practices: the use of social media throughout emergency & disaster relief. Retrieved from https://www.american.edu/soc/communication/upload/Erica-Goldfine.pdf

[49] Lindsay B.R (2011, September 6) Social media and disasters: current uses, future options, and policy considerations. Retrieved from http://fas.org/sgp/crs/homesec/R41987.pdf

[50] Merchant RM, Elmer S, Lurie N (2011, July 28) Integrating social media into emergency preparedness efforts. New England Journal of Medicine 365(4):289–291.

[51] Resnyansky L (2014, March 10) Social media, disaster studies, and human communication. IEEE Technology and Society Magazine 33(1):54–65. doi: 10.1109/MTS.2014.2301857.

[52] White CM (2012) Social media, crisis communication, and emergency management: leveraging Web 2.0 technologies. CRC Press, Boca Raton.

References

Coleman A (2013) Managing a crisis in the era of social communication: how Greater Manchester Police is developing community engagement and communication. J Brand Strategy 2(2):129–133

Covello VT (2003) Best practices in public health risk and crisis communication. J Health Commun: International Perspectives 8(1):5–8. doi:10.1080/10810730390224802

FedEx and Ketchum benchmark best practices in social media (2011) Strateg Commun Manag 15(1):9

Heath RL (2006) Best practices in crisis communication: evolution of practices through research. J Appl Commun Res 34(3):245–248. doi:10.1080/00909880600771577

Hornick L (2011) Summit sheds light on social media best practices. Comm World 28(5):8–9

Gibbons CM, Fleisher L, Slamon RE, Bass S, Kandadai V, Beck JR (2011) Exploring the potential of Web 2.0 to address health disparities. J Health Commun 16(1):77–89. doi:10.1080/10810730.2011.596916

González-Herrero A, Smith S (2008) Crisis communications management on the Web: how Internet-based technologies are changing the way public relations professionals handle business crises. J Conting Crisis Manag 16(3):143–153

Hobbs R (2009) Media literacy gets fair use: the best practices model in teaching. Paper presented at the International Communication Association annual meeting, Chicago, 21–25 May 2009

Janoske ML, Liu BF, Madden S (2013) Congress report: experts' recommendations on enacting best practices in risk and crisis communication. J Conting Crisis Manag 21(4):231–235. doi:10.1111/1468-5973.12031

Korbas Magal D, Reich Z (2010) The 'best practices' in risk and crisis communication: past, present and future. Paper presented at the International Communication Association annual meeting, Singapore, 22–26 Jun 2010

Leavesley J, Brannan C, Hemus J (n.d.) Braving the social media crisis. White paper produced by Insignia—Reputation Management and Communications Consultancy. Downloadable from http://insigniacomms.com/resources_w_papers/braving-the-social-media-crisis/

Mckintosh A (2012) Empowering employees to social ambassadors at Pepsi Co. Strateg Commun Manag 16(5):26–29

Moran E, Gossieaux F (2010) Marketing in a hyper-social world. J Advert Res 50(3):232–239. doi:10.2501/S0021849910091397

Palen L, Vieweg S, Liu SB, Hughes AL (2009) Crisis in a networked world: features of communication in the April 16, 2007 Virginia Tech event. Soc Sci Comput Rev 000(00):1–14. doi:10.1177/08944393093323

Reynolds B (2006) Response to best practices. J Appl Commun Res 34(3):249–252

Ropeik D (2006) Best practices response. J Appl Commun Res 34(3):253–256. doi:10.1080/00909880600771601

Sandman PM (2006) Crisis communication best practices: some quibbles and additions. J Appl Commun Res 34(3):257–262. doi:10.1080/00909880600771619

Seeger MW (2006) Best practices in crisis communication: an expert panel process. J Appl Commun Res 34(3):232–244. doi:10.1080/00909880600769944

Veil SR, Buehner T, Palenchar MJ (2011) A work-in-process literature review: incorporating social media in risk and crisis communication. J Conting Crisis Manag 19(2):110–122. doi:10.1111/J.1468-5973.2011.00639.x

Veil SR, Husted RA (2012) Best practices as an assessment for crisis communication. J of Commun Manag 16(2):131–145. doi:10.1108/13632541211217560

Venette SJ (2006) Special section introduction: best practices in risk and crisis communication. J Appl Commun Res 34(3):229–231. doi:10.1080/00909880600769464

Wendling C, Radisch J, Jacobzone S (2013) working paper, The use of social media in risk and crisis communication, OECD working papers on Public Governance, No. 24, OECD Publishing, Paris. doi:http://dx.doi.org/10.1787/5k3v01fskp9s-en

Chapter 3
Polluted Discourse: Communication and Myths in a Climate of Denial

Peter H. Jacobs, Ari Jokimäki, Ken Rice, Sarah A. Green, and Bärbel Winkler

Abstract Human activities, principally the burning of fossil fuels, are changing the climate. Despite widespread scientific consensus on this fact, communicating the risks posed by climate change to the public remains challenging. We examine the role of contrarian narratives in climate communication, focusing on two terminological claims—(1) that scientists abandoned the term *global warming* in favor of *climate change* in response to a change in temperature evolution, and (2) that *catastrophic anthropogenic global warming* is the mainstream scientific position—and find them to be without merit. We discuss how scientists and communicators can neutralize these myths while informing the public. Finally, we summarize the existing literature on word choice in climate communications and suggest best practices based on target audiences.

Keywords Climate change • Global warming • Terminology • Science communication

P.H. Jacobs, B.A. (✉)
Department of Environmental Science and Policy, George Mason University, MS 5F2, 4400 University Drive, Fairfax 22030, VA, USA
e-mail: Pjacobs4@gmu.edu

A. Jokimäki, B.Sc. • B. Winkler, B.A.
Skeptical Science, Brisbane, Australia

K. Rice, Ph.D.
Institute for Astronomy, School of Physics and Astronomy, The University of Edinburgh, Edinburgh, UK

S.A. Green, Ph.D.
Department of Chemistry, Michigan Technological University, Houghton, MI, USA

J.L. Drake et al. (eds.), *Communicating Climate-Change and Natural Hazard Risk and Cultivating Resilience*, Advances in Natural and Technological Hazards Research 45, DOI 10.1007/978-3-319-20161-0_3

3.1 Introduction

3.1.1 Climate Communication

The scientific community overwhelmingly agrees that humans are changing the climate (Anderegg et al. 2010; Cook et al. 2013; Doran and Zimmerman 2009; Oreskes 2004). However, significant obstacles to effective public understanding of the problem currently prevent mitigating and adaptive actions (Gifford 2011). There has been a great deal of interest in the role of denial or contrarianism as one such obstacle, be it the role of industry-funded *think tanks*, which serve as well-moneyed front groups for anti-regulatory campaigns and narratives (Brulle 2014; Dunlap and Jacques 2013; Oreskes and Conway 2010), or the dimensions and causes of the partisan divide in public opinion (Dunlap and McCright 2008; Guber 2013; Malka et al. 2009).

We focus on the role of contrarian myths, which can be both a contributing source to and an amplifier of this partisan divide and the scientist-public disconnect. By their very nature, contrarian myths often bypass the normal societal filters against misinformation, such as journalistic fact checking or academic peer review (Elsasser and Dunlap 2013). While addressing the underlying drivers of climate denialism is invaluable, addressing myths directly is worthwhile in its own right. Rebutting myths may not itself be sufficient to cause committed partisans to change their position, but it has the potential to prevent propagation of myths to those who are undecided or disengaged. We are hopeful that by reducing the persuasive efficacy of myths, rebuttals can lead to a reduction in their use. Effective refutation of these myths may necessitate multichannel rebuttals, including rebuttals in scholarly, mainstream, and social media (Cook et al. 2014). Although social media outlets (e.g., SkepticalScience.com) have begun to address the problem, journalistic and scholarly refutations remain the exception (e.g., Peterson et al. 2008) rather than the norm.

3.2 The Terminology of Environmental Change

A potential source of misconception and misinformation in communicating environmental risk is the often overlapping, but not precisely equivalent, meanings of different terms used to discuss aspects of environmental change. *Global warming*, *climate change*, and *global environmental change* can all refer to the present human-caused warming of the planet, though each can also refer to specific aspects of environmental change that the others may not encompass (see Fig. 3.1). Thus, it is possible to have climate change that is neither global nor warming, global warming that is not anthropogenic, or anthropogenic global environmental change that is neither warming nor climatic. Scientists and communicators who wish to reach lay audiences should be mindful that terminology, especially terminology with technical as well as general meanings, can confuse instead of clarify (Somerville and Hassol 2011).

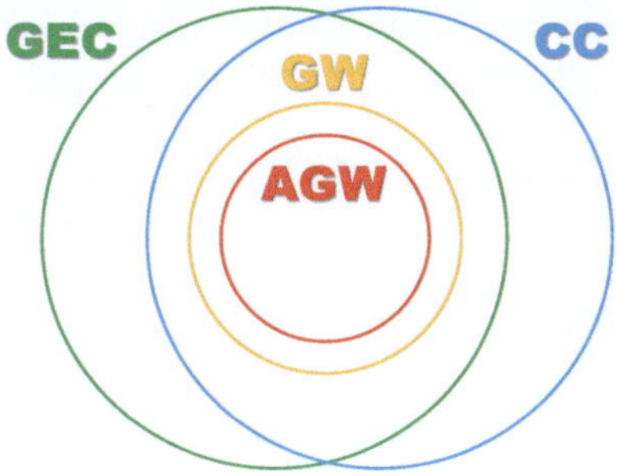

Fig. 3.1 Euler diagram of overlapping terms for environmental change. Overlapping but noninterchangeable terms for environmental change: *GEC* global environmental change, *CC* climate change, *GW* global warming, *AGW* anthropogenic global warming

Global Warming *Global warming* (GW) is a relatively narrowly circumscribed term, typically referring to a sustained increase in the mean surface temperature of a planet. GW can refer to changes caused by human activities, such as in the present warming of the planet through our increases in greenhouse gases and other changes in radiative forcings. GW may also occur in response to natural changes in radiative forcings, such as an increase in solar activity (Lockwood 2012) or an increase in greenhouse gases due to volcanism (as during the Permian-Triassic mass extinction; Cui and Kump 2014; Joachimski et al. 2012). Anthropogenic global warming (AGW) refers specifically to human-caused warming and is therefore a subset of GW. Although the phrase GW is primarily used in reference to Earth, it is also applicable to other planetary objects (e.g., Fenton et al. 2007).

Climate Change *Climate change* (CC) is a more encompassing term and may refer spatially to national or smaller scales (e.g., Salnikov et al. 2014; Coulson et al. 1993), or to scales as large as planets (e.g., Solomon et al. 1999). Unlike GW, CC need not refer only to an increase in the mean temperature of an area, nor necessarily to a change in any average at all; CC can refer to changes in the statistical properties of a wide range of climatological features or processes, such as the prevalence of extreme events (Meehl et al. 2000). In addition to changes in temperatures, CC can also refer to changes in the mean, extreme, or spatial distribution of precipitation (Wigley and Jones 1985), changes to the persistence and distribution of droughts (Shaw 2003; Touchan et al. 2011), changes in the frequency and intensity of tropical cyclones (e.g., Elsner et al. 2008), or other changes to the typical properties of climate.

Global Environmental Change An overlapping term is *global change* or *global environmental change* (GEC). While GEC can refer to global warming or climate change, it is also frequently, and sometimes preferentially, used to describe global-scale environmental problems unrelated to climate change, such as habitat fragmentation, invasive species, biodiversity loss, or freshwater usage and contamination (Stern et al. 1992; Tilman et al. 2001; World Health Organization 2005). GEC can also encompass geochemical changes resulting from the increase of greenhouse gases such as ocean acidification (Beman et al. 2011) or other industrial emissions such as acid rain or stratospheric ozone depletion (Mazur 1998; The Social Learning Group 2001) that are not strictly considered climatic changes in and of themselves.

3.2.1 Myths and Implications

Several contrarian myths have arisen around the terminology of human-driven climate change. Here, we focus on two: (1) that GW was the preferred terminology of scientists, but was recently abandoned in favor of CC, and (2) that *catastrophic anthropogenic global warming* (CAGW) is the mainstream prediction of the scientific community and thus forms the justification for policy. Both of these myths implicitly call into question the credibility of the scientific community and those who convey its results. The GW vs. CC myth portrays scientists as both incompetent (incorrect in their predictions of temperature evolution) and dishonest (engaging in revisionism by "moving the goalposts" in order to encompass previously unanticipated events). The CAGW myth portrays the mainstream as doomsayers who are constantly rebuffed by study after study within climate scientists' own field.

3.2.2 Global Warming Versus Climate Change Versus Temperature

Both CC and GW have been in use by the scientific community for many decades. Usage of CC can be traced back at least to the 1920s (Joffe 1929; Willis 1925), and the similar term *climatic change* can be traced back at least to the 1850s (Anderson 1857; Mayer 1856). The first use of the term GW has been popularly attributed to Broecker (1975), but we found that it had been used more than ten years earlier. Although Broecker was one of the first to use it in the context of the current warming, Mitchell (1961) used GW in a similar context.

Yet a frequent contrarian claim is that the term GW had been the preferred terminology, but that it was recently abandoned in favor of CC. This myth is often made in conjunction with the allegation that global warming has stopped, or that it is presently cooling as in the following example:

> Human-caused global warming' has now officially been re-named 'climate change' to explain the inconvenient truth that the winter of 2007–8 was the coldest in a century, in spite of all those tons of 'greenhouse gas' being spewed into the air from all the new factories in China and India. Worldwide temps dropped 0.6 of a degree C in one year. (Lewis 2008, para. 3)

Likewise, Littlejohn (2012) stated, "When it became apparent that temperatures were actually falling, they simply changed the name of their religion from 'global warming' to 'climate change'" (para. 9). Given the more encompassing meaning of CC relative to GW, we believe the contrarian myth can be debunked by examining the scientific literature and comparing the two terms' relative use. Further, we expect there to be no relationship between the preference of GW over CC and observed warming.

3.2.3 The Scientific Mainstream Versus So-Called "Catastrophic Anthropogenic Global Warming"

Another claim advanced by those who reject the mainstream scientific agreement on climate is that the consensus position consists of a claim of *catastrophic anthropogenic global warming* or the frequently used acronym CAGW (e.g., Hickey 2014; The Hockey Schtick 2012; Milloy 2012; Starck 2012). However, CAGW is rarely, if ever, defined or sourced to a mainstream scientific organization or study. Any scientific study's result, or statement by a researcher, that does not fit a contrarian's personal, flexible definition of CAGW can therefore be adopted as ostensibly supporting their view and refuting the mainstream, even when such results are actually consistent with the mainstream position on climate (e.g., The Hockey Schtick 2014).

3.3 Materials and Methods

3.3.1 Academic Databases

To address the claim of CC vs. GW in the scientific literature, we used two established academic reference databases (Thomson Reuters' *Web of Science*™ and Elsevier's *Scopus*) to record the returned number of papers for each year between 1950 and 2013 using each search term. The search terms we used were *climate change* and *global warming*. Additionally, we searched the term *catastrophic anthropogenic global warming* in both *Scopus* and *Web of Science*. To assess paper counts that used only GW or only CC, we also recorded the number of papers that used both terms. We then adjusted the GW and CC values by subtracting the results of papers using both terms.

Thomson Reuters' *Web of Science™* The *Web of Science* Core Collection contains information about peer-reviewed journals in the science and the social sciences from 1900 till the present. It also contains information about conference proceedings and books; however, we restricted our *Web of Science* search to articles only, ensuring these works would not be included in our search results. A review of *Web of Science* performance can be found in Jacsó (2011a).

Elsevier's *Scopus* *Scopus* is an abstract and citation database covering the fields of science, technology, medicine, social sciences, and arts and humanities. Like *Web of Science*, *Scopus* also contains information about books and conference proceedings, but here we also restricted our search to articles only. A review of *Scopus*' performance can be found in Jacsó (2011a, b).

3.3.2 Newspaper Databases

ProQuest *Historical Newspapers™* The ProQuest *Historical Newspapers™* database provides a full text archive of 36 newspapers spanning 1764–2011. The database covers so-called *prestige press* US papers like the Los Angeles Times, New York Times, Wall Street Journal, and Washington Post, a selection of international papers, as well as collections of traditionally *Black* and *American Jewish* newspapers. We searched for articles including the term *catastrophic anthropogenic global warming* or *CAGW*, but excluding the phrase *Citizens Against Government Waste*, which shares the same acronym and thus distorts queries by returning a prohibitively high number of false positives. We restricted the returned results to exclude the following items: birth/marriage announcements, obituaries, classified advertisements, credit/acknowledgments, illustrations/images/photographs, military/war news, stock quotes, and tables of contents.

LexisNexis® *Academic* The LexisNexis® *Academic* database was used to supplement the results of the historical newspaper search, with larger coverage, particularly over recent decades when the myth is most relevant. The *Academic* newspaper database contains a full text archive of over 3000 newspapers. We searched for articles including the term *catastrophic anthropogenic global warming* or *CAGW*, but excluding the phrase *Citizens Against Government Waste*. We restricted the source type to newspapers only.

3.3.3 Activist Websites and Blogs

We analyzed 72 organizational websites, of which we classified 32 as activist organizations. From each of the 32 websites, we analyzed three pages: the site's home page, its *About Us* page, and a page that discussed the issue of climate change (and/or global warming), if such a page was available. We also searched and recorded the number of hits for GW and CC for each site and calculated a GW percentage from the resulting hits ($P_{GW} = N_{GW} / (N_{GW} + N_{CC})\ 100$, where P is percentage and N is number of hits). We also checked 27 activist blogs in a similar manner.

3.3.4 Global Temperature Data

Time series for the global mean surface temperature evolution are produced by a number of scientific organizations studying climate (see Fig. 3.2a; Brohan et al. 2006; Cowtan and Way 2014; Hansen et al. 2010; Morice et al. 2012; Muller et al. 2013; Smith et al. 2008). To be as generous to the myth as possible, we present our

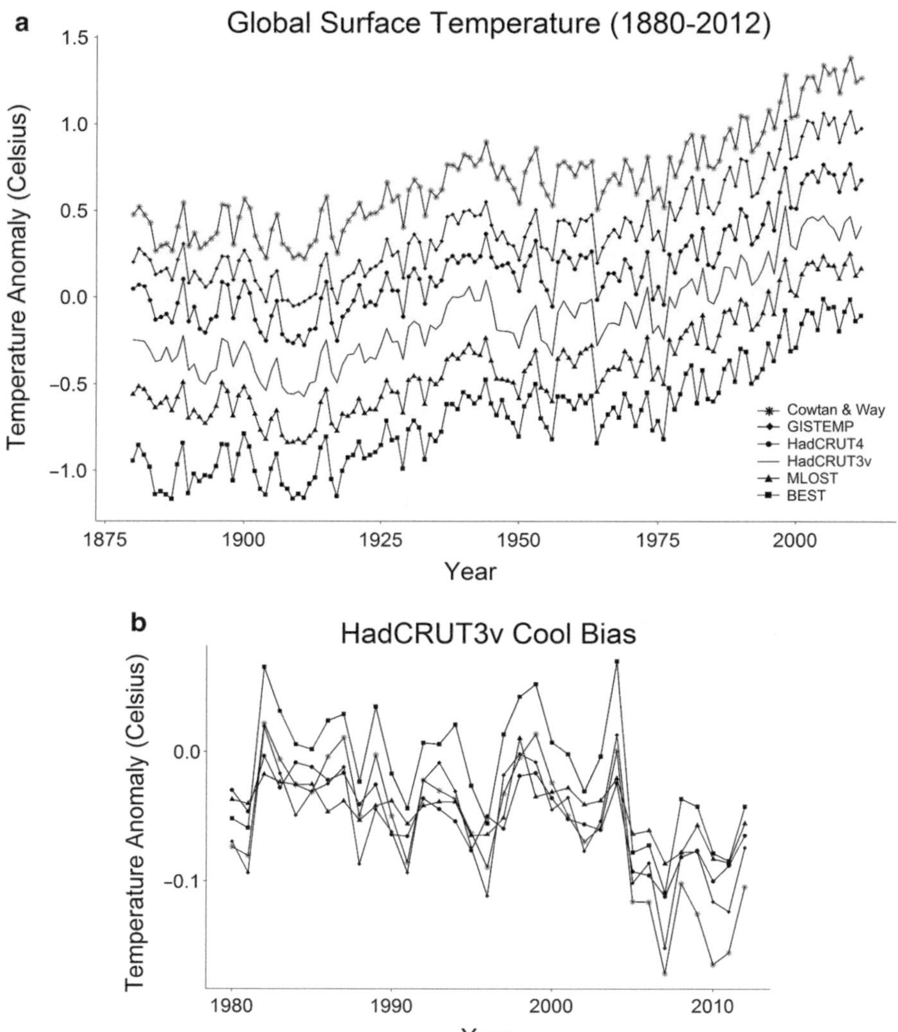

Fig. 3.2 Global temperature records. (**a**) Globally averaged surface instrumental records from various groups; HadCRUT3v anomalies shown relative to 1961–1990 baseline; other records are offset for clarity. (**b**) Difference between HadCRUT3v and other records, plotted on a common baseline

analysis using the HadCRUT3v dataset, which suffers from known coverage biases and shows markedly less warming in recent years compared to other surface temperature analyses (see Fig. 3.2b). HadCRUT3v temperature was annualized and downloaded from KNMI Climate Explorer (http://climexp.knmi.nl/).

3.4 Analyses and Results

3.4.1 Use of CC and GW in the Scientific Literature

The total number of papers found using each term in each year during the period 1950–2013 is presented in Table 3.1. Papers using CC outnumbered those using GW by around 5–9 times, depending on the database. The evolution of each term over time is presented in Fig. 3.3. Neither CC nor GW was used with much

Table 3.1 Total results for all terms

	Scopus	Web of science
Climate change	75,163	173,647
Global warming	15,672	18,804
Both	8,564	7,136

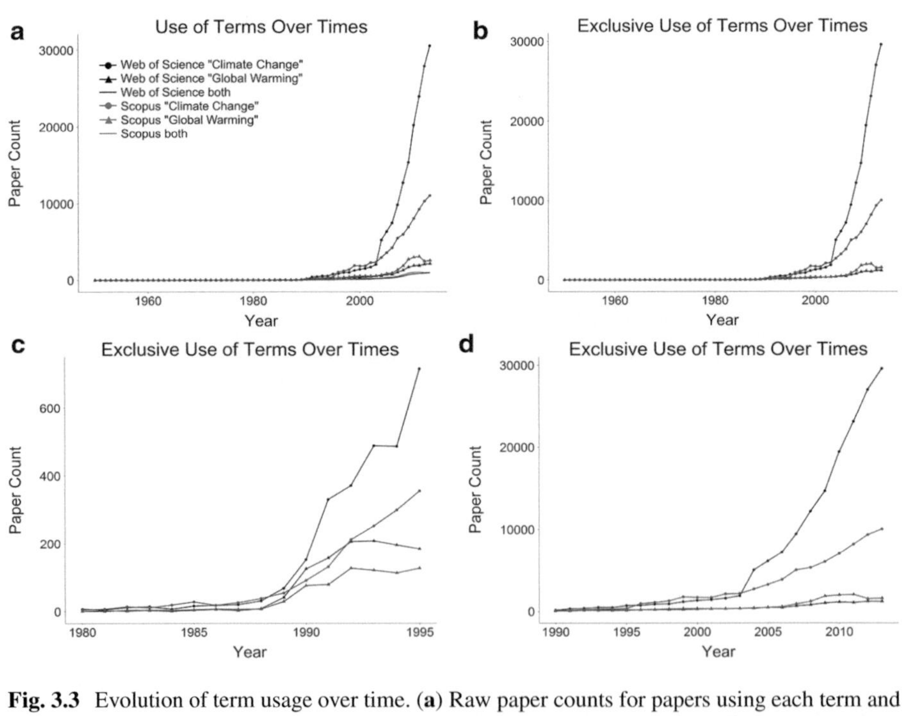

Fig. 3.3 Evolution of term usage over time. (**a**) Raw paper counts for papers using each term and both terms; (**b**) adjusted (papers using both terms removed) paper counts; (**c**) adjusted values from 1980 to 1995; (**d**) adjusted values from 1990s to 2000s. Annual paper counts in Web of Science (*black*) and in Scopus (*gray*) database for CC (*solid line*), GW (*dashed line*), and both (*dotted line*)

frequency until the 1980s. The numbers of CC and GW articles were somewhat similar until the early 1990s, when the increase in the number of CC papers occurs earlier than the increase in GW papers in both *Web of Science* and *Scopus* databases. From the early to mid-1990s, the number of CC papers continues to grow steadily while the number of GW papers remains flat during this period (Fig. 3.3c). An upward trend in GW papers does not resume until the 2000s (Fig. 3.3d).

3.4.2 Change in Usage Relative to Temperature Change

We find that CC has been the more prevalent term at least since the early 1990s, which precludes the claim that it was recently adopted in favor of GW in response to purportedly cooling temperatures. The ongoing warming of the climate system also precludes the possibility of this particular myth being true (Fig. 3.2a). Nor does there appear to be any relationship between the ratio of GW to CC papers and the evolution of global temperatures over time (Fig. 3.4). Nevertheless, we tested for a relationship between GW's preferential use and global temperatures. We performed linear correlation tests (Pearson product-moment correlation coefficient) with annual ratios reflecting the relative usage of CC vs. GW against annual temperature data. Calculations were performed using the *R* statistical programming language (R

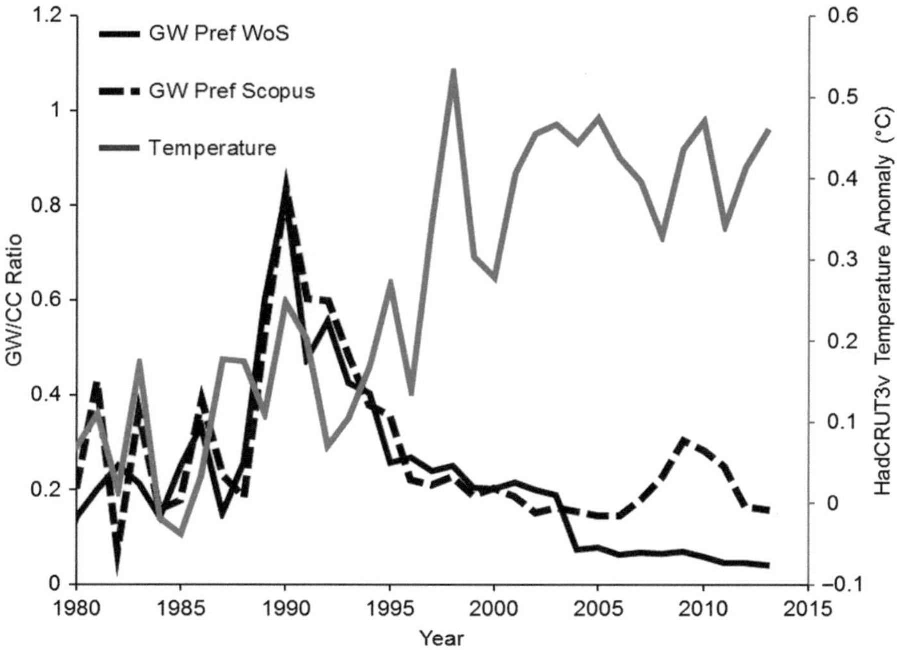

Fig. 3.4 Preferential use of GW vs. CC and temperature change. Change in the preference (ratio) of GW over CC over time compared to the global temperature evolution

Core Team, 2014). For the contrarian claim to be supported, we would expect a higher ratio of GW to CC papers during increasing temperatures and a decline in the ratio when temperature decreases. The correlation test was performed again with lagged responses for the scholarly results (ranging from 1 to 5 years, to allow for the delay between a change in temperature and the writing and publication of a paper). We found correlations in the opposite direction as posited by the myth, and these negative correlations were for the most part statistically significant (Fig. 3.5).

3.4.3 Activist Websites

According to our results, about two-thirds of all activist organizations sampled use only CC on their home pages. Most of the remaining third of home pages use neither CC nor GW. Few homepages use GW or both GW and CC. *About Us* pages of activist organization websites use only CC (~70 %) or neither CC nor GW (~30 %). In the activist organization websites, the page that discusses the issue of climate change and/or global warming included both CC and GW in most of the sites (~91 %) and only CC in the rest of the sites (~9 %). Search engines find both GW and CC somewhere within almost all websites of the sampled organizations. Of the number of hits for either CC or GW, on average about one-fifth is GW (GW percentage ~22 %).

Home pages of activist blogs typically use both terms. About 30 % use only CC and 7 % use only GW. *About Us* pages show a similar distribution to home pages, although none used only GW (however, only 15 blogs in our sample had an *About Us* page, so the sample is small here). Only one blog in our sample had a page that describes CC and GW, and this page included both terms. Search engine results show that activist blogs use GW and CC almost in similar proportions. Mean GW percentage is 48 %. None of the blogs use only CC. Of the sources we analyzed, activist blogs clearly have the highest percentage of GW (48 %). Activist organization websites had GW percentage of 22 %. Peer-reviewed papers have GW percentage of 11.3 % when calculated for the years from 2000 to 2013, averaged from Web of Science and Scopus data.

3.4.4 Usage of CAGW

The phrase *catastrophic anthropogenic global warming* and its acronym (*CAGW*) were found only once in our query of the scientific databases (Carlin 2011). It is used by an author taking a contrarian position to the mainstream scientific community, and the phrase is neither defined nor sourced to a mainstream scientific publication or organization. Further, we find the phrase is rarely used in the mainstream press and in the results of our query appears only in opinion editorials, letters to the

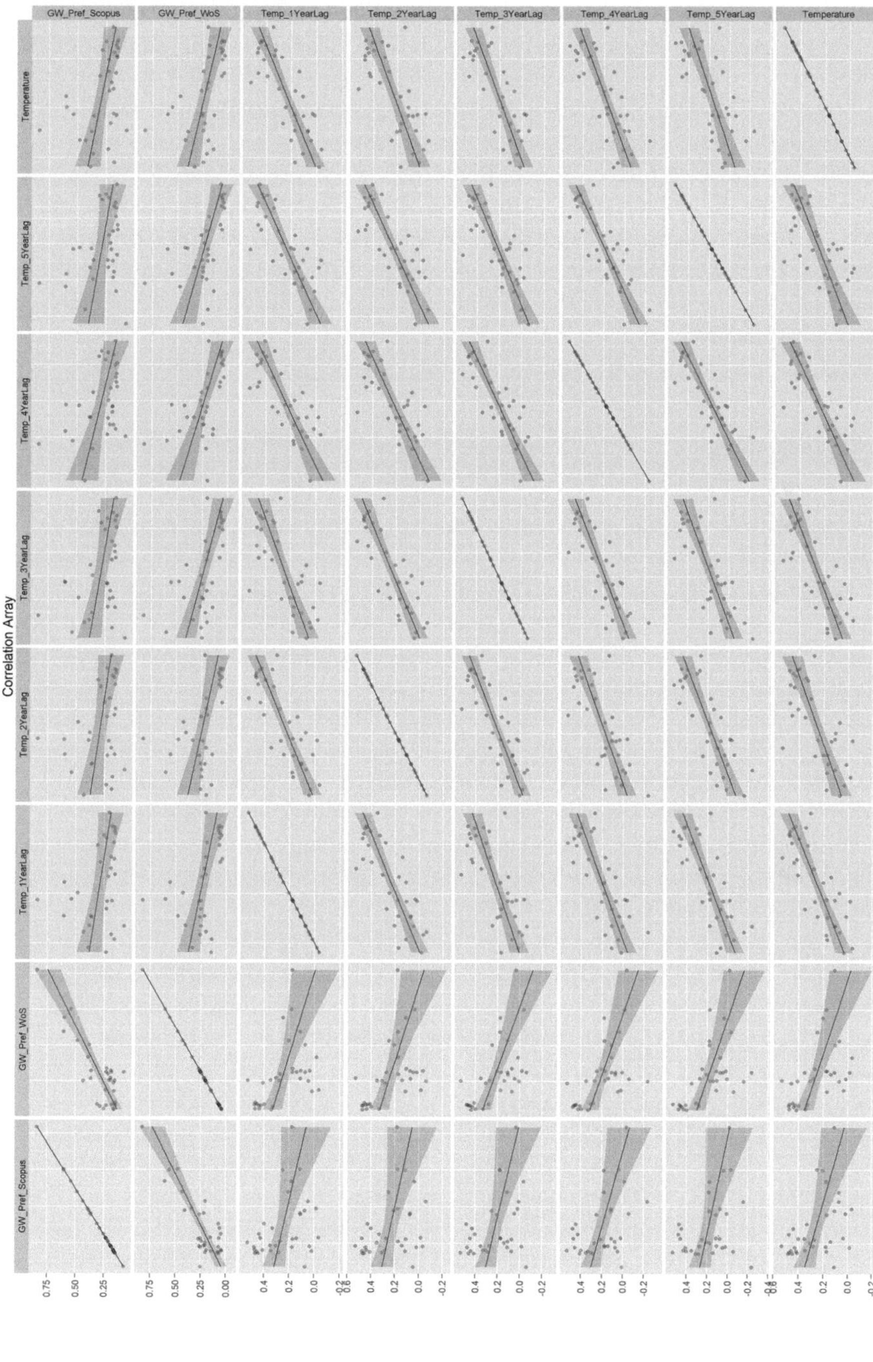

Fig. 3.5 Correlation test results for global warming preference and temperature change. Correlations between the relative preference of GW over CC and global temperature, with lags of 1–5 years

editor, blog posts, or other nonnews items. Again, in all cases, CAGW is used by authors taking a contrarian position to the mainstream scientific view on climate.

Although the databases used in our analyses have their strengths and weaknesses (Jacsó 2011a, b), we believe our results to be robust. Other scientific publications databases exist, but most are either more limited in scope (e.g. PubMed, focused on medical research; BioOne, focused on biological research since 1998) or too broad to be useful in this study (e.g., Google Scholar, or ProQuest, which includes sources far beyond peer-reviewed papers). A number of other terms could be included in future analyses (e.g., global cooling, climate crisis, climate disruption, etc.) but are beyond the scope of this work. Expanding our search to include non-peer-reviewed, nonphysical science scholarly material could return different results; however, the relevance of such queries in the assessment of the position of the scientific mainstream is questionable.

3.5 Discussion

The myth that GW was the preferred term until recently and CC was adopted in response to a decline in temperatures is groundless. In our analysis, CC was used earlier and has been used more prevalently—both overall and in all but a handful of years over the last several decades. Use of GW in lieu of CC does not appear to be related to changes in global temperature; rather, preference for GW is negatively correlated with temperature change. Nor has there been cooling to prompt the purported switch. Changes in solar irradiance, volcanic eruptions, and tropical Pacific variability can give the appearance of periods of flat or even declining temperatures during periods when the overall trend is unquestionably one of warming (Easterling and Wehner 2009; Lean and Rind 2009; Rahmstorf et al. 2012; Thompson et al. 2009). After accounting for such factors, the overall warming trend continues in-line with expectations (e.g., Kosaka and Xie 2013; Huber and Knutti 2014; Schmidt et al. 2014).

Amusingly, the period during the early 1990s–2000s actually saw the sharpest decline in preference for GW relative to CC, even as global temperature was increasing more rapidly than in previous years. This increased warming rate was likely attributable to a rebound from the cooling of the Mount Pinatubo eruption and positive El Niño Southern Oscillation (ENSO) conditions, superimposed on top of the anthropogenically forced trend (Rahmstorf et al. 2007). If the myth was at all based in fact, preference of GW over CC should have climbed in concert with this rapid warming, when in fact it plummeted. The second myth we examined is similarly unfounded. So-called CAGW is not a concept discussed by the mainstream scientific community, let alone its consensus position. Our analysis was restricted to the particular concept of CAGW—other terms would likely return different results. We do not doubt that climate change has been discussed in the context of catastrophe, be it in paleoecological contexts, or regarding the present, human-driven change. However, the specificity of the phrase CAGW, along with its frequent contrarian use and presentation as the mainstream scientific position, lent itself more readily to analysis.

Mainstream goals such as avoiding "dangerous interference" with the climate system (Anderson and Bows 2011; Hansen 2005; Keller et al. 2005; Kriegler 2007; Oppenheimer 2005; Mann 2009; Ramanathan and Feng 2008; Schneider and Lane 2006; UNFCCC 1992) and specific numerical targets such as 2 °C (UNFCCC 2009, 2010) are obviously much lower criteria to satisfy than the undefined CAGW. By its nature, *dangerous* is a conditional descriptor, while *catastrophic* is both higher in stakes as well as more certain. Such a dramatically higher and ever-shifting standard, in combination with its lack of attribution to mainstream sources, leads us to conclude that CAGW exists solely as a straw man of the mainstream position and is used almost exclusively by contrarians themselves.

Given that the examined myths are unfounded, communicators might wonder how to successfully rebut them. Effective myth refutation techniques are explained in depth in Cook and Lewandosky's *Debunking Handbook* (2011). Briefly, when communicators are debunking myths, they should: emphasize the core facts of the issue and avoid the tendency to begin by restating the myth; warn the audience that the myth is false before restating it; provide an alternative explanation to fill the cognitive gap left by debunking the myth (e.g., why misinformers promote the myth or what the actual facts regarding the issue are); and, when possible, present rebutting information in the form of, or along with, easily understood graphical representations. Inoculation theory offers a powerful method of myth refutation (Banas and Rains 2010; Compton and Pfau 2009; McGuire 1964) but requires a scenario in which the communicator can introduce the myth to his or her audience for the first time. Inoculation theory suggests that motivating an audience to perceive a myth as threatening (e.g., those perpetuating the myth are trying to take advantage of the audience) and presenting the audience with a preemptive refutation can combine to create resistance to future exposure to the myth, analogous to biological immunity through exposure to weakened or inactive viruses. A final method of myth refutation is agnotology-based learning (Bedford 2010; Cook et al. 2014). Agnotology-based learning is a pedagogical tool that seeks to teach audiences about a subject by having them critically compare a source of misinformation with a refutational text. Directing the audience to identify the flaws in myths themselves can lead to greater and longer-lasting increases in knowledge than passive learning activities like listening to lectures.

Communicators may also wonder which phrase (CC or GW) is the most effective to use themselves. A growing body of academic work has arisen in recent years, investigating the impact of using the term CC or GW with different audiences. According to Schuldt et al. (2011), the magnitude of the partisan divide on climate in the United States is somewhat dependent on the term used. Only 44 % of Republicans believe GW is occurring, while 60 % believe CC is happening. At the same time, Democrats are as equally likely to accept the existence of GW (87 %) as CC (86 %). Likewise, Villar and Krosnick (2011) found that Republicans tend to see CC as a more serious issue than GW (whereas Democrats see GW more serious than CC). Interestingly, Schuldt et al. (2011) find that conservative websites prefer to use GW over CC. Given the previous findings (that conservatives respond more negatively to GW relative to CC), such a propensity for using GW would have the effect of reinforcing contrarian messages on such sites. These results are consistent with Akerlof and Maibach (2011), who found that those who reject the mainstream position on climate prefer to use GW over CC. Assessing to what extent this GW preference by contrarians is a deliberate strategy is a question that might be examined in future work. Other research has begun to examine the interaction between terminology choice and meteorological events (Schuldt and Roh 2014a, b) and the communications efficacy of terms beyond CC and GW, such as *climate crisis* and *climate disruption* (Jaskulsky and Besel 2013).

Taken together, these findings suggest that while GW might have more of an impact when communicating with known Liberal or Democratic-leaning American audiences, CC might be the optimal term when communicating to a Republican-leaning audience or audience of unknown composition. Although GW may be the more familiar term (Whitmarsh 2009), CC is both the more widely used term in the scientific literature and appears to have fewer negative connotations.

3.6 Conclusions

In this study, we examine two myths centered on the use of terminology in communicating climate change. Our results indicate that "climate change" has been used more often than *global warming* in the scientific literature and continues to be the more prevalent term in recent years. There also appears to be no relationship between the greater use of the term "global warming" (relative to *climate change*) and higher temperatures. Activist websites also seem to be continuing to use the term *global warming* into the present. Our analysis therefore leads us to conclude that the claim that the term *global warming* was abandoned in favor of the term *climate change*, in response to stalling temperature, is baseless.

Additionally, we find that *catastrophic anthropogenic global warming* is essentially a term that is never used in the relevant scientific literature by mainstream sources. Furthermore, in the press it appears to be used exclusively by climate contrarians. The term is typically neither defined nor attributed to a mainstream scientific source. Our conclusion is therefore that CAGW is simply a straw man used by climate contrarians to criticize the mainstream position. Evidence-based debunking strategies, inoculation theory, and agnotology-based learning can be employed to neutralize potential misconceptions created by such myths. Finally, we urge scientists and communicators to be mindful of the potential for confusion when using various terms relating to environmental change and to tailor their language to maximize the communications' impact for their audiences.

Acknowledgments The authors would like to thank Anne-Marie Blackburn, John Cook, David Kirtley, Bob Lacatena, Dana Nuccitelli, Mark Richardson, Andy Skuce, Glenn Tamblyn, and an anonymous reviewer for their helpful suggestions. The authors would also like to thank the editors, especially Katia Kontar, for their guidance. The authors declare no conflicts of interest.

References

Akerlof K, Maibach EW (2011) A rose by any other name …?: What members of the general public prefer to call "climate change". Clim Change 106(4):699–710. doi:10.1007/s10584-011-0070-4

Anderegg WR, Prall JW, Harold J, Schneider SH (2010) Expert credibility in climate change. Proc Natl Acad Sci U S A 107(27):12107–12109. doi:10.1073/pnas.1003187107

Anderson T (1857) Notes on the Indian species of Lycium. Ann Mag Nat Hist: Series 2 20(116):126–127. doi:10.1080/00222935709487885

Anderson K, Bows A (2011) Beyond 'dangerous' climate change: emission scenarios for a new world. Phil Trans R Soc A 13, 369(1934):20–44. doi:10.1098/rsta.2010.0290

Banas JA, Rains SA (2010) A meta-analysis of research on inoculation theory. Commun Monogr 77(3):281–311. doi:10.1080/03637751003758193

Bedford D (2010) Agnotology as a teaching tool: learning climate science by studying misinformation. J Geogr 109(4):159–165. doi:10.1080/00221341.2010.498121

Beman JM, Chow C-E, King AL, Feng Y, Fuhrman JA, Andersson A, Bates NR, Popp BN, Hutchins DA (2011) Global declines in oceanic nitrification rates as a consequence of ocean acidification. Proc Natl Acad Sci U S A 108(1):208–213. doi:10.1073/pnas.1011053108

Broecker WS (1975) Climatic change: are we on the brink of a pronounced global warming? Science 189(4201):460–463. doi:10.1126/science.189.4201.460

Brohan P, Kennedy JJ, Harris I, Tett SFB, Jones PD (2006) Uncertainty estimates in regional and global observed temperature changes: a new data set from 1850. J Geophys Res 111, D12106. doi:10.1029/2005JD006548

Brulle RJ (2014) Institutionalizing delay: foundation funding and the creation of US climate change counter-movement organizations. Clim Change 122(4):681–694. doi:10.1007/s10584-013-1018-7

Carlin A (2011) A multidisciplinary, science-based approach to the economics of climate change. Int J Environ Res Public Health 8(4):985–1031. doi:10.3390/ijerph8040985

Compton J, Pfau M (2009) Spreading inoculation: inoculation, resistance to influence, and word-of-mouth communication. Commun Theory 19(1):9–28. doi:10.1111/j.1468-2885.2008.01330.x

Cook J, Lewandowsky S (2011) The debunking handbook. University of Queensland, St. Lucia, 5 Nov 2011. http://sks.to/debunk. Accessed 14 Oct 2014

Cook J, Nuccitelli D, Green SA, Richardson M, Winkler B, Painting R, Way R, Jacobs P, Skuce A (2013) Quantifying the consensus on anthropogenic global warming in the scientific literature. Environ Res Lett 8(2):024024. doi:10.1088/1748-9326/8/2/024024

Cook J, Bedford D, Mandia S (2014) Raising climate literacy through addressing misinformation: case studies in agnotology-based learning. J Geosci Educ 62(3):296–306. doi:10.5408/13-071.1

Coulson S, Hodkinson ID, Strathdee A, Bale JS, Block W, Worland MR, Webb NR (1993) Simulated climate change: the interaction between vegetation type and microhabitat temperatures at Ny Ålesund, Svalbard. Polar Biol 13(1):67–70. doi:10.1007/BF00236585

Cowtan K, Way RG (2014) Coverage bias in the HadCRUT4 temperature series and its impact on recent temperature trends. Q J Roy Meteorol Soc 140(683):1935–1944. doi:10.1002/qj.2297

Cui Y, Kump LR (2014) Global warming and the end-Permian extinction event: proxy and modeling perspectives. Earth Sci Rev (in press). doi:10.1016/j.earscirev.2014.04.007

Doran PT, Zimmerman MK (2009) Examining the scientific consensus on climate change. Eos Trans AGU 90(3):22–23. doi:10.1029/2009EO030002

Dunlap RE, Jacques PJ (2013) Climate change denial books and conservative think tanks: exploring the connection. Am Behav Sci 57(6):699–731. doi:10.1177/0002764213477096

Dunlap RE, McCright AM (2008) A widening gap: Republican and Democratic views on climate change. Environ Sci Policy Sustain Dev 50(5):26–35. doi:10.3200/ENVT.50.5.26-35

Easterling DR, Wehner MF (2009) Is the climate warming or cooling? Geophys Res Lett 36, L08706. doi:10.1029/2009GL037810

Elsasser SW, Dunlap RE (2013) Leading voices in the denier choir: conservative columnists' dismissal of global warming and denigration of climate science. Am Behav Sci 57(6):754–776. doi:10.1177/0002764212469800

Elsner JB, Kossin JP, Jagger TH (2008) The increasing intensity of the strongest tropical cyclones. Nature 50:92–95. doi:10.1038/nature07234

Fenton LK, Geissler PE, Haberle RM (2007) Global warming and climate forcing by recent albedo changes on Mars. Nature 446:646–649. doi:10.1038/nature05718

Gifford R (2011) The dragons of inaction: psychological barriers that limit climate change mitigation and adaptation. Am Psychol 66(4):290. doi:10.1037/a0023566

Guber DL (2013) A cooling climate for change? Party polarization and the politics of global warming. Am Behav Sci 57(1):93–115. doi:10.1177/0002764212463361

Hansen JE (2005) A slippery slope: how much global warming constitutes "dangerous anthropogenic interference"? Clim Change 68(3):269–279. doi:10.1007/s10584-005-4135-0

Hansen J, Ruedy R, Sato M, Lo K (2010) Global surface temperature change. Rev Geophys 48, RG4004. doi:10.1029/2010RG000345

Hickey JG (2014) Scientists rebut White House global warming claims. Newsmax. http://www.newsmax.com/Newsfront/scientists-rebut-global-warming/2014/05/18/id/571987/. Accessed 15 Oct 2014

Huber M, Knutti R (2014) Natural variability, radiative forcing and climate response in the recent hiatus reconciled. Nat Geosci 7:651–656. doi:10.1038/ngeo2228

Jacsó P (2011a) The h-index, h-core citation rate and the bibliometric profile of the Web of Science database in three configurations. Online Inf Rev 35(5):821–833

Jacsó P (2011b) The h-index, h-core citation rate and the bibliometric profile of the Scopus database. Online Inf Rev 35(3):492–501

Jaskulsky L, Besel R (2013) Words that (don't) matter: an exploratory study of four climate change names in environmental discourse. Appl Environ Educ Commun 12(1):38–45. doi:10.1080/1533015X.2013.795836

Joachimski MM, Lai X, Shen S, Jiang H, Luo G, Chen B, Chen J, Sun Y (2012) Climate warming in the latest Permian and the Permian–Triassic mass extinction. Geology 40(3):195–198. doi:10.1130/G32707.1

Joffe JS (1929) Soil profile studies: I. Soil as an independent body and soil morphology. Soil Sci 28(1):39–54

Keller K, Hall M, Kim S-R, Bradford DF, Oppenheimer M (2005) Avoiding dangerous anthropogenic interference with the climate system. Clim Change 73(3):227–238. doi:10.1007/s10584-005-0426-8

Kosaka Y, Xie S-P (2013) Recent global-warming hiatus tied to equatorial Pacific surface cooling. Nature 501:403–407. doi:10.1038/nature12534

Kriegler E (2007) On the verge of dangerous anthropogenic interference with the climate system? Environ Res Lett 2, 011001. doi:10.1088/1748-9326/2/1/011001

Lean JL, Rind DH (2009) How will Earth's surface temperature change in future decades? Geophys Res Lett 36(15). doi:10.1029/2009GL038932

Lewis J (2008) The epicycles of Global Warming. The American Thinker. From http://www.americanthinker.com/2008/03/the_epicycles_of_global_warmin.html. Accessed 16 Oct 2014

Littlejohn R (2012) It's raining, it's pouring, that'll be the global warming…, 12 Jun 2012. http://www.dailymail.co.uk/debate/article-2157831/Its-raining-pouring-thatll-global-warming-.html. Accessed 15 Oct 2014

Lockwood M (2012) Solar influence on global and regional climates. Surv Geophys 33(3–4):503–534. doi:10.1007/s10712-012-9181-3

Malka A, Krosnick JA, Langer G (2009) The association of knowledge with concern about global warming: trusted information sources shape public thinking. Risk Anal 29(5):633–647. doi:10.1111/j.1539-6924.2009.01220.x

Mann ME (2009) Defining dangerous anthropogenic interference. Proc Natl Acad Sci U S A 106(11):4065–4066. doi:10.1073/pnas.0901303106

Mayer B (1856) Observations on Mexican history and archaeology: with a special notice of Zapotec remains, as delineated in JG Sawkins's drawings of Mitla 9(4). Smithsonian Institute, Washington DC

Mazur A (1998) Global environmental change in the news: 1987–90 vs 1992–6. Int Sociol 13(4):457–472. doi:10.1177/026858098013004003

McGuire WJ (1964) Some contemporary approaches. Adv Exp Soc Psychol 1:191–229. doi:10.1016/S0065-2601(08)60052-0

Meehl GA, Zwiers F, Evans J, Knutson T, Mearns L, Whetton P (2000) Trends in extreme weather and climate events: issues related to modeling extremes in projections of future climate change*. Bull Am Meteorol Soc 81(3):427–436

Milloy S (2012) "Skewered! William D. Nordhaus says we're wrong!" JunkScience.com, http://junkscience.com/2012/02/29/skewered-william-d-nordhaus-says-were-wrong/. Accessed 15 Oct 2014

Mitchell JM (1961) Recent secular changes of global temperature. Ann N Y Acad Sci 95(1):235–250. doi:10.1111/j.1749-6632.1961.tb50036.x

Morice CP, Kennedy JJ, Rayner NA, Jones PD (2012) Quantifying uncertainties in global and regional temperature change using an ensemble of observational estimates: the HadCRUT4 data set. J Geophys Res 117, D08101. doi:10.1029/2011JD017187

Muller RA, Curry J, Groom D, Jacobsen R, Perlmutter S, Rohde R, Rosenfeld A, Wickham C, Wurtele J (2013) Decadal variations in the global atmospheric land temperatures. J Geophys Res Atmos 118(11):5280–5286. doi:10.1002/jgrd.50458

Oppenheimer M (2005) Defining dangerous anthropogenic interference: the role of science, the limits of science. Risk Anal 25(6):1399–1407. doi:10.1111/j.1539-6924.2005.00687.x

Oreskes N (2004) The scientific consensus on climate change. Science 306(5702):1686–1686. doi:10.1126/science.1103618

Oreskes N, Conway EM (2010) Merchants of doubt: how a handful of scientists obscured the truth on issues from tobacco smoke to global warming. Bloomsbury Publishing, New York

Peterson TC, Connolley WM, Fleck J (2008) The myth of the 1970s global cooling scientific consensus. Bull Am Meteorol Soc 89(9):1325–1337. doi:10.1175/2008BAMS2370.1

R Core Team (2014). R: a language and environment for statistical computing. R Foundation for Statistical Computing, Vienna. http://www.R-project.org/. Accessed 14 Oct 2014

Rahmstorf S, Cazenave A, Church JA, Hansen JE, Keeling RF, Parker DE, Somerville RCJ (2007) Recent climate observations compared to projections. Science 316(5825):709. doi:10.1126/science.1136843

Rahmstorf S, Foster G, Cazenave A (2012) Comparing climate projections to observations up to 2011. Environ Res Lett 7:044035. doi:10.1088/1748-9326/7/4/044035

Ramanathan V, Feng Y (2008) On avoiding dangerous anthropogenic interference with the climate system: formidable challenges ahead. Proc Natl Acad Sci U S A 105(38):14245–14250. doi:10.1073/pnas.0803838105

Salnikov V, Turulina G, Polyakova S, Petrova, Y, Skakov A (2014) Climate change in Kazakhstan during the past 70 years. Quat Int 358: 77–82. Accessed 9 Feb 2015. http://www.sciencedirect.com/science/article/pii/S1040618214006557

Schmidt GA, Shindell DT, Tsigaridis K (2014) Reconciling warming trends. Nat Geosci 7(3):158–160

Schneider SH, Lane J (2006) An overview of 'dangerous' climate change. In: Schellnhuber HJ, Cramer WP (eds) Avoiding dangerous climate change. Cambridge University Press, Cambridge

Schuldt JP, Roh S (2014a) Of accessibility and applicability: how heat-related cues affect belief in "global warming" versus "climate change". Soc Cogn 32(3):217–238. doi:10.1521/soco.2014.32.3.217

Schuldt JP, Roh S (2014b) Media frames and cognitive accessibility: what do "global warming" and "climate change" evoke in partisan minds? Environ Comm (ahead of print). doi:10.1080/17524032.2014.909510

Schuldt JP, Konrath SH, Schwarz N (2011) "Global warming" or "climate change"? Whether the planet is warming depends on question wording. Public Opin Q 75(1):115–124. doi:10.1093/poq/nfq073

Shaw JM (2003) Climate change and deforestation: implications for the Maya collapse. Anc Mesoam 14(1):157–167. doi:10.1017/S0956536103132063

Smith TM, Reynolds RW, Peterson TC, Lawrimore J (2008) Improvements to NOAA's historical merged land–ocean surface temperature analysis (1880–2006). J Climate 21(10):2283–2296. doi:10.1175/2007JCLI2100.1

Social Learning Group, Clark WC, Jaeger J, van Eijndhoven J, Dickson N (2001) Learning to manage global environmental risks (Vol. 1): A comparative history of social responses to climate change, ozone depletion and acid rain. MIT Press

Solomon SC, Bullock MA, Grinspoon DH (1999) Climate change as a regulator of tectonics on Venus. Science 286(5437):87–90. doi:10.1126/science.286.5437.87

Somerville RC, Hassol SJ (2011) Communicating the science of climate change. Phys Today 64(10):48. doi:10.1063/PT.3.1296

Starck W (2012) "Speak loudly and carry a busted hockey stick". Quadrant Online. http://quadrant.org.au/opinion/doomed-planet/2012/11/speak-loudly-and-carry-a-busted-hockey-stick/. Accessed 15 Oct 2014

Stern PC, Young OR, Druckman D (eds) (1992) Global environmental change: understanding the human dimensions. Committee on the human dimensions of global change, National Research Council

The Hockey Schtick (2012) Why belief in CAGW (Catastrophic Anthropogenic Global Warming) is not currently justified by the standards of the scientific method. http://hockeyschtick.blogspot.com/2012/05/why-belief-in-cagw-catastrophic.html. Accessed 15 Oct 2014

The Hockey Schtick (2014) New paper is a huge blow to CAGW: 'Missing heat' NOT found in the deep oceans, 6 Oct 2014. http://hockeyschtick.blogspot.fi/2014/10/new-paper-is-huge-blow-to-cagw-missing.html. Accessed 15 Oct 2014

Thompson DW, Wallace JM, Jones PD, Kennedy JJ (2009) Identifying signatures of natural climate variability in time series of global-mean surface temperature: methodology and insights. J Climate 22(22):6120–6141

Tilman D, Fargione J, Wolff B, D'Antonio C, Dobson A, Howarth R, Schindler D, Schlesinger WH, Simberloff D, Swackhamer D (2001) Forecasting agriculturally driven global environmental change. Science 292(5515):281–284. doi:10.1126/science.1057544

Touchan R, Anchukaitis KJ, Meko DM, Sabir M, Attalah S, Aloui A (2011) Spatiotemporal drought variability in northwestern Africa over the last nine centuries. Climate Dynam 37(1–2):237–252. doi:10.1007/s00382-010-0804-4

United Nations Framework Convention on Climate Change (1992) Article 2 of the Convention. http://unfccc.int/essential_background/convention/background/items/1353.php. Accessed 15 Oct 2014

United Nations Framework Convention on Climate Change (2009) Copenhagen Accord. http://unfccc.int/resource/docs/2009/cop15/eng/11a01.pdf. Accessed 15 Oct 2014

United Nations Framework Convention on Climate Change (2010) Decision 1/CP.16. The Cancun Agreements: Outcome of the work of the ad hoc working group on long-term cooperative action under the convention. http://unfccc.int/resource/docs/2010/cop16/eng/07a01.pdf#page=2. Accessed 15 Oct 2014

Villar A, Krosnick JA (2011) Global warming vs. climate change, taxes vs. prices: does word choice matter? Clim Change 105(1–2):1–12. doi:10.1007/s10584-010-9882-x

Whitmarsh L (2009) What's in a name? Commonalities and differences in public understanding of "climate change" and "global warming". Public Underst Sci 18(4):401–420. doi:10.1177/0963662506073088

Wigley TM, Jones PD (1985) Influences of precipitation changes and direct CO2 effects on streamflow. Nature 314:149–152. doi:10.1038/314149a0

Willis R (1925) Physiography of the California coast ranges. GSA Bull 36(4):641–678. doi:10.1130/GSAB-36-641

World Health Organization (2005) http://www.who.int/globalchange/environment/en/. Accessed 14 Oct 2014

Chapter 4
Public Perceptions of Global Warming: Understanding Survey Differences

Kelly Klima

Abstract Since 2007, no scientific body of national, or international, standing rejects the findings of human-induced climate change. Yet in the United States, public opinion and public policy remain deeply divided on the issue. I review five longitudinal surveys from Yale/George Mason, Stanford/Resources for the Future, University of Michigan/Brookings, Gallup, and the Pew Research Center to understand different surveys of Americans have different results to the question "Does global warming exist?" I find that question wording makes a difference, and researchers may want to focus their efforts on answering the questions that lead to the "Don't know" responses.

Keywords Climate change • Global warming • Public opinion • Polls

4.1 Introduction

The vast majority of scientists and researchers agree that man-made emissions are likely exacerbating climate change; since 2007 no scientific body has disagreed with this position (Oreskes 2007). However, it is clear from US news articles to the contrary (e.g., Taylor 2013) that Americans remain deeply divided on whether anthropogenic climate change exists. As this makes many forms of climate change legislation nearly impossible to achieve a bipartisan consensus, this begs further study into how and why respondents answer "Is global warming happening?"

K. Klima, Ph.D. (✉)
Department of Engineering and Public Policy, Carnegie Mellon University, Pittsburgh, PA, USA
e-mail: kklima@andrew.cmu.edu

J.L. Drake et al. (eds.), *Communicating Climate-Change and Natural Hazard Risk and Cultivating Resilience*, Advances in Natural and Technological Hazards Research 45, DOI 10.1007/978-3-319-20161-0_4

4.2 Background: A Snapshot of Current Beliefs

To begin to understand the breadth of the disagreement, we examined news headlines from June 2014. The following three articles presumably were generated in response to the US Environmental Protection Agency's (EPA) June 2, 2014, proposal to regulate utilities' greenhouse gas emissions (EPA 2014). First, in New York Times, "Is Global Warming Real? Most Americans Say Yes" suggests that most Americans believe in global warming (Kopicki 2014). Second, the Washington Post/ABC News in "Broad Concern about Global Warming Boosts Support for New EPA Regulations" is more lukewarm, but still suggests that most Americans are in favor (Langer 2014). Third, the Pittsburgh Post-Gazette's "Pennsylvania voters favor EPA greenhouse gas curbs, poll shows" also shows that that Americans believe in global warming (Hopy 2014). Based on these headlines, it would seem clear that the majority of Americans believe global warming is real and support regulation of greenhouse gas emissions.

Then let's take a longer-term perspective to try and better understand what Americans really think. Figure 4.1 shows America's response to Stanford University,

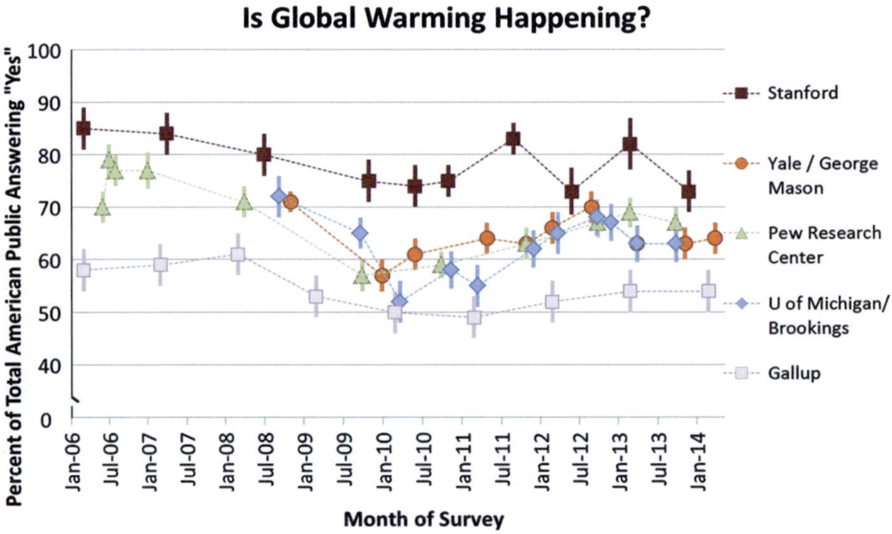

Fig. 4.1 Percent of total American public answering "Yes" to the question "Is global warming happening?" *Top*: Where possible, interpret *loosely*, as in "Temperature may already have or may in the future warm; no causal attribution." *Bottom*: Where possible, interpret *strictly*, as in "Temperatures have been increasing; this is at least partially caused by man." The *vertical bars* give the polling error indicated by the survey authors (Sources: Yale (Leiserowitz et al. 2012a, b, c, 2013, 2014), U Michigan/Brookings (Borick and Rabe 2012a, b, 2013, 2014), Stanford (Krosnick and MacInnis 2011; Krosnick 2012; Jordan 2013; Harrison 2013; Resources for the Future 2014), Gallup (Saad 2012, 2013, 2014), and Pew (2011, 2012, 2013, 2014))

Yale University/George Mason, the Pew Research Center, the University of Michigan (previously Brookings), and Gallup polls asking basically the same question: "Is global warming happening?" While the differences among polls likely occur due to question wording, one stark realization stands out. Since 2006, a majority (50–85 %) of Americans have agreed that global warming is happening, but there is also variance among those numbers. This variance is likely sufficient to influence policymakers' actions, particularly when broken out on a state, regional, or congressional district level.

4.3 Grounding the Problem: Understanding the Responses

To better understand why the different polls fail to agree, I decided to look more closely at the wording in each survey. Table 4.1 gives a breakdown of the slightly different slices of "Is global warming happening?" that each survey asks for, and the subsections below give more detail about each question. For each survey, I highlight the answer to the question "Is global warming happening?" in the text and in Fig. 4.1. If the option is given, I interpret this in two ways:

1. *Loosely* Where possible, interpret as "Temperature may already have or may in the future warm; no causal attribution."
2. *Strictly* Where possible, interpret as "Temperatures have been increasing; this is at least partially caused by man."

Table 4.1 A breakdown of the slightly different slices of "Is global warming happening?" that each survey asks for

Research group	Time frame	Certainty level	Type of change	Attribution
Stanford	100 years ago to present	May have been going up	Slow change in temperature	None specified
Yale/George Mason	150 years ago to unspecified future time	Has been increasing/may be increasing	Change in temperature may affect climate	None specified
Pew Research Center	Unspecified historic time to present	Solid evidence	Earth is warming	Human activity, natural patterns, don't know
University of Michigan/Brookings	40 years ago to present	Solid evidence	Earth is warming	None specified
Gallup	Unspecified historic time to different future ranges	Will or has already happened	Effects of global warming are happening	None specified

4.3.1 Stanford

Stanford Woods Institute Senior Fellow Jon Krosnick has been conducting public opinion studies on climate change and related efforts since at least 2006 (Krosnick and MacInnis 2011; Krosnick 2012; Stanford 2013; Harrison 2013; Jordan 2013). Now run by Stanford University's Political Psychology Research Group (Resources for the Future 2014), this survey question continues to be run roughly every year:

Q: You may have heard about the idea that the world's temperature may have been going up slowly over the past 100 years. What is your personal opinion on this—do you think this has probably been happening, or do you think it probably has not been happening?
A: Has been happening, Has not been happening, Don't know/refuse?

The values we report in Fig. 4.1 are the "*Has been happening*" (73–85 %); approximately 2–3 % of respondents typically answer "*Don't Know/ refuse.*" The percentage point error is between ±3 and ±5 depending on the number of respondents for each survey. Likely, this has the highest value of "yes" for all the polls because it emphasizes a slow change over the last century and uses the word choice "probably."

4.3.2 Yale/George Mason

The Yale Project on Climate Change Communication and the George Mason University Center for Climate Change Communication have successfully collaborated for several years on this question (Leiserowitz et al. 2012a, b, c, 2013, 2014). The 2014 PIs, Anthony Leiserowitz, Geoff Feinberg, Seth Rosenthal, Edward Maibach, and Connie Roser-Renouf, have asked roughly twice a year:

Q: Recently, you may have noticed that global warming has been getting some attention in the news. Global warming refers to the idea that the world's average temperature has been increasing over the past 150 years, may be increasing more in the future, and that the world's climate may change as a result. What do you think? Do you think that global warming is happening?
A: Yes, No, Don't Know.

The values we report are the "*Yes*" (61–71 %); approximately 10–20 % of respondents typically answer "*Don't Know.*" The percentage point error is between ±2 and ±3 depending on the number of respondents each survey. Similarly to the Stanford poll, this poll adds uncertainty by using "*may*," so we would expect similar results. Unfortunately, this poll had a much higher response of "*Don't Know*," possibly due to the final phrase "*the world's climate may change as a result.*" To compare the two, if you assume that half of all "Don't Knows" would have answered "Yes," this polling will then agree quite strongly with the Stanford finding.

4.3.3 Pew Research Center

The Pew Research Center based in Washington, D.C., conducts nonpartisan, non-advocacy public opinion polling and demographic research. Regarding global warming (Pew 2011, 2012, 2013, 2014), it asks:

Q: Is there solid evidence the earth is warming?
A: Yes (because of human activity), Yes (because of natural patterns), Yes (don't know), No, Mixed evidence/Don't Know

The values we report are those answering *loosely*, "*Yes (any answer)*" (57–77 %), and those answering *strictly*, "*Yes (because of human activity)*" (34–47 %); approximately 6–10 % of respondents typically answer "*Mixed evidence/Don't Know.*" The percentage point error is ±3. Likely, this falls in the middle of the pack because they ask respondents about "solid evidence," as opposed to the first two polls that express high levels of uncertainty.

4.3.4 University of Michigan/Brookings

This poll was initially housed at Brookings as the National Survey of American Public Opinion on Climate Change (NSAPOCC) and is now run through University of Michigan's Center for Local, State, and Urban Policy's National Surveys on Energy and Environment (Borick and Rabe 2012a, b, 2013, 2014). There was no noticeable change when switch occurred in December 2011. They ask:

Q: Is there solid evidence that the average temperature on Earth has been getting warmer over the past four decades?
A: Yes, No, Don't know

The values we report are the "*Yes*" (52–72 %); here again the "*Don't knows*" are quite high at approximately 10–20 %. The percentage point error is between ±3 and ±4 depending on the number of respondents each survey. Likely, this falls lower because they ask respondents about "*solid evidence*" in the "*past four decades*," as opposed to the first two polls that express high levels of uncertainty over a century or longer.

4.4 Gallup

Gallup has extensive experience running polls on just about any US policy question that exists. Of our surveys, they are the only ones to have consistently asked their question the same month (March) of every year since at least 2006 (Saad 2012, 2013, 2014). Specifically,

Q: Which of the following statements reflects your view of when the effects of global warming will begin to happen:

A: They have already begun to happen, they will start happening within a few years, they will start happening within your lifetime, they will not happen within your lifetime, but they will affect future generations, or they will never happen?

We report those who answered *loosely,* everything except "*They will never happen*" (81–92 %), and those answering *strictly*, "*They have already begun to happen*" (49–61 %). The percentage point error is ±4. We suspect there is a low response rate in this instance because the question didn't ask about "*global warming*"; it asked about "*the effects of global warming.*" Also, this question has a much more specific timeline than the other questions.

4.5 Discussing the Effect of Demographics

Prior to interpreting these results, we must first ask whether the differences in response might be due to demographics of the respondents. It is a common practice in surveys to obtain a large enough respondent pool to include the views of the entire population and then weight particular responses to adjust the convenience sample into a true representative sample. In each of the surveys reported here, the researchers obtained sufficient participants (typically 1,000 or more) and then adjusted their results to reflect the entire United States. The uncertainty caused by this analysis is typically +/− 2 to 5 percentage points per response and reported in both the text and in Fig. 4.1. Since all of the surveys included have interpreted their results in this manner, the differences should not be due to demographics.

However, the question of demographics remains interesting. Consequently, many studies, including each of the five longitudinal studies referenced in this chapter, have examined respondent demographics. For instance, the Yale group has split their respondents into "Six Americas" with differing responses to whether global warming was real: alarmed, concerned, cautious, disengaged, doubtful, and dismissive (Leiserowitz et al. 2014). Each group has different demographics, including age, sex, race, and wealth. For instance, the "alarmed" group believes in global warming and is highly worried about the consequences. This group tends to be moderate to liberal Democrats who are women, older to middle aged, college educated, and of upper income levels. On the opposite end of the spectrum, the "dismissive" group tends to not believe global warming exists. This group tends to be high-income, well-educated, white men who are very conservative Republicans.

John Ramos (2014) examines whether education level affects the response. He used the 2010 Cooperative Congressional Election Study (CCES) asked over 50,000 respondents (CCES 2010):

Q: "From what you know about global climate change or global warming, which one of the following statements comes closest to your opinion?"

A: "Global climate change has been established as a serious problem, and immediate action is necessary," "There is enough evidence that climate change is taking place and some action should be taken," "We don't know enough about global climate change, and more research is necessary before taking any actions," "Concern about global climate change is exaggerated. No action is necessary," or "Global climate change is not occurring; this is not a real issue."

For this framing, education has little effect on views on climate change.

4.6 Lessons Learned

Given that each survey samples a demographically similar group of US citizens and therefore demographics are not causing the differences, the above analysis illustrates two points. First, *question wording makes a difference in the polling results*. For instance, recall we interpreted the question "Is global warming happening?" in the strictest sense possible: temperatures have been rising in the past and this is due to mankind's activities. This may not be the same interpretation that others might use. For instance, the Stanford question is only about temperatures; someone who believes that temperatures are increasing, regardless of the reason, would answer "yes." However, the Yale survey mentions global warming; a reader who believes temperatures are increasing but not because of human activity may answer "yes" or "no," depending on their understanding of whether "global warming" means warming caused by greenhouse gases in the atmosphere. Other differences between the questions arise; for instance, respondents are much more likely to agree with survey questions that ask about more gradual change over longer time periods and mention uncertainty, as opposed to questions that ask about fast and certain change over shorter time periods. Second, *researchers may want to focus their efforts on answering the questions that lead to the "Don't know" responses.* While we cannot infer how the 10–20 % of respondents in the Stanford and Michigan polls might have answered if the question was clearer, it may be that with clearer questions and answers to them, these reports would show very different "Yes" results.

4.7 Conclusions

This chapter discussed five longitudinal studies using various wording to ask Americans whether they believe global warming exists. After controlling for demographics, I find that question wording makes a difference, and researchers may want to focus their efforts on answering the questions that lead to the "Don't know" responses. As scientists and engineers, we have an obligation to help those who "Don't know" better understand climate change and the methods that can be used to mitigate emissions and adapt to the changes that may be in our future. Although it's not very easy to take a step outside of our normal comfort zone and explore the policy implications of our work, it's important that we do so. Yet, to truly prepare for our future, we will need to understand not only the science of climate change and engineering to mitigate greenhouse gas emissions and adapt to climate change but also the policy implications for the challenges of our time.

Acknowledgments I thank the Carnegie Mellon Electricity Industry Center and the Center for Climate and Energy Decision Making (created through a cooperative agreement between the NSF (SES-0345798) and Carnegie Mellon) for funding this work. Also, thanks for the significant support and input from the Center for Clean Air Policy (2012), the American Geophysical Union (2013), and the Scott Institute for Energy Innovation (2014) for publishing blogs leading up to the development of this work. Finally, I thank one anonymous reviewer who provided additional insights and wording to improve the chapter.

References

Borick C, Rabe B (2012a) The fall 2011 National survey of American public opinion on climate change. Issues Gov Stud 44:8

Borick C, Rabe B (2012b) Continued rebound in American belief in climate change: Spring 2012 NSAPOCC findings. Governance Studies at Brookings

Borick C, Rabe B (2013) The fall 2012 national surveys on energy and environment findings report for belief-related questions. The Center for Local, State, and Urban Policy, March, p. 9

Borick C, Rabe B (2014) The fall 2011 national survey of American public opinion on climate change. Issues Energ Environ Pol 7:8

CCES (2010) Cooperative congressional election study. http://projects.iq.harvard.edu/cces/home. Accessed 5 Mar 2015

Environmental Protection Agency (EPA n.d.) Carbon pollution emission guidelines for existing stationary sources: electric utility generating units. Fed Regist: Dl J United States Gov. 79 FR 67406. Proposed Rule, 13 Nov 2014

Harrison C (2013) Survey results: US views on climate adaptation. Report from Stanford Woods Institute for the Environment. https://woods.stanford.edu/sites/default/files/documents/Climate_Survey_Exec_Summ_US.pdf. Accessed 27 Feb 2015

Hopy D (2014) Pennsylvania voters favor EPA greenhouse gas curbs, poll shows. Pittsburgh Post-Gazette, 5 Jun 2014. http://www.post-gazette.com/news/politics-state/2014/06/05/Pennsylvania-voters-favor-EPA-greenhouse-gas-curbs-polls-show/stories/201406050309. Accessed 5 Mar 2015

Jordan R (2013) Americans back preparation for extreme weather and sea level rise. Stanford Woods Institute for the Environment, 28 Mar 2013. https://woods.stanford.edu/news-events/news/americans-back-preparation-extreme-weather-and-sea-level-rise. Accessed 27 Feb 2015

Kopicki A (2014) Is global warming real? Most Americans say yes. The New York Times, 1 Jun 2014. http://www.nytimes.com/2014/06/02/upshot/is-global-warming-real-most-in-US-believe-in-climate-change.html?partner=rss&emc=rss&smid=tw-nytimes&_r=1&abt=0002&abg=1. Accessed 27 Feb 2015

Krosnick J (2012) Washington Post-Stanford University survey data, June. http://www.washingtonpost.com/wp-srv/national/documents/global-warming-poll-2.pdf. Accessed 27 Feb 2015

Krosnick J, MacInnis B (2011) National survey of American public opinion on global warming: Stanford University with Ipsos and Reuters, September

Langer G (2014) Broad concern about global warming boosts support for new EPA regulations. ABC News, 2 Jun 2014

Leiserowitz A, Maibach E, Roser-Renouf C (2012a) Climate change in the American mind: public support for climate and energy policies in March 2012. Report from Yale project on climate change communication. George Mason University Center for Climate Change Communication

Leiserowitz A, Maibach E, Roser-Renouf C (2012b) Climate change in the American mind: Americans' global warming beliefs and attitudes in March 2012. Report from Yale project on climate change communication. George Mason University Center for Climate Change Communication

Leiserowitz A, Maibach E, Roser-Renouf C, Feinberg G, Howe P (2012c) Extreme weather and climate change in the American mind: September 2012. Report from Yale project on climate change communication. George Mason University Center for Climate Change Communication

Leiserowitz A, Maibach E, Roser-Renouf C (2013) Climate change in the American mind: Americans' global warming beliefs and attitudes in April 2013. Report from Yale project on climate change communication. George Mason University Center for Climate Change Communication

Leiserowitz A, Maibach E, Roser-Renouf C (2014) Climate change in the American mind: April 2014. Report from Yale project on climate change communication. George Mason University Center for Climate Change Communication

Oreskes N (2007) The scientific consensus on climate change: how do we know we're not wrong? In: DiMento JF, Doughman P (eds) Climate change: what it means for us, our children, and our grandchildren. MIT Press, Cambridge, pp 65–71

Pew Research Center for People and the Press (2011) Modest rise in number saying there is 'solid evidence' of global warming: more moderate republicans see evidence of warming, 1 Dec 2011. http://www.people-press.org/2011/12/01/modest-rise-in-number-saying-there-is-solid-evidence-of-global-warming/?src=prc-headline. Accessed 27 Feb 2015

Pew Research Center for People and the Press (2012) More say there is solid evidence of global warming, 15 Oct 2012. http://www.people-press.org/2012/10/15/more-say-there-is-solid-evidence-of-global-warming/2/. Accessed 27 Feb 2015

Pew Research Center for People and the Press (2013) Keystone XL pipeline draws broad support: Continuing partisan divide in views of global warming, 2 Apr 2013. http://www.people-press.org/2013/04/02/keystone-xl-pipeline-draws-broad-support/. Accessed 27 Feb 2015

Pew Research Center for People and the Press (2014) Climate change: key data points from Pew Research, 27 Jan 2014. http://www.people-press.org/2013/04/02/keystone-xl-pipeline-draws-broad-support/. Accessed 27 Feb 2015

Ramos JL (2014) Education does not explain views on climate change well, and is not why the two parties disagree, 24 Jul 2014. http://www.insidesources.com/education-does-not-explain-views-on-climate-change-well-and-is-not-why-the-two-parties-disagree/. Accessed 27 Feb 2015

Resources for the Future (2014) Surveying American attitudes toward climate change. http://www.rff.org/RFF/Documents/RFF-Resources-186_Infographic.pdf. Accessed 27 Feb 2015

Saad L (2012) In US, global warming views steady despite warm winter. Gallup Politics, 30 Mar 2012. http://www.gallup.com/poll/153608/global-warming-views-steady-despite-warm-winter.aspx. Accessed 27 Feb 2015

Saad L (2013) Americans' concerns about global warming on the rise. Gallup politics, 8 Apr 2013. http://www.gallup.com/poll/161645/americans-concerns-global-warming-rise.aspx. Accessed 27 Feb 2015

Saad L (2014) One in four in US are solidly skeptical of global warming. Gallup politics, 22 Apr 2014. http://www.gallup.com/poll/168620/one-four-solidly-skeptical-global-warming.aspx. Accessed 27 Feb 2015

Stanford University (2013) Stanford University climate adaptation national poll conducted by GfK Custom Research North America, March. https://woods.stanford.edu/sites/default/files/documents/Climate-Adaptation-Results-TOPLINE.pdf. Accessed 27 Feb 2015

Taylor J (2013) Peer-reviewed survey finds majority of scientists skeptical of global warming crisis. Forbes, 13 Feb 2013. http://www.forbes.com/sites/jamestaylor/2013/02/13/peer-reviewed-survey-finds-majority-of-scientists-skeptical-of-global-warming-crisis/. Accessed 5 Mar 2015

Chapter 5
Building Interfaces That Work: A Multi-stakeholder Approach to Air Pollution and Climate Change Mitigation

Julia Schmale, Erika von Schneidemesser, Ilan Chabay, Achim Maas, and Mark G. Lawrence

Abstract Air pollution and climate change are two major environmental problems. These issues are not only inextricably linked with regard to their effects and mitigation options but also through their causes that include human behavior, infrastructures, technology, and other factors. This implies that societal transformation to a sustainable human-atmosphere relationship will require the involvement of the many facets of society for discussions of normative and value-related issues for the codesign of salient and legitimate solutions. We describe the ClimPol project and a specific subproject *Mobility and Climate*, which create integrated and long-lasting strategies by applying a transdisciplinary approach together with the framework of *coupled-human-atmosphere-systems-thinking*.

Keywords Transdisciplinary • Air pollution • Climate change • Coupled thinking

5.1 Introduction

Humankind has been changing the face of the Earth and the composition of the atmosphere to ever-increasing degrees so that the current geological epoch is suggested to be termed the *Anthropocene* (Crutzen 2002). Human activities related to transport, energy, food production, and industry, for example, emit large quantities of a variety of gases and particles to the atmosphere that lead to air pollution and

J. Schmale, Ph.D. (✉)
Institute for Advanced Sustainability Studies, Potsdam, Germany

Paul Scherrer Institute, Villigen, Switzerland
e-mail: julia.schmale@gmail.com

E. von Schneidemesser, Ph.D. • I. Chabay, Ph.D. • A. Maas, M.A. • M.G. Lawrence, Ph.D.
Institute for Advanced Sustainability Studies, Potsdam, Germany

J.L. Drake et al. (eds.), *Communicating Climate-Change and Natural Hazard Risk and Cultivating Resilience*, Advances in Natural and Technological Hazards Research 45, DOI 10.1007/978-3-319-20161-0_5

climate change, which are currently critical environmental problems. Air pollution has now been classified as the single most important global environmental cause of premature death (OECD 2014). And according to the fifth IPCC assessment report, it is more likely than not that the global mean surface temperature will exceed the 2 °C threshold by 2100 (IPCC 2013). This has been recognized as a threshold above which dangerous anthropogenic interference with the climate system becomes increasingly likely.

The obstacles to sustainable solutions to these issues are manifold. Here, we address three of the obstacles together: (a) the failure of applying knowledge regarding links between air pollution and climate change, (b) the structure of environmental policies on related topics, and (c) the communication between science, policy, and society. We do so by applying a framework of *coupled-human-atmosphere-systems-thinking* (CHAST) in combination with a transdisciplinary research approach developed in our project, Short-lived *Clim*ate-forcing *Pol*lutants: Research Needs and Pathways to *Pol*icy Implementation (ClimPol). We will first elaborate on the challenges to be overcome and introduce the transdisciplinary research approach and CHAST. Then we present and discuss results from their combined application in the sub-project *Mobility and Climate* carried out with the city administration of Potsdam, Germany. The objective of this sub-project was to prioritize measures fostering long-lasting solutions that would jointly reduce air pollution and climate change impacts from the local traffic sector.

5.2 Challenges

5.2.1 Applying Knowledge on the Linkages Between Air Pollution and Climate Change

Air pollution and climate change are inextricably linked with regard to their causes, effects, and mitigation options (e.g., Jacob and Winner 2009). While the general linkage might be apparent to decision-makers, the relevant details which are fundamental to develop adequate mitigation strategies are often not known. One crucial reason for this is that science provides few, if any, comparable metrics for measuring the effect of these two issues (Schmale et al. 2014a). Some of the main scientific points that would have to be worked out more specifically to adequately support decision-making at multiple levels are elaborated in the following text.

Many air pollutants and greenhouse gases are emitted from the same sources, especially through combustion processes for energy production, transport, heating, and cooking, among others. In addition to their detrimental effects on health and ecosystems, certain air pollutants have warming or cooling effects on the climate (Ramanathan and Xu 2010). Benefits for climate change mitigation can be achieved through reducing black carbon and ozone precursors, such as methane, resulting in roughly 0.5 °C less global mean warming by 2050 (Shindell et al. 2012). Furthermore, by 2030 the global annual premature deaths due to $PM_{2.5}$ and ozone could be reduced by 0.6–4.4 and 0.04–0.52 million persons, respectively, if the 14 most effective, currently available measures as identified by Shindell et al., were to be implemented (Anenberg et al. 2012).

Conversely, greenhouse gases leading to climate change have the potential to aggravate air pollution, e.g., through more frequent high-pressure blocking leading to stagnant weather patterns or increased emissions owing to higher temperatures (e.g., Brasseur 2009). It has also been shown that more ambitious climate mitigation strategies are beneficial to air quality (e.g., Williams 2012). Especially given current urbanization trends, treating air quality, and climate change challenges jointly have the potential to be highly beneficial for a large number of people.

Cities are subject to confounding effects from industry and transport, as well as the built environment, population density, and other added stresses that lead to feedbacks, making them and their citizens particularly vulnerable (Harlan and Ruddell 2011). However, no specific jobs or task forces generally exist in city or regional administrations to coordinate efforts and expenditures for air quality and climate change mitigation to make use of synergistic effects (Schmale et al. 2014b).

5.2.2 Policy and Regulatory Environments

In the regulatory and policy environments, air quality management and climate change mitigation are commonly treated separately across all levels from local to global scale. There are a number of reasons for this thematic separation. For example, due to the short atmospheric lifetime of air pollutants (days/weeks to months), air quality is a local to regional problem, and mitigation generally leads to immediate and local benefits. Air pollution is covered by national and regional legislation (e.g., the Air Quality Directive 2008/50/EC in Europe, the Clean Air Act in the United States). It has been recognized as a problem since the late nineteenth century and has been regulated since the early twentieth century owing to strong pollution episodes that caused significant numbers of premature deaths, for example, in 1952 in London (Bell et al. 2004).

In comparison, climate change has only become an issue in the past few decades and has been addressed primarily at an international level since the establishment of the Intergovernmental Panel on Climate Change in 1988 and the 1992 United Nations Framework Convention on Climate Change. In contrast to air pollution, the actual location of long-lived greenhouse gas emission sources is less relevant for global climate change, and reducing emissions has no immediate local effect. International negotiations for substantial, binding emission reductions have shown limited progress so far. Only recently have some countries introduced binding climate legislation or targets on a national level (e.g., the UK Climate Change Act from 2008).

There has been a hesitancy to integrate the two areas, especially because of the concern that the slow progress in climate change negotiations might hinder improvements in air quality, leading to immediate detrimental effects on health and ecosystems that could otherwise be prevented. Nevertheless, this separation leads to trade-offs, meaning that some policies that focus on mitigation of either air pollution or climate change will simultaneously exacerbate the other. For example, reducing sulfur dioxide is directed to the protection of human health and ecosystems. However, sulfur dioxide is oxidized to form sulfate-containing particles in the atmosphere that reflect sunlight and have a cooling effect by masking the anthropogenic greenhouse effect (Ramanathan and Xu 2010).

Emission mitigation (or transformation) technologies that are added at the point of emission, also known as *end-of-pipe* measures, can yield mixed results. For example, scrubbers on power plants specifically reduce sulfur dioxide, but do not remove any other species that might be climate-warming substances, which means that these scrubbers will be beneficial for air quality but will contribute to warming by their removal of cooling substances from the atmosphere. Conversely, some *green* technologies, so dubbed because of their reduction of CO_2 emissions relative to conventional technologies, can actually have adverse impacts on air quality. An example of this is wood-fueled stoves for residential heating. While these stoves have lower CO_2 emissions relative to heating with fossil fuel combustion, the emissions of particulate matter, including black carbon, are typically substantially greater, thus reducing air quality.

In addition to the connections through their impacts, there are also socioeconomic benefits to air quality from action on climate change, which have been estimated to be 35€ per ton of CO_2 avoided on average, which is tremendous given the current global emissions of over 30 billion tons of CO_2 (EPA 2014; Nemet et al. 2010,). Generally, more ambitious climate change mitigation strategies will lead to infrastructure changes in source sectors such as energy and transportation resulting in solutions that are long-lasting and simultaneously more beneficial to air quality than nearsighted, end-of-pipe technology deployment (McCollum et al. 2013). Making use of the potential for synergies will, however, require integrated approaches across multiple sectors, such as agriculture, transport, the built environment, conservation, and energy policies. Applying multi-criteria assessments across different policy options and specific implementation measures can help to explicitly identify benefits and trade-offs, which can support informed choices for integrated policies (Schmale et al. 2014b).

5.2.3 Collaboration Among Science, Policy, and Civil Society Communities

Integrated responses to the environmental challenges of the Anthropocene are needed. Conventionally trained scientists and societal actors have difficulty in grappling with such challenges (Hirsch Hadorn et al. 2006), mainly because traditional science often delivers only meticulously separated information (Tàbara and Chabay 2013). In many instances, there are sufficient incentives and knowledge to act, yet there remains an obvious divide between the available knowledge and actions taken (Cash et al. 2003).

Much of science, including air quality and climate change impact research, is still conducted in the *mode 1* production of knowledge (Gibbons et al. 1994): research that happens in a highly academic context among disciplinary experts. Knowledge is delivered to a wider audience unidirectionally after the production of results, rather than incorporating stakeholders in the knowledge generation process from the beginning (Roux et al. 2006). Approaches to managing the boundary between knowledge and action include *mode 2* science, transdisciplinary research, the continuous engagement model, or the cogeneration of usable knowledge (Dilling and Lemos 2011; Gibbons et al. 1994; Hessels and van Lente 2008; Hirsch Hadorn et al. 2006). All these aim at jointly engaging stakeholders and scientists throughout the process of knowledge generation. Knowledge generation in this context means that all relevant information, data, research results, and value judgments, among others, provided by the stakeholders—including scientists—are integrated and used to develop a problem-specific knowledge that can lead to solution strategies. The transdisciplinary approach, as applied here (see *The Transdisciplinary Approach* in Sect. 5.3.1), is especially appropriate for identifying integrated policy solutions for local to regional air pollution problems and the global climate change challenge, because it can deliver context-specific solutions for the local level that simultaneously serve the global community (Blok et al. 2012; Pohl 2008).

5.3 Methods

In the ClimPol project (http://climpol.iass-potsdam.de), in order to jointly develop integrated solution options for the simultaneous mitigation of air pollution and climate change with stakeholders, we applied a transdisciplinary research approach, as well as *coupled-human-atmosphere-systems-thinking* (CHAST). While CHAST provides the theoretical framework to create solution options, the transdisciplinary research approach is a methodology facilitating its application in multi-stakeholder projects.

5.3.1 The Transdisciplinary Approach

Generally, transdisciplinary (TD) research involves different disciplinary scientist as well as collaborators outside of science. The purpose is to cocreate solution options for sustainability challenges involving complex societal concerns (Funtowicz and Ravetz 1993; Hirsch Hadorn et al. 2006). We followed the basic process as outlined by Lang et al. (2012). The first phase is dedicated to problem framing and team building (see Fig. 5.1 stage A). Here, it is important to define a common *real-world* problem among the partners from policy, society, and science. In addition, *boundary objects*—which are objects of common concern between science and

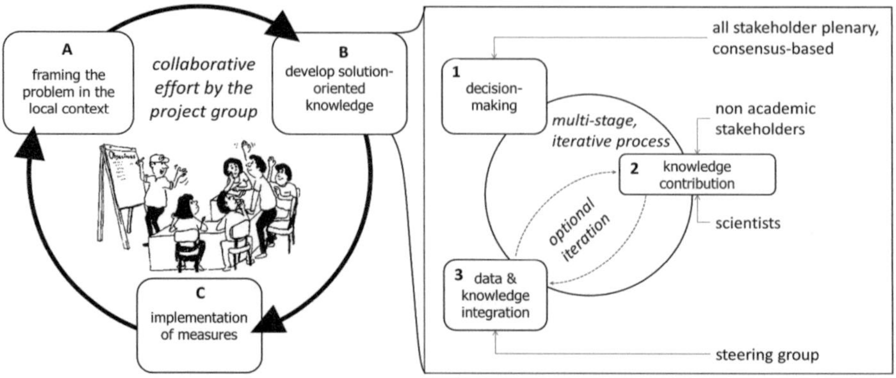

Fig. 5.1 The iterative and multistage work cycle. *Left*: overall cycle. *Right*: the procedure applied here for solution-oriented knowledge development (Adapted from Schmale et al. 2015)

practice, for example, an emissions inventory—are defined at this stage that are both researchable and advance science and allow for implementation in the local context (Clark et al. 2011). In the second phase (see Fig. 5.1, box B), solution options are developed jointly with all partners. This is an iterative process (see right box in Fig. 5.1), which is performed as many times as necessary in the overall project. For each cycle, the objectives and deliverables of each subphase are defined with as many project partners as appropriate. Thereafter, all partners gather their relevant knowledge or conduct targeted research (see box 2 in Fig. 5.1), which is *collected* by the project leadership and integrated for the purpose of finding solution strategies (see box 3 in Fig. 5.1).

The integrated knowledge is presented to the partners, in a form that is understandable for all, and discussed to codevelop the solution (see box 1 in Fig. 5.1). The ClimPol project aims at creating environments in which mutual learning can happen to increase all partners' capabilities to address the challenge, as well as to foster consensus-based decisions before the project enters the next phase. This increases the buy-in and ownership among all partners (Hirsch Hadorn et al. 2006). The different types of knowledge most often included in the context of finding integrated solutions for air quality and climate change mitigation are natural and social sciences, practical and locally relevant knowledge from city administrations and environment agencies, policy and regulatory information from decision-makers, and normative perspectives from NGOs.

Typically, the cycle will be applied at least three times in order to create integrated systems, target, and transformative knowledge (ProClim 1997). Systems knowledge is used to describe the status quo of the problem as comprehensively as possible. Target knowledge is needed to describe the possible desirable states of the system in the future. This step often includes scenario building (Stauffacher and Scholz 2013). In order to choose which of the scenarios is most desirable, within ClimPol, multi-criteria assessments are developed among all partners to allow for objective, as well as subjective and value-based judgments. Transformative knowledge is developed to enable change and lead into the third phase according to Lang et al. (2012, p. 28): "(re)integration and application of integrated knowledge" (see box C in Fig. 5.1).

5.3.2 A Framework for Integrated Solution Options*:* *Coupled-Human-Atmosphere-Systems-Thinking*

Coupled-human-atmosphere-systems-thinking (CHAST), developed by the authors of this chapter, provides a theoretical framework that connects two main challenges. First, it considers all anthropogenic emissions to the air at the same time, meaning that there is no isolated consideration of emissions and mitigation options of air pollutants or greenhouse gases. Second, solution options for mitigation are sought that directly address human behavior and its socio-technological context and the built environment responsible for the emissions.

CHAST requires collecting knowledge specific to the problem context that relates the current state of the system (system knowledge) to impacts through air pollution and climate change, as well as knowledge of the desired state of the system (target knowledge). Therefore, it is necessary to bridge the disciplinary sciences, including natural and social sciences and humanities. Regarding solution options, CHAST asks for context-specific solutions for the local problem while simultaneously thinking of implications for the global community to bridge the scales of impacts of air pollution and climate change.

Many of the current solutions regarding emissions to the atmosphere are first-order technical approaches using end-of-pipe technologies like diesel particulate filters for cars. Second-order solutions often involve energy efficiency, including the promotion of enhanced devices, e.g., low emission vehicles. Both types of solutions accommodate current patterns of behavior, reinforcing the status quo. For example, new car technology allows individuals to continue using their private car instead of switching to public transport or other alternatives. This is linked to the challenge that a transition to sustainability is coupled with and aggravated by the strong path dependencies and lock-ins in the existing sectors (Åhman and Nilsson 2008).

Established technologies and their improvements are highly intertwined with user practices and life styles, complementary technologies, business models, regulations, and even political structures (Rip and Kemp 1998). As a consequence, established socio-technological systems undergo incremental rather than radical changes (Markard et al. 2012). In addition, environmental protection, air quality regulation, and climate change mitigation are usually reactive responses, while society's attention is much more attracted by catastrophic one-time events, rather than the dangers of underlying fundamental structures and infrastructures (Wiek et al. 2012). For that reason, CHAST aims to foster long-term solutions that involve changes in infrastructure and behavior. Here, stakeholder involvement is important, because societal transformations for the sake of a sustainable human-atmosphere relationship are highly context dependent and will require discussions of normative and value-related issues, which cannot be led by scientists alone but need to include many facets of society for the codesign of salient and legitimate solutions (Seidl et al. 2013).

5.4 Results and Discussion

Here, we present the basic design of the ClimPol project and lessons learned from the first full application of TD and CHAST in the sub-project *Mobility and Climate*. The project was set up to function at the interface between the air quality and climate change problems, contributing to science, policy, and society. ClimPol focuses on creating solutions with stakeholders in mutual learning environments at multiple levels and identifying pathways to policy implementation for integrated and socio-technological solutions. It aims to improve understanding and awareness that air pollution and climate change problems are inexorably linked and that changes in human behavior, the built environment, and infrastructure are directly relevant to reducing any kind of emissions in the long term. The project is built on three main pillars:

1. Codesigning usable knowledge and solution options to find pathways to policy implementation through collaboration with decision-making and decision-informing stakeholders at different administrative and policy levels
2. Setting salient research agendas by communicating informational needs from the nonacademic world to the scientific community
3. Developing narratives that is communication that is not only informational but also emotionally resonant and memorable, to improve understanding of the issues and motivate action

In the sub-project *Mobility and Climate*, the ClimPol project collaborated with the city administration of Potsdam, Germany, to develop a priority list of measures in the urban transport sector to be implemented in the following 6 years. A new method was developed for integrated assessment of impacts from traffic-related measures on air quality, climate change, noise pollution, safety, eco-mobility, and quality of life.

In brief, the city had identified more than 100 measures based on a variety of regulatory and strategic planning documents, such as an air quality management plan, a local climate strategy, and a city development concept, among others. Since the identified measures served different primary objectives, it was unclear to which extent they were complementary, contradictory, or overlapping. As the implementation of all measures could not be financially supported, the mayor commissioned a project group to identify priority measures. The group consisted of ten representatives from the relevant departments in the municipality. In addition, one person each from the state province's environmental protection agency and the ministry for environment, as well as two natural scientists from the ClimPol project, joined the group. The areas of expertise covered road traffic control, city and traffic development, traffic infrastructure, traffic management, air quality, climate change, and public relations.

The work of the project group followed the three stages of the transdisciplinary process as outlined (see *The Transdisciplinary Approach* in Sect. 5.3.1). First, the problem was collaboratively framed and the to-be-developed integrated assessment approach was identified as the *boundary object* between science and practice. In the second phase, the multi-criteria assessment method was jointly developed integrating all represented knowledge types, following the iterative process shown in Fig. 5.1 (right box). All measures were assessed by the multi-criteria method that considered impacts on six categories: air quality, climate change, noise emissions, road safety, eco-mobility, and quality of life. Subsequently, the results were averaged and different weighting factors attributed to the categories, to give greater weight to those categories responding to legal obligations and city targets and less weight to more subjective categories. Based on this, an average index value per measure was calculated, resulting in a ranking where measures with the highest synergies received higher priorities. In the third phase, the proposed priority list was discussed in the city financial planning committee with the result that a budget was fixed for the next 6 years to implement the measures generally following their ranking. A full description of the work can be found elsewhere (Schmale et al. 2015).

This project clearly profited from the transdisciplinary approach, because the prioritization of measures would not have been grounded on sound science and practical knowledge if practitioners and scientists had not worked together. The involvement of the scientists gave the results higher credibility, which in turn generated higher buy-in from all affected city departments. This credibility and buy-in is important for the provision of financial means to guarantee the implementation of the priority measures. In addition, the process was also successful with regard to ClimPol objectives.

The project started under the headline of *Mobility and Climate* but developed a more holistic vision through applying the CHAST framework in the multi-criteria assessment that clearly highlighted co-benefits and trade-offs of individual measures for air quality, climate change, and beyond. For example, enlarging and incentivizing the park and ride facilities outside of the city center would simultaneously reduce emissions of air pollutants, greenhouse gases, and noise, while increasing the share of eco-mobility passengers within the city. Linking the city's objective of increasing eco-mobility to the integrated assessment highlighted the importance of *soft* measures, such as the creation of a mobility agency, which would motivate and help citizens to change their mobility behavior over the long term.

5.5 Conclusions and Outlook

The impacts owing to anthropogenic emissions to the atmosphere result in significant environmental problems that humanity faces. To identify strategies, a number of challenges need to be overcome, including the lack of awareness regarding linkages of air quality and climate change, the separation of the two issues in thematic policies and regulations, and the collaboration between society, policy makers, and scientists. Here, we have shown how the combination of a transdisciplinary approach together with the framework of *coupled-human-atmosphere-systems-thinking*, such as in the context of the ClimPol project, can successfully address these underlying problems. The ClimPol project aims to codevelop with stakeholders practically relevant solutions for an integrated approach to mitigating air pollution and climate change, with an emphasis on the need for changes in infrastructure and behavior. In the Potsdam project group on *Mobility and Climate*, we demonstrated the successful application of the combined TD and CHAST approach for the first time.

The approach applied by ClimPol is not a silver bullet, but it will yield benefits primarily for those who implement the codeveloped solutions at the local to national scale. Currently, in the atmospheric science community, as in many natural science disciplines, transdisciplinary research is still a novel method, and more projects at the leading edge of the development and application of the transdisciplinary approach are needed to reach a critical mass. Developments at the international level will be supported by the formation of Future Earth (http://www.icsu.org/future-earth) and its emphasis on socially relevant solutions through transdisciplinarity, as well as initiatives such as the International and Global Atmospheric Chemistry project that fosters integration of air pollution and climate change mitigation (IGBP/IGAC 2012). On the level of education and career opportunities, however, further development is urgently needed to prepare the next generation of scientists for the challenge and also to develop adequate mechanisms for recognition of the value of experience with the transdisciplinary approach for future career pathways.

References

Åhman M, Nilsson LJ (2008) Path dependency and the future of advanced vehicles and biofuels. Util Polic 16(2):80–89

Anenberg SC, Schwartz J, Shindell D, Amann M, Faluvegi G, Klimont Z, Ramanathan V (2012) Global air quality and health co-benefits of mitigating near-term climate change through methane and black carbon emission controls. Environ Health Perspect 120(6):831–839

Bell ML, Davis DL, Fletcher T (2004) A retrospective assessment of mortality from the London smog episode of 1952: the role of influenza and pollution. Environ Health Perspect 112(1):6

Blok K, Höhne N, Van Der Leun K, Harrison N (2012) Bridging the greenhouse-gas emissions gap. Nat Clim Chang 2(7):471–474

Brasseur G (2009) Implications of climate change for air quality. WMO Bull 58(1):10–15

Cash DW, Clark WC, Alcock F, Dickson NM, Eckley N, Guston DH, Mitchell RB (2003) Knowledge systems for sustainable development. Proc Natl Acad Sci U S A 100(14):8086–8091

Clark WC, Tomich TP, van Noordwijk M, Guston D, Catacutan D, Dickson NM, McNie E (2011) Boundary work for sustainable development: natural resource management at the Consultative Group on International Agricultural Research (CGIAR). Proc Natl Acad Sci. doi:10.1073/pnas.0900231108

Crutzen PJ (2002) Geology of mankind: the anthropocene. Nature 415:23

Dilling L, Lemos MC (2011) Creating usable science: opportunities and constraints for climate knowledge use and their implications for science policy. Glob Environ Chang 21(2):680–689

EPA (2014) Global greenhouse gas emissions data. http://www.epa.gov/climatechange/ghgemissions/global.html. Accessed 14 Jul 2014

Funtowicz SO, Ravetz JR (1993) Science for the post-normal age. Futures 25(7):739–755

Gibbons M, Limoges C, Nowotny H, Schwartzmann S, Scott P, Trow M (1994) The new production of knowledge: the dynamics of science and research in contemporary societies. Sage, London

Harlan SL, Ruddell DM (2011) Climate change and health in cities: impacts of heat and air pollution and potential co-benefits from mitigation and adaptation. Curr Opin Environ Sustain 3(3):126–134

Hessels LK, van Lente H (2008) Re-thinking new knowledge production: a literature review and a research agenda. Res Polic 37(4):740–760. doi:10.1016/j.respol.2008.01.008

Hirsch Hadorn G, Bradley D, Pohl C, Rist S, Wiesmann U (2006) Implications of transdisciplinarity for sustainability research. Ecol Econ 60(1):119–128. doi:10.1016/j.ecolecon.2005.12.002

IGBP/IGAC (2012) Time to act: the opportunity to simultaneously mitigate air pollution and climate change, International Geosphere-Biosphere Programme (IGBP) and International Global Atmospheric Chemistry (IGAC) Project, p 6. http://www.accent-network.org/accent_documents/IGBP_IGAC_AirPolClim_Statement_FINAL.pdf. Accessed 28 Jul 2013

IPCC (2013) Summary for policymakers. In: Climate change 2013: the physical science Basis. Contribution of working group I to the fifth assessment report of the intergovernmental panel on climate change. Stocker TF, Qin D, Plattner G-K, Tignor M, Allen SK, Boschung J, Nauels A, Xia Y, Bex V and Midgley PM (eds). Cambridge University Press, Cambridge, UK/New York. http://www.ipcc.ch/pdf/assessment-report/ar5/wg1/WGIAR5_SPM_brochure_en.pdf. Accessed 30 Jun 2014

Jacob DJ, Winner DA (2009) Effect of climate change on air quality. Atmos Environ 43(1):51–63

Lang DJ, Wiek A, Bergmann M, Stauffacher M, Martens P, Moll P, Thomas CJ (2012) Transdisciplinary research in sustainability science: practice, principles, and challenges. Sustain Sci 7(1):25–43

Markard J, Raven R, Truffer B (2012) Sustainability transitions: an emerging field of research and its prospects. Res Polic 41(6):955–967

McCollum DL, Krey V, Riahi K, Kolp P, Grubler A, Makowski M, Nakicenovic N (2013) Climate policies can help resolve energy security and air pollution challenges. Clim Change 119(2):479–494

Nemet GF, Holloway T, Meier P (2010) Implications of incorporating air-quality co-benefits into climate change policymaking. Environ Res Lett 5(1):014007–014015

Organisation for Economic Co-operation and Development (2014) The cost of air pollution. OECD Publishing, Paris

Pohl C (2008) From science to policy through transdisciplinary research. Environ Sci Polic 11(1):46–53. doi:10.1016/j.envsci.2007.06.001

ProClim (1997) Research on sustainability and global change—visions in science policy by Swiss researchers. Sciences Academy of Sciences, Bern

Ramanathan V, Xu YY (2010) The Copenhagen accord for limiting global warming: criteria, constraints, and available avenues. Proc Natl Acad Sci U S A 107(18):8055–8062. doi:10.1073/pnas.1002293107

Rip A, Kemp R (1998) Technological change. In: Rayner S, Malone EL (eds) Human choice and climate change resources and technology. Battelle, Columbus, pp 327–399

Roux DJ, Rogers KH, Biggs HC, Ashton PJ, Sergeant A (2006) Bridging the science-management divide: moving from unidirectional knowledge transfer to knowledge interfacing and sharing. Ecol Soc 11(1):4 [online]. http://www.ecologyandsociety.org/vol11/iss1/art4/

Schmale J, von Schneidemesser E, Van Aardenne J (2014a) New directions: support for integrated decision-making in air and climate policies—development of a metrics-based information portal. Atmos Environ 90:146–148

Schmale J, Shindell D, von Schneidemesser E, Chabay I, Lawrence MG (2014b) Clean up our skies. Nature 515:335–337. doi:10.1038/515335a

Schmale J, von Schneidemesser E, Dörrie A (2015) An integrated assessment method for sustainable transport system planning in a middle sized German city. Sustainability 7(2):1329–1354. doi:10.3390/su7021329

Seidl R, Brand F, Stauffacher M, Krütli P, Le Q, Spörri A, Scholz R (2013) Science with society in the anthropocene. AMBIO 42(1):5–12. doi:10.1007/s13280-012-0363-5

Shindell D, Kuylenstierna JCI, Vignati E, van Dingenen R, Amann M, Klimont Z, Fowler D (2012) Simultaneously mitigating near-term climate change and improving human health and food security. Science 335(6065):183–189. doi:10.1126/science.1210026

Stauffacher M, Scholz RW (2013) HES based transdisciplinary case studies: the example of sustainable transformation of leisure traffic in the city of Basel. In: Mieg HA, Toepfer K (eds) Institutional and social innovation for sustainable urban development. Routledge, London/New York, pp 25–43

Tàbara JD, Chabay I (2013) Coupling human information and knowledge systems with social-ecological systems change: reframing research, education, and policy for sustainability. Environ Sci Polic 28:71–81

Wiek A, Farioli F, Fukushi K, Yarime M (2012) Sustainability science: bridging the gap between science and society. Sustain Sci 7(1):1–4

Williams M (2012) Tackling climate change: what is the impact on air pollution? Carbon Manag 3(5):511–519

Part II
Before Disaster: New Methods to Predict, Prepare, and Prevent the Worst

Chapter 6
Fostering Resilience in the Face of an Uncertain Future: Using Scenario Planning to Communicate Climate Change Risks and Collaboratively Develop Adaptation Strategies

Nancy Fresco and Kristin Timm

Abstract Communicating about climate change is a serious challenge. There is an urgent need to develop effective processes at the local level to engage, inform, and support decision-makers in their efforts to plan for the impacts of climate change. This is particularly urgent in Alaska, where the impacts of climate change are already being felt. The Scenarios Network for Alaska and Arctic Planning (SNAP) at the University of Alaska Fairbanks has spent the past 7 years facilitating the use of scenarios for climate change planning. In this case study highlighting SNAP's collaboration with the National Park Service (NPS), we explore how the scenario planning process can be applied to support effective communication of climate risks. The scenario planning process is a promising approach for engaging diverse individuals in a dialogue that promotes a new way of thinking about uncertainty while facilitating the coproduction of knowledge.

Keywords Scenarios planning • Narrative theory • Risk communication • Climate change • Uncertainty • Adaptation • National Park Service • Alaska • Scenarios Network for Alaska and Arctic Planning

6.1 Introduction and Background

Climate change can be a challenging subject to communicate (Moser 2010; Pidgeon and Fischhoff 2011). Nonetheless, effective communication is urgently needed to engage, inform, and support decision-makers in their efforts to respond to and plan

N. Fresco (✉) • K. Timm
International Arctic Research Center (IARC), Scenarios Network for Alaska and Arctic Planning (SNAP), University of Alaska Fairbanks, Fairbanks, AK, USA
e-mail: nlfresco@alaska.edu

J.L. Drake et al. (eds.), *Communicating Climate-Change and Natural Hazard Risk and Cultivating Resilience*, Advances in Natural and Technological Hazards Research 45, DOI 10.1007/978-3-319-20161-0_6

for the impacts of climate change (Knapp and Trainor 2013; Moss et al. 2013; National Research Council 2009). The inherent complexity of the science, its politically charged nature, and the global scope of the problem are just a few of the reasons that climate change communication efforts often fail (Moser 2010; Weber and Stern 2011). Successful approaches to climate change communication must find ways to overcome these common barriers. The scenario planning process is one technique that holds particular promise (Eisenack et al. 2014; Moser 2014).

Scenario planning is a process for developing narratives for a range of possible futures, rather than attempting to predict the probability of a given future (Mahmoud et al. 2009; Schwartz 1996; Thompson et al. 2012; Xiang and Clarke 2003). Scenarios are growing in popularity as an approach to plan for the impacts of climate change. The specific tactics of scenario development have varied from highly quantitative assessments to interpretive narratives. For example, the Intergovernmental Panel on Climate Change (IPCC) generates scenarios of future global temperatures using global circulation models and measurements of greenhouse gas emissions (IPCC 2013). Shaw and colleagues used computer visualizations to generate scenarios of future sea level rise to encourage stakeholders to discuss adaptation options in their community (Shaw et al. 2009). Scenarios can also be written or spoken narratives that qualitatively describe possible futures (Schwartz 1996).

Since its formation in 2007, the Scenarios Network for Alaska and Arctic Planning (SNAP) at the University of Alaska Fairbanks (UAF) has worked with dozens of different groups—including government agencies, individuals, academic researchers, private industry, and communities—to develop scenarios that aid in the climate change planning process. In this chapter, we describe the Climate Change Scenario Planning Project (CCSPP) recently undertaken by SNAP and the Alaska Region of the National Park Service (NPS). Used in government, private industry, and resource management for years, scenario planning is a proven approach for planning under uncertainty (Ernst and van Riemsdijk 2013; Lawler et al. 2010; Peterson et al. 2003; Schwartz 1996). The approach emphasizes diverse participation, bridging different types of knowledge, and managing uncertainty—making this an effective method for climate change communication. In analyzing this case study we ask, how does the scenario planning process promote effective communication of climate change risks and uncertainty with diverse audiences?

6.2 Literature Review

6.2.1 Narrative Theory

The use of narrative as a means of expressing future possibilities—scenarios—is in keeping with the narrative paradigm first posited in the 1980s by Walter Fisher (Fisher 1984, 1985) and further developed by many other social scientists. The theory suggests that all human communication is a form of storytelling, involving characters, conflicts, and chronological reporting. It relies on the idea that in order to be persuasive and believable, narratives must adhere to rules of coherence and fidelity (Griffin 2009). Narrative theory has been widely applied in the field of communication. William Kirkwood's work (1992) suggests that Fisher's theories can be extended beyond the dominant values or beliefs of the time, in order to promote social change. This idea is supported by communication strategies used in a business context (Jameson 2001). It is in this vein that narrative theory seems most applicable to scenarios planning.

6.2.2 Climate Change in Alaska

Alaska is one of the most rapidly warming places on earth. In the past 50 years, the average annual temperature of Alaska has increased by approximately 4 °F (Markon et al. 2012). Increases in temperature have led to reductions in sea ice extent and thickness (Stroeve et al. 2011) and glacier mass (Arendt et al. 2009). Permafrost is thawing, leading to land subsidence, changes in hydrologic systems, and changes to ecosystem processes (Romanovsky et al. 2010). Increases in temperatures have made the ground dryer and more susceptible to fire (Wolken et al. 2011). Plant, animal, and insect species are moving into new areas of the landscape (Wolken et al. 2011). In the waters surrounding Alaska, temperatures are increasing and the water is becoming more acidic (Steinacher et al. 2009). By the middle of this century, the average annual mean temperature of Alaska is expected to warm by an additional 4–6 °F (Markon et al. 2012).

6.2.3 Climate Impacts on People

Changes in the biophysical environment are impacting Alaska's people and communities. Fall storms—common in the Bering Sea—are posing greater threats to communities that no longer have the protection of coastal sea ice and permafrost to stabilize their coastlines (Markon et al. 2012). In addition to the risks of erosion, changes in seasonality are leading to more dangerous travel conditions on rivers and sea ice (Markon et al. 2012). Many indigenous rural communities have relied on certain animal and plant species for centuries, and now the availability, locations, and habitats of some of these species are shifting and creating food security issues (Markon et al. 2012). Across Alaska communities and other organizations are responding to these impacts and developing climate change adaptation plans (Knapp and Trainor 2013). Many of these plans call for a more collaborative approach to climate change research and planning that involves the integration of local knowledge, the formation of knowledge-sharing networks, and in general, the transformation of the role for climate science in decision-making and adaptation planning (Knapp and Trainor 2013).

6.2.4 Risk Communication and Adaptation

Planning for and adapting to climate change requires learning, experimenting, deliberation, and communication (Adger et al. 2008; Chapin et al. 2009). Effective communication and information are ongoing needs in the climate change adaptation process. However, climate change can be a challenging subject to communicate for several reasons (Hulme 2009; Moser 2010, 2014). Among these reasons is the nature of the phenomenon itself. Climate is not a single phenomenon, but rather a long-term pattern of weather events subject to statistical analysis. The causes of climate change are largely invisible, and for many Americans the signals and impacts are difficult to attribute (Moser 2010; Weber and Stern 2011). The impacts of climate change are difficult to predict and possess layers of uncertainty related to future population size, emission levels, probability of extreme events, and regional impacts of these processes on ecosystems and communities (Pidgeon and Fischhoff 2011).

Noting the public's limited or maladaptive response to climate change, scientists, politicians, and other people have assumed that this lack of action stems from poor scientific understanding of the issue. Thus, they have tried to remedy the problem by conveying scientific facts about climate change. This is an example of the deficit model of science communication, in which the audience is perceived to be lacking a specific type of knowledge, and the purpose of communication is to convey information to fill this deficit (Brossard and Lewenstein 2010; Nisbet and Scheufele 2009). However, this model often fails. Although striving for greater education and increases in scientific literacy are valuable endeavors, it is overly simplistic to assume that transferring information is all that is needed to create motivation and change behavior (Moser 2010).

Humans, despite our seemingly logical and forward-looking minds, are not always adept at judging risks or weighing the relative costs and benefits of various actions (or inaction). Underlying worldviews and beliefs shape how people understand climate change information, evaluate the risks, and choose to respond (Hulme 2009; Kahan et al. 2012; Lorenzoni et al. 2007). When it comes to risk, evidence is mounting that people most often rely on their values to identify issues of importance and to choose ideologically congruent interpretations of those issues (Kahan et al. 2012; Nisbet 2010). In risk communication, special effort and attention must be made to communicate in ways that do not threaten deeply held, individual values.

A more user-centered, public engagement model of science communication can be more effective in risk communication (Brossard and Lewenstein 2010). Research shows that when subject matter is made personally relevant, even people with limited scientific backgrounds or formal education can grasp complex concepts (Brossard and Lewenstein 2010). Not only do attempts at top-down transfer of scientific information undermine people's need to have their own viewpoints heard and credited, they can also result in a failure by scientists and policy-makers to learn about important social and cultural factors that strongly influence opinion and behavior (Moser 2014).

6.3 The Scenario Planning Process

Natural systems—including global climate—are often too complex and uncontrollable to be reliably predicted via models (Peterson et al. 2003; Schwartz 1996). Thus, in the case of climate change adaptation planning, actions based on a single predictive forecast can be extremely risky (Schoemaker 1991). To that end, the product of the scenario planning process is not a prediction about the future, but a set of possibilities (see Fig. 6.1) (IPCC 2008; Mahmoud et al. 2009). During the scenario planning process, these divergent possible futures are explored, analyzed, and discussed in advance, so that stakeholders can weigh adaptive strategies that apply to multiple futures, rather than merely adapting reactively as one of these futures ultimately plays out over time.

6.3.1 Stages of the Scenario Planning Process

The scenario planning process (see Fig. 6.2) fosters engagement by bringing together diverse people to orient on a single stated purpose or question (Mahmoud et al. 2009; Peterson et al. 2003). Participants then explore the issue in question through a systemic process of collecting, discussing, analyzing, and synthesizing a wide variety of quantitative and qualitative information sources (Shaw et al. 2009; Shearer 2005). From this information, the group generates potential scenarios and creates narratives of these scenarios that are plausible, relevant, and challenging.

Developing the scenarios is a multistep process in which participants select two key drivers of change that are both important (likely to cause multiple significant effects) and uncertain (in terms of the magnitude or direction of the change). These drivers,

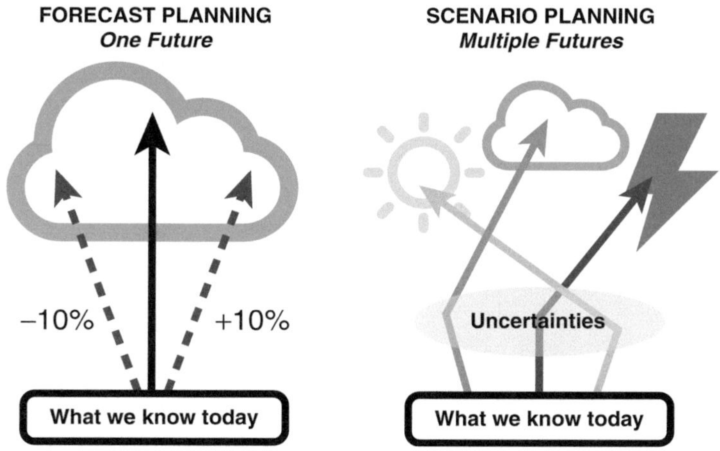

Fig. 6.1 The difference between forecasting and scenario planning is that scenarios are possibilities and not probabilistic predictions about the future (Adapted from the Global Business Network)

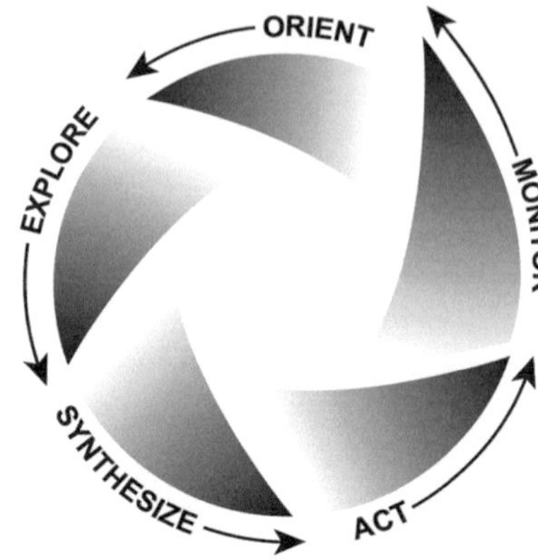

Fig. 6.2 The five stages of the scenario planning process (Adapted from the Global Business Network)

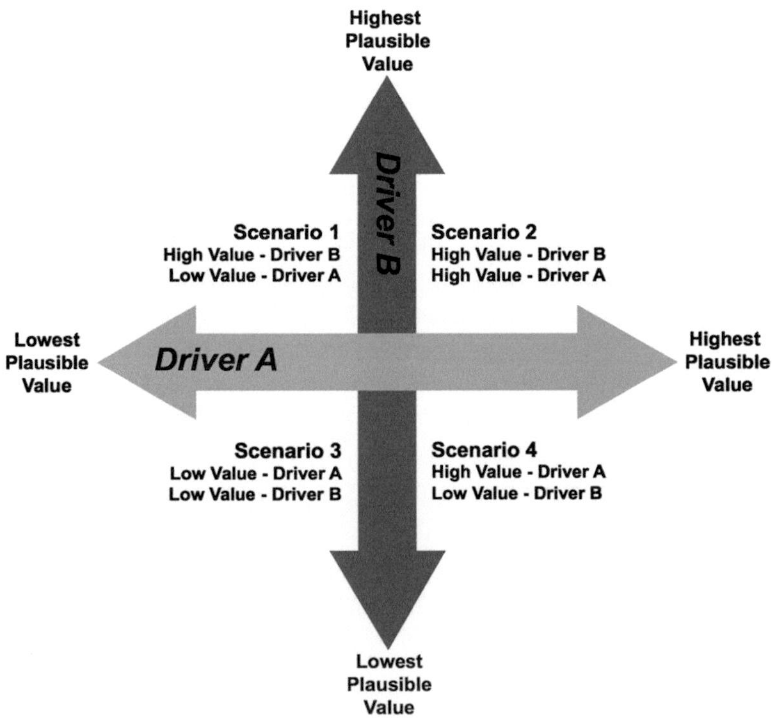

Fig. 6.3 Combining two selected drivers (*arrows*) creates four possible futures (each quadrant), and each future combines the endpoints of the drivers in a novel manner. These endpoints often, but not always represent the "low" and "high" end of plausible values that are estimated with the best available science. Endpoints values are estimated based on the best available science (Adapted from the Global Business Network)

when intersected, yield four possible futures (see Fig. 6.3). By selecting the drivers with the greatest importance and uncertainty, the futures represent highly divergent scenarios that approximate the full range of possibilities. The original characteristics of the selected drivers—plausibility, relevance, divergence, and challenge—are also qualities that participants are asked to keep in mind during the development of scenarios and the narratives to describe those scenarios.

Once the narratives are fleshed out, participants can identify what actions are appropriate and possible to address the potential outcomes of each scenario (see Fig. 6.4). Scenario planning offers participants the opportunity to search for actions that perform well under all scenarios (often called *no-regrets* or *robust* actions) (Carvalho et al. 2011, p. 2024; Peterson et al. 2003, p. 362). These actions are often among the most immediate and powerful scenario outcomes, and they often appeal to managers, because they tend to be noncontroversial (Lempert and Schlesinger 2000). After actions have been selected, everyone can work to ensure that the results of these actions are carried out and monitored.

6.3.2 The Scenarios Network for Alaska and Arctic Planning

The idea to foster integration and develop a scenario planning process for Alaska emerged in 2006 from discussions by an interdisciplinary group of about a dozen faculty members at the University of Alaska. The consensus of that group was that such a process would be one of the most useful ways that researchers could share

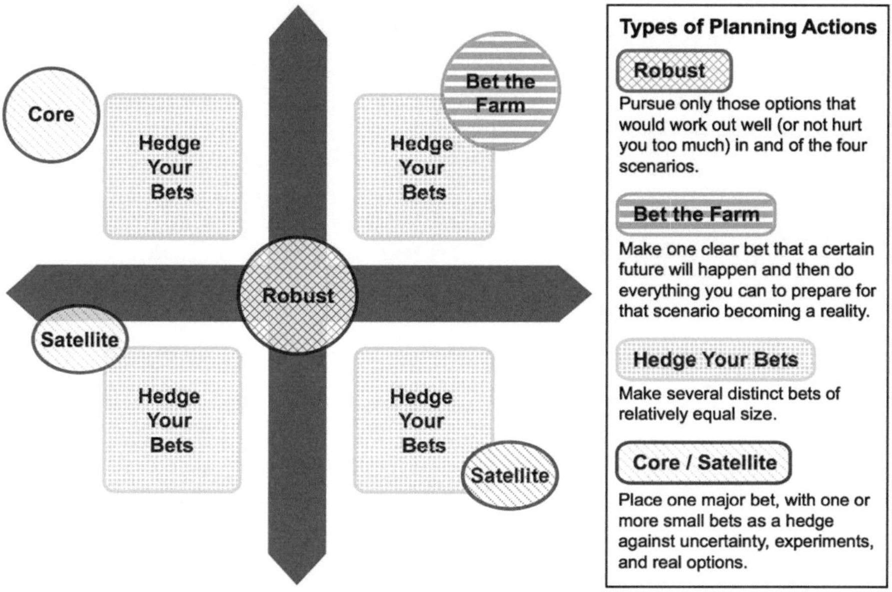

Fig. 6.4 If one imagines that each of these four quadrants represents a divergent scenario, planning resources must be allocated based on the costs and benefits associated with these four possible futures. Choosing an optimal planning action depends on weighing choices based on their short-term and long-term outcomes (Adapted from the Global Business Network)

their expertise with decision-makers and other stakeholders across Alaska and the Arctic. Starting in 2007, SNAP's work philosophy embraces interdisciplinary collaboration, efficiency, and innovation.

Central to SNAP's work is the development of climate projections. Climate data are derived from global circulation models (GCMs) created at scientific centers around the world. SNAP downscales these data, making GCMs more practical for assessing how climate change may impact an area the size of a community or land management unit. In partnership with its collaborators, SNAP also links climate projections to variables such as fire behavior, soil temperature, permafrost dynamics, biome shift, and landscape and hydrologic change. Not only does SNAP produce sets of data, maps, and data visualization tools, but they also provide support for the integration of these projections. This begins with making the methods and results of their work as transparent and as accessible as possible. Their model of collaboration involves iteration and communication throughout the process—from identifying data needs to helping their collaborators share the results.

6.3.3 The Climate Change Scenario Planning Project

In 2010, the NPS national and Alaska regional offices released climate change response strategies for the National Park System and the Alaska Region, respectively (National Park Service 2010a, b). Scenario planning was identified in both strategies as a high priority for understanding potential climate change impacts to park resources, assets, and operations (National Park 2010a, b). In August 2010, SNAP and NPS embarked on a collaborative effort to assess the potential effects of climate change in Alaska's National Parks.

The impetus for the Climate Change Scenario Planning Project (CCSPP) was a growing sense of urgency and a feeling that traditional management tools were not adequate to meet the challenges of climate change. Although a scenarios approach had been part of SNAP since its inception, this marked the research group's first formal approach in the process, using methods outlined by third-party experts in the field.

Key aspects and components of the Climate Change Scenario Planning Project included:

- Regional focus, with emphasis on issues of concern at the local level
- Participation by a broad range of stakeholders
- Integration of scientific information and local knowledge
- Immersive participation
- Focus on narrative rather than data outputs
- Open-ended process
- Outputs intended to be advisory rather than prescriptive

The CCSPP began with a training workshop, led by Jonathan Star of the Global Business Network (GBN), at which a team of NPS Alaska Region and SNAP employees learned how to develop scenarios based on nested frameworks of critical uncertainties. The ultimate goal of the workshops was to envision and describe scenarios that would help NPS managers, cooperating personnel, and key stakeholders consider potential consequences of climate change for all of Alaska's NPS areas. Analyzing and weighing the potential consequences of these scenarios would help prepare park managers and others for impending changes, gather diverse input, and make informed decisions.

The National Park Service manages over 50 million acres of land in Alaska. Spanning several distinct regions and ecosystems—and experiencing very different sets of climate change impacts—the workshops were organized around each of the four regional Inventory and Monitoring (I&M) networks. Over the course of 2 years, SNAP and NPS organized workshops in five regions of the state: interior, southeast, southwest, western Arctic, and interior Arctic.

One of the goals of the CCSPP was to include a broad range of participants with a wide range of diverse perspectives and expertise. First, a list of key stakeholders and community members was generated by the NPS and SNAP. From that list, potential participants were contacted via e-mail and/or phone and asked for their time, expertise, and assistance. No one was paid to take part, although some travel funds were provided, particularly for members of rural communities. For each regional network, the workshop participants included about 50 % representation from the region's parks and NPS staff from the Alaska Regional Office. Another 25 % of participants included SNAP personnel, key individuals from other agencies, and nongovernment organizations. Another 25 % of participants were members of communities near the parks or with a stake in the region (Ernst and van Riemsdijk 2013).

Although climate was always at the heart of the discussion, the drivers selected by each group of workshop participants differed, and the key issues for each region were widely divergent (see Fig. 6.5). For example, in the Interior Alaska workshop, forest fire and permafrost thaw were of greatest concern, while Western Alaska workshop participants were worried about coastal erosion. In the Southeast Alaska workshop, fisheries were at the top of the list. At each workshop, the preliminary scenarios derived from the matrix were then nested in a second matrix that took into account sociopolitical variables and uncertainty. By adding these variables, each group broadened their discussion from biophysical impacts to the complex social ramifications of climate change. Details of the scenarios tended to be radically different from region to region, with some groups focusing almost entirely on oceanic effects (e.g., fisheries) and others focusing on coastal erosion, increased forest fire, or changes in the length of the summer season.

The outcomes of the scenarios were communicated to the rest of the workshop participants using a narrative format. An important component of the scenarios process is turning bullet-point summaries of the potential effects of climate change into stories that come alive. As a result, the narrative styles varied wildly between groups; different people found very different ways to make future stories speak to them in a way that mattered. Just as the key concerns varied, so too did the manner in which the narratives or the *stories of the future* were expressed. They ranged from hypothetical transcripts of future meetings with government officials to interactive maps of future park resources to imagined social media chats between affected individuals. One story took the form of a rhyming children's storybook. Entitled *Gussaq*

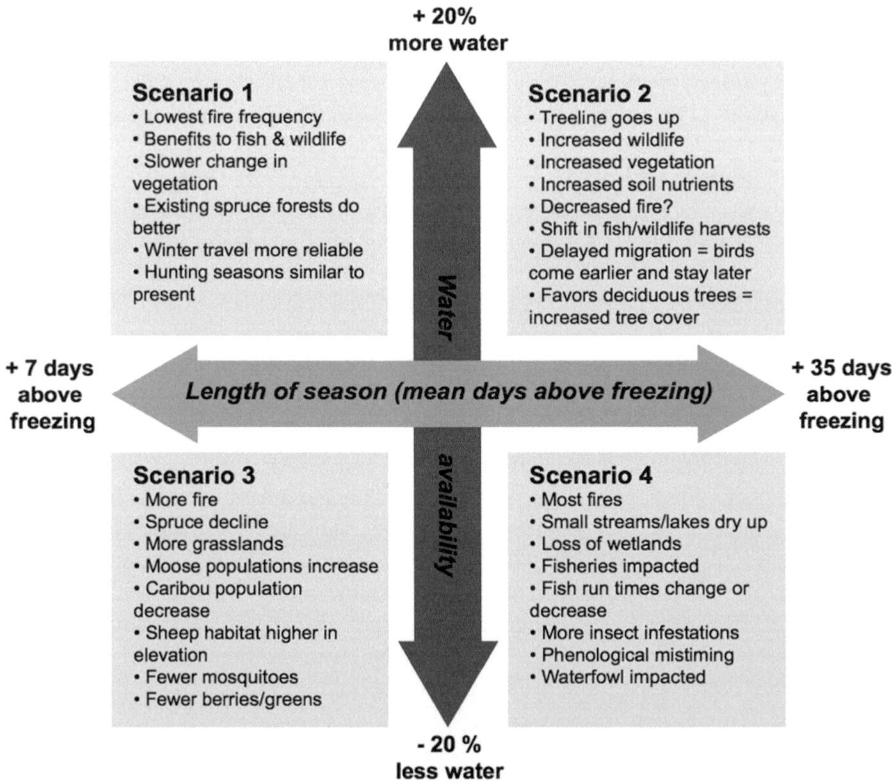

Fig. 6.5 CCSPP participants selected key drivers and developed scenarios for the systems in which they live and work. This example, from the Central Alaska workshop, shows the first stage of scenario development using the drivers of warm season length (mean temperatures above freezing) and water availability (soil moisture during the growing season). The combination of these drivers generates four different, but plausible, futures

the Tussock, it begins, "Far to the north, where the grasses still grow/Though the winds blow with smoke and soot when they blow/And the rivers are silty and slow when they flow…" It goes on, through 15 stanzas, to humorously yet vividly describe an interior Alaska future with seasonal shifts; fewer caribou, fish, and waterfowl; thawing permafrost; increased development; and little governmental oversight.

While such outputs might seem frivolous, the opposite seemed to be the case. It was in creating these stories that participants became most engaged with the very real possibilities represented by the relevant, challenging, divergent, yet plausible futures they had mapped out. The CCSPP was intended to stimulate creative thinking, and the outputs of the project included not only tangible reports but also the new mind-set, increased awareness, and new communication and education tools of project participants.

The climate change scenario planning process was not intended to end at the completion of the workshops or at the issuing of the associated reports and presentations. It was designed to be a living process. In addition to putting together workshop summary reports, SNAP and NPS collaborated on a 24-page outreach booklet and digital iBook. This booklet is entitled *State of Change: Climate Change in Alaska's National Park Areas* and shares information about climate impacts and actions in Alaska's National Parks from a narrative perspective that mirrors the communication tools used in the scenario planning workshops. Ultimately, the suggested management choices that came from each group were relatively similar. General findings from all five workshops included the need for increased cooperation, trust building, and information sharing among diverse groups; increased monitoring of resources; and improved emergency planning. More specifically, findings suggested building flexibility and cooperation by using traditional ecological knowledge (TEK) (e.g., site-specific knowledge of wildlife habits or memories of how extreme events played out in the past) and comanagement in planning and hiring staff with multidisciplinary skills. Workshop participants further suggested the need for specific monitoring programs and using the parks as living laboratories. They also urged managers to assess the vulnerability and risks to key resources and infrastructure. Workshop participants admitted, however, that budget constraints might limit these efforts.

6.4 Discussion

At both the local and national level, the National Park Service is becoming increasingly aware that changing climatic conditions are already impacting environmental, social, and economic conditions in and around National Park Service lands (National Park Service 2010a, b). The NPS and SNAP collaborated to carry out the CCSPP—a scenario planning process aimed at exploring climate change impacts and adaptation options in Alaska's National Park areas. Of all SNAP's projects since its inception, its collaboration with the National Park Service is the most comprehensive effort to use all the tools that scenario planning offers. Several previous studies have documented the ways that scenario planning and the similar processes of vulnerability assessment, adaptive management, and adaptive risk management facilitate stakeholder engagement to more effectively plan for an uncertain future (Kettle et al. 2014; Lawler et al. 2010; Shaw et al. 2009). Rarely have these efforts been analyzed specifically for their contributions to improving climate change communication.

6.4.1 Toward an Engagement Model of Communication

The CCSPP facilitated a more engaging approach to climate changecommunication than the top-down approach to science communication typical of many research projects. The participants of the scenario planning workshops were highly diverse, with varying levels of knowledge and experience with climate impacts. The 4-day workshops began with fairly broad scientific information about possible climate change impacts. When subgroup members were asked to embark upon the scenarios process by selecting key elements driving change, local issues of concern were identified by the participants that were not particularly familiar to NPS or SNAP researchers. In some cases, the scientists participating in the workshops had to seek additional sources of information to answer participants' questions that were well outside of their own disciplines, such as ocean acidification. Thus, participants not only contributed local knowledge and decided what was most important; they sometimes drove the scientific process as well. In this way, the scenario planning process helped move climate change communication and planning toward the more collaborative approach to climate science and planning that Knapp and Trainor (2013) recommend in their analysis of climate change adaptation plans.

6.4.2 Drawing from What We Know: Integrating Knowledge

Ernst and van Riemsdijk (2013), who used these workshops as the focus of their study of stakeholder involvement in scenarios planning, described how many non-NPS stakeholders felt like their opinions were valued and heard through the CCSPP. Integrating local knowledge and values are essential for culturally responsive approaches to climate change planning, and by including community members, these voices and perspectives became an important element of the CCSPP. There is evidence that the scenario planning process provides a collaborative atmosphere that sets participants as equals in discussion, which helps fulfill a commonly articulated need to integrate local knowledge (Knapp and Trainor 2013; Moser 2014). Because everyone is invited to contribute and presumably wants to speak and be heard, the CCSPP also eliminated the power dynamics that come with one-way science communication and provided an incentive to avoid unnecessarily technical information.

6.4.3 Using Narrative to Integrate Science into Climate Change Planning

In addition to local and traditional ecological knowledge, climate change information from several disciplines—atmospheric, geophysical, biological, and social sciences—was also included in the scenario planning process. The scenarios that were developed were nested in information about political and economic futures. In academia, where specialization has historically been the pinnacle of success, it can be difficult to integrate across disciplines and include local knowledge. The CCSPP helped facilitate the integration of many types and sources of knowledge and provided a safe forum for this complexity to be explored through the use of narratives.

The CCSPP concluded with the articulation of likely futures—including the risks, uncertainty, knowledge, and values—in a narrative format. Articulating possible futures in narrative format makes the abstract issue of climate change into concrete examples that are place-based and meaningful to the participants. The narrative format allows for a wide variety of information—from highly technical models to personal experiences and stories—to be integrated in one likely future that can be easily shared with others (Dahlstrom 2014). It also makes it easier to consider both climatic and non-climatic elements of the system in question (Dahlstrom 2014; Peterson et al. 2003). There is also increasing evidence that humans are wired for narrative. When listening to a set of bullet points, we use some analytical parts of our brain to decode the meaning, but when listening to a story, the sensory parts of our brain are also activated, making the experience more memorable (Heath and Heath 2008). The narratives also give participants the opportunity to explore the ideas that don't always show up in empirical data, like extreme events or the interrelationship between different types of natural and social events.

6.4.4 Facing Uncertainty Head-On

We consider uncertainty in many of the decisions we make every day, yet many individuals and institutions use uncertainty as a justification for not responding or preparing for climate change (Bidwell et al. 2013). While climate change is not without manageable types of uncertainty, there has also been a well-funded effort to create undue uncertainty around climate change (Hulme 2009; Moser 2010). Rather than avoiding complexity and uncertainty, the scenario planning process embraces it and sets out to uncover it (Schwartz 1996). In the process of developing scenarios, the workshop participants looked for management actions that could perform well in each scenario. Accepting uncertainty and looking for these kinds of solutions build resilience in systems by avoiding rigidity traps and making institutions strong enough to bend when the wind blows. This approach also helps avoid a common problem in climate change communication—overwhelming people's finite pool of worry—by enabling participants to learn about the risks while also exploring realistic and reasonable actions that can be taken in response (Moser 2010; Pidgeon and Fischhoff 2011).

The scenario planning process also adopts an intuitive approach to dealing with uncertainty. In our daily lives, we all routinely make judgments about risk and reward and apportion our time, energy, and resources accordingly. For example, we might put forth the effort needed to apply for a new job, knowing that our chances of getting it were low, but still plausible. Conversely, we might purchase insurance against fire or theft, knowing that these calamities, while not likely, are possible—and potentially devastating. Whether we also purchase tornado insurance or hurricane insurance depends on our judgment of whether these phenomena are possible in our area. Such judgment, of course, depends on what information is available, whether we have access to that information and whether we truly understand what it means, in terms of risk, uncertainty, and probable effects.

6.5 Conclusion

Increasingly, diverse groups—including land managers and community members with intimate ties to the land and its resources—are using the idea of resilience to discuss desirable futures. Linked to the concept of resilience is the understanding that uncertainty is always present and that the future is not a known quantity nor is it predictable. At the same time, new understandings of effective risk and science communication are suggesting a movement toward more engaging approaches to communication. Given this shift in both management philosophies and communication techniques, the scenario planning process is well situated to be an ideal tool for helping diverse groups of people engage with each other to explore uncertainty, integrate knowledge, and plan for a range of possible futures.

During the Climate Change Scenario Planning Project, SNAP and the Alaska Region of the National Park Service collaborated to host five scenario planning workshops with NPS employees, other resource stakeholders, and local community members. While the climate impacts and issues varied greatly among different regions of the state, the management suggestions that came from the CCSPP participants contained many similar recommendations. Because each group of participants arrived at these ideas independently, without the influence of the scientists or organizers, they had a sense of ownership to the stories and scenarios that were created. This sense of ownership—of coming to conclusions based on personal knowledge and group brainstorming—offers a very different, more collaborative form of resource management that has been sought by climate change adaptation and communication researchers and practitioners alike.

As elucidated by Ernst and van Riemsdijk (2013), flaws in the scenario planning process, including imbalanced participation, can impede communication and efficacy. The scenario planning process is ripe for more rigorous analysis as a communication tool. Scenario planning offers novel opportunities to link science with management that is meaningful to stakeholders. These aspects include creating an environment in which true involvement and engagement can take place among diverse groups; integrating scientific and place-based or traditional knowledge; addressing uncertainty, particularly high levels of uncertainty that result in divergent possible futures; and using humans' innate penchant for considering the future in narrative rather than data-driven terms.

References

Adger WN, Dessai S, Goulden M, Hulme M, Lorenzoni I, Nelson DR, Wreford A (2008) Are there social limits to adaptation to climate change? Clim Change 93(3–4):335–354. doi:10.1007/s10584-008-9520-z

Arendt AA, Walsh J, Harrison W (2009) Changes of glaciers and climate in northwestern North America during the late twentieth century. J Climate 22(15):4117–4134. doi:10.1175/2009jcli2784.1

Bidwell D, Dietz T, Scavia D (2013) Fostering knowledge networks for climate adaptation. Nat Clim Chang 3(7):610–611. doi:10.1038/nclimate1931

Brossard D, Lewenstein BV (2010) A critical appraisal of models of public understanding of science. In: Kahlor L, Stout PA (eds) Communicating science: new agendas in communication. Routledge, New York, pp 11–39

Carvalho SB, Brito JC, Crespo EG, Watts ME, Possingham HP (2011) Conservation planning under climate change: toward accounting for uncertainty in predicted species distributions to increase confidence in conservation investments in space and time. Biol Conserv 144(7):2020–2030. doi:10.1016/j.biocon.2011.04.024

Chapin FS, Kofinas GP, Folke C (2009) Principles of ecosystem stewardship. Springer, New York

Dahlstrom MF (2014) Using narratives and storytelling to communicate science with nonexpert audiences. Proc Natl Acad Sci U S A 111(Suppl 4):13614–13620. doi:10.1073/pnas.1320645111

Eisenack I, Moser SC, Hoffmann E, Klein RJT, Oberlack C, Pechan A, Termeer CJAM (2014) Explaining and overcoming barriers to climate change adaptation. Nat Clim Chang 4:867–872. doi:10.1038/NCLIMATE2350

Ernst KM, van Riemsdijk M (2013) Climate change scenario planning in Alaska's National Parks: stakeholder involvement in the decision-making process. Appl Geogr 45:22–28. doi:10.1016/j.apgeog.2013.08.004

Fisher WR (1984) Narration as human communication paradigm: the case of public moral argument. Commun Monogr 51:1–22

Fisher WR (1985) The narrative paradigm: an elaboration. Commun Monogr 52:347–367

Griffin EA (2009) A first look at communication theory, 7th edn. McGraw-Hill, Boston

Heath C, Heath D (2008) Made to stick: why some ideas survive and others die. Random House, New York

Hulme M (2009) Why we disagree about climate change. Cambridge University Press, New York

IPCC (2008) Definitions. Retrieved from http://www.ipcc-data.org/ddc_definitions.html. Accessed 2 Feb 2014

IPCC (2013) Climate change 2013: the physical science basis. Contribution of working group I to the fifth assessment, report of the intergovernmental panel on climate change. In: Strocker TF, Qin D, Plattner G-K, Tignor M, Allen SK, Boschung J, Nauels A, Xia Y, Bex V, Midgley PM (eds) Summary for policymakers. Cambridge University Press, Cambridge/New York, p 23

Jameson DE (2001) Narrative discourse and management action. Int J Bus Comm 38(4):476–511

Kahan D, Peters E, Wittlin M, Slovic P, Ouellette LL, Braman D, Mandel G (2012) The polarizing impact of science literacy and numeracy on perceived climate change risks. Nat Clim Change 2:732–735. doi:10.1038/nclimate154710.1038/NCLIMATE1547

Kettle NP, Dow K, Tuler S, Webler T, Whitehead J, Miller KM (2014) Integrating scientific and local knowledge to inform risk-based management approaches for climate adaptation. Clin Risk Manag 4–5:17–31. doi:10.1016/j.crm.2014.07.001

Kirkwood WG (1992) Narrative and the rhetoric of possibility. Commun Monogr 59(1):30–47

Knapp CN, Trainor SF (2013) Adapting science to a warming world. Glob Environ Chang 23(5):1296–1306. doi:10.1016/j.gloenvcha.2013.07.007

Lawler JJ, Tear TH, Pyke C, Shaw MR, Gonzalez P, Kareiva P, Pearsall S (2010) Resource management in a changing and uncertain climate. Front Ecol Environ 8(1):35–43

Lempert RJ, Schlesinger ME (2000) Robust strategies for abating climate change – an editorial essay. Clim Change 45(3–4):387–401. doi:10.1023/A:1005698407365

Lorenzoni I, Nicholson-Cole S, Whitmarsh L (2007) Barriers perceived to engaging with climate change among the UK public and their policy implications. Glob Environ Chang 17(3–4):445–459. doi:10.1016/j.gloenvcha.2007.01.004

Mahmoud M, Liu Y, Hartmann H, Stewart S, Wagener T, Semmens D, Winter L (2009) A formal framework for scenario development in support of environmental decision-making. Environ Model Software 24:798–808. doi:10.1016/j.envsoft.2008.11.010

Markon CJ, Trainor SF, Chapin III FS (2012) The United States national climate assessment—Alaska technical regional report: U.S. Geological Survey Circular 1379, 148. http://pubs.usgs.gov/circ/1379/pdf/circ1379.pdf
Moser SC (2010) Communicating climate change: history, challenges, process and future directions. WIREs Clim Change 1:31–53. doi:10.1002/wcc.011
Moser SC (2014) Communicating adaptation to climate change: the art and science of public engagement when climate change comes home. Wiley Interdiscip Rev Clim Chang 5(3):337–358. doi:10.1002/wcc.276
Moss RH, Meehl GA, Lemos MC, Smith JB, Arnold JR, Arnott JC, Wilbanks TJ (2013) Hell and high water: practice-relevant adaptation science. Science 342:696–698
National Park Service (2010a) Alaska region climate change strategy. http://www.nps.gov/akso/docs/AKCCRS.pdf. Accessed 18 Jan 2015
National Park Service (2010b) National Park Service climate change response strategy. Fort Collins Co. http://www.nature.nps.gov/climatechange/docs/NPS_CCRS.pdf. Accessed 18 Jan 2015
National Research Council (2009) Informing decisions in a changing climate. The National Academies Press, Washington, DC. http://www.nap.edu/catalog/12626/informing-decisions-in-a-changing-climate. Accessed 5 Jul 2014
Nisbet MC (2010) Framing science a new paradigm in public engagement. In: Kahlor L, Stout PA (eds) Communicating science. Routledge, New York, pp 40–67
Nisbet MC, Scheufele DA (2009) What's next for science communication? Promising directions and lingering distractions. Am J Bot 96(10):1767–1778. doi:10.3732/ajb.0900041
Peterson GD, Cumming GS, Carpenter SR (2003) Scenario planning: a tool for conservation in an uncertain world. Conserv Biol 17(2):358–366
Pidgeon N, Fischhoff B (2011) The role of social and decision sciences in communicating uncertain climate risks. Nat Clim Chang 1:35–41. doi:10.1038/nclimate108010.1038/NCLIMATE1080
Romanovsky VE, Smith SL, Christiansen HH (2010) Permafrost thermal state in the polar Northern Hemisphere during the international polar year 2007–2009: a synthesis. Permafr Periglac Process 21(2):106–116. doi:10.1002/ppp.689
Schoemaker PJH (1991) When and how to use scenario planning: aheuristic approach with illustration. J Forecast 10(6):549–564. doi:10.1002/for.3980100602
Schwartz P (1996) The art of the long view: planning for the future in an uncertain world. Doubleday, New York
Shaw A, Sheppard S, Burch S, Flanders D, Wiek A, Carmichael J, Cohen S (2009) Making local futures tangible—synthesizing, downscaling, and visualizing climate change scenarios for participatory capacity building. Glob Environ Chang 19:447–463. doi:10.1016/j.gloenvcha.2009.04.002
Shearer AW (2005) Approaching scenario-based studies: three perceptions about the future and considerations for landscape planning. Environment and Planning B: Planning and Design 32(1):67–87. doi:10.1068/B3116
Steinacher M, Joos F, Frolicher TL, Plattner G-K, Doney SC (2009) Imminent ocean acidification in the Arctic projected with the NCAR global coupled carbon cycle-climate model. Biogeosciences 6:515–533
Stroeve JC, Serreze MC, Holland MM, Kay JE, Malanik J, Barrett AP (2011) The Arctic's rapidly shrinking sea ice cover: a research synthesis. Clim Change 110(3–4):1005–1027. doi:10.1007/s10584-011-0101-1
Thompson JR, Wiek A, Swanson FJ, Carpenter SR, Fresco N, Hollingsworth T, Foster DR (2012) Scenario studies as a synthetic and integrative research activity for long-term ecological research. BioScience 62(4):367–376. doi:10.1525/bio.2012.62.4.8
Weber EU, Stern PC (2011) Public understanding of climate change in the United States. Am Psychol 66(4):315–328. doi:10.1037/a0023253
Wolken JM, Hollingsworth TN, Rupp TS, Chapin FS, Trainor SF, Barrett TM, Yarie J (2011) Evidence and implications of recent and projected climate change in Alaska's forest ecosystems. Ecosphere 2(11):1–35. doi:10.1890/es11-00288.1
Xiang N, Clarke KC (2003) The use of scenarios in land-use planning. Environ Plan B 30(6):885–909

Chapter 7
Barriers to Using Climate Information: Challenges in Communicating Probabilistic Forecasts to Decision-Makers

Melanie Davis, Rachel Lowe, Sophie Steffen, Francisco Doblas-Reyes, and Xavier Rodó

Abstract Despite the strong dependence of certain sectors (e.g. energy, health, agriculture, tourism and insurance) on weather and climate variability, and several initiatives towards demonstrating the added benefits of integrating probabilistic climate forecasts into decision-making processes, such information is still underutilised. Improved communication is fundamental to stimulate the use of climate products by end users. This chapter evaluates current approaches to the visual communication of probabilistic seasonal climate forecast information. The overall aim of this study is to establish a visual communication protocol for such forecasts, which does not currently exist. Global Producing Centres (GPCs) show their own probabilistic forecasts with limited consistency in communication between different centres, which complicates how end users understand and interpret the products. A communication protocol that encompasses both the visualisation and description of climate forecasts can help to introduce a standard format and message to end users across different climate-sensitive sectors. It is hoped that this work will facilitate the improvement of decision-making processes that rely on climate forecast information and enable their wide-range dissemination via new climate services.

M. Davis, B.S. (✉) • R. Lowe, Ph.D.
Institut Català de Ciències del Clima (IC3), Barcelona, Spain
e-mail: melanie.davis@ic3.cat

S. Steffen, M.A.
University of Barcelona, Barcelona, Spain

F. Doblas-Reyes, Ph.D.
Institut Català de Ciències del Clima (IC3), Barcelona, Spain

Institució Catalana de Recerca i Estudis Avançats (ICREA), Barcelona, Spain

Barcelona Supercomputing Center-Centro Nacional de Supercomputación (BSC-CNS), Barcelona, Spain

X. Rodó, Ph.D.
Institut Català de Ciències del Clima (IC3), Barcelona, Spain

Institució Catalana de Recerca i Estudis Avançats (ICREA), Barcelona, Spain

J.L. Drake et al. (eds.), *Communicating Climate-Change and Natural Hazard Risk and Cultivating Resilience*, Advances in Natural and Technological Hazards Research 45, DOI 10.1007/978-3-319-20161-0_7

Keywords Climate • Forecast • Probability • Communication • Visualisation • Climate services

7.1 Introduction

Climate is defined as the average weather over relatively long periods of time (Solomon et al. 2007). A climate projection is an estimate of the future, whereas a climate forecast enables some level of confidence to be attached to a projection that is branded with *most likely* (IPCC 2013), where a range of outcomes is provided, with the probability of each one to occur. For climate-sensitive sectors (e.g. energy industry, health, agriculture, water resource management, tourism, insurances, etc.) climate forecasts are useful tools to better plan for the future (Thomson et al. 2006; García-Morales and Dubus 2007; Gámiz-Fortis et al. 2008; Lowe et al. 2014).

Seasonal climate forecasts occupy an intermediate zone between weather forecasting and climate projections and intend to issue a statement of the expected climate conditions for the next 1–6 months (Doblas-Reyes et al. 2013). Despite the recent effort to develop underpinning science for seasonal to decadal (s2d) climate forecasts, there has been relatively little uptake and use of these new tools by users for decision-making in Europe (Falloon et al. 2013). Improved communication is fundamental to stimulate the use of climate forecast information within decision-making processes, so that highly climate-dependent industries are able to adapt to climate variability and change.

The Global Framework for Climate Services (GFCS) initiative (Hewitt et al. 2012), developing under the United Nations system and spearheaded by the World Meteorological Organization (WMO), aims to "mainstream" climate information into decision-making at all levels in society via new climate services. Following the objectives of the GFCS and subsequent European projects, Seasonal-to-Decadal Climate Prediction for the Improvement of European Climate Service (SPECS) and the European Provision of Regional Impacts Assessment in Seasonal and Decadal Timescales (EUPORIAS: Hewitt et al. 2013), this study investigates the challenges associated with the visualisation and description of seasonal climate forecasts.

The overall objective is to establish a protocol for the visual communication of probabilistic climate forecasts, which can promote the wide-range dissemination and exchange of *actionable* seasonal forecasts from Global Producing Centres (GPCs). This chapter examines the existing barriers and deficits in the visual communication of climate forecasts to end users. Although the importance of forecast visualisation has gained more weight and some progress has been made to advise visualisation techniques (e.g. Jupp et al. 2012; Slingsby et al. 2009), there is still much work to be done. Therefore, we investigate the current state of the art in visualising seasonal climate forecasts in a decision-making context.

Current approaches to produce seasonal climate forecasts include the use of physically based dynamical global climate models and/or empirically based statistical models. These approaches can be combined in a multi-model ensemble approach to produce probabilistic forecasts that adequately express the uncertainties of the future climate statement (Rodrigues et al. 2014). Probabilistic forecasts can be visualised in different ways. Often, they are presented as categories, such as terciles (three categories: lower, middle and upper thirds of a distribution) or quintiles (five categories). In some cases, the probability of the most likely category is provided, where the categories are formulated with respect to a climatological reference, such as the climatological upper and lower terciles. These define equiprobable categories that contain on average 1/3 of the events over the reference period.

Any probabilistic forecast should be accompanied by an estimate of its past performance, known as *forecast verification*. This should address the *accuracy* (how close the forecast probabilities are to the observed frequencies), the *skill* (how the probabilistic forecasts compare with some reference system) and the *utility* (the economic or other advantages of the probabilistic forecasts) (Jolliffe and Stephenson 2012). A number of different verification methods exist (e.g. relative operating characteristic (ROC), Brier skill score, Gerrity score, reliability diagram, correlation, etc.).

In this study, examples of probabilistic forecasts and their associated verifications were analysed from different forecasting centres and used to make a catalogue of current approaches, to assess their advantages and limitations and, finally, to recommend better alternatives. The variable and time period considered was precipitation for the December-January-February (DJF) season for the year 2009/2010. However, the analysis could also be extended to examine other climate variables, seasons and timescales. This time period was selected because an El Niño event was underway, which is characterised by a large-scale abnormal warming in the equatorial Pacific. During El Niño episodes, tropical precipitation becomes disrupted, resulting in persistent temperature and precipitation anomalies in many regions (Halpert and Ropelewski 1992; Ropelewski and Halpert 1987; Trenberth and Stepaniak 2001). For example, during El Niño wetter than normal conditions are observed over northern Brazil in the summer season (Halpert and Ropelewski 1992; Ropelewski and Halpert 1987, 1989).

During El Niño (and La Niña) events, climate forecasts are shown to be more accurate (Goddard and Dilley 2005). When these events occur, there is a clear opportunity to incorporate climate information into decision-making processes for climate-sensitive sectors. GPCs show their own probabilistic forecasts with limited consistency in the visual communication techniques across different centres. The recommended communication protocol for both the visualisation and description of climate forecasts can help to introduce a standard format and message to end users in climate-sensitive sectors. This information does not yet exist and is, therefore, very timely as new climate services rapidly develop.

7.2 Methods

First, a case study to evaluate the availability, accessibility (exercise 1), visualisation and description (exercise 2) of seasonal forecasts from different forecasting centres was undertaken. This was followed by qualitative participatory exercises, conducted with potential decision-makers to receive feedback on how to improve current forecast visualisation and description techniques (exercise 3). Finally, in another qualitative participatory exercise, a simplified decision-making case study for Brazilian hydropower management was used to illustrate the range of possible decision outcomes, based on precipitation forecasts issued by various forecasting centres for the DJF 2009/2010 period (exercise 4). Thirteen forecasting centres were compared that are either officially designated World Meteorological Organization (WMO) GPCs for long-range forecasts or are identified by the WMO as lead or major centres providing global seasonal forecasts (WMO 2013). As not all of the centres analysed are official GPCs, they are herein referred to as forecast centres (FCs) and include:

1. Met Office (UKMO)
2. International Research Institute for Climate and Society (IRI)
3. European Centre for Medium-Range Weather Forecasts (ECMWF)
4. Asia-Pacific Economic Cooperation Climate Center (APCC)
5. Centro de Previsão de Tempo e Estudos Climáticos (CPTEC)
6. Bureau of Meteorology (BoM)
7. Beijing Climate Center (BCC)
8. Climate Prediction Center (CPC) National Oceanic and Atmospheric Administration (NOAA)
9. Tokyo Climate Center (TCC)
10. Korea Meteorological Administration (KMA)
11. Météo-France (MF)
12. Meteorological Service of Canada (MSC)
13. South African Weather Service (SAWS)

Exercise 1: Forecast Availability and Accessibility The availability and accessibility of climate forecasts and their corresponding verification for each of the FC websites were analysed. Firstly, general aspects were compared, such as if a login is required, how many clicks away the required information is from the homepage and whether the site is available in English. The second aspects focused on the forecast product and type available, the time periods available and how far in advance the information is issued (i.e. lead time). The geographical region and seasonal climate variables that are forecast were also analysed. Finally, the corresponding forecast verification products and types were assessed in a similar way to the forecasts.

Exercise 2: Forecast Visualisation and Description For each FC, the probabilistic precipitation forecast for December-January-February (DJF) 2009/2010 season (and the corresponding verification map for the DJF season) was chosen to assess the different visualisation techniques adopted. Precipitation was the only variable available from all FCs. As previously mentioned, this period was marked by a strong El Niño event. An assessment was made of the descriptive information that appeared in the title/lettering, such as the forecast product/verification type, the probability of categories, the region, the variable, the forecast period and when it was issued. The appearance of a legend, an indication of units, type of labels, colours used and the number of different colours in the legend were also examined.

Exercise 3: Forecast Evaluation by End Users A workshop "Building Two-Way Communication: A Week of Climate Services" took place at the International Centre for Theoretical Physics (ICTP) from 2 to 6 December 2013, organised by the CLIMRUN (Climate Local Information in the Mediterranean Region Responding to User Needs) project. During the workshop, individual and group participatory interviews/exercises were conducted to evaluate the potential end-user perception of the seasonal climate forecast information and its communication from several FCs. In total, approximately 40 participants were involved, with a range of professional backgrounds from industry, policy and academia. For the first task, participants were divided into 12 groups and each group was assigned an FC. They were introduced to the FC's homepage to access the probabilistic forecasts and verification information. Participants had to find the seasonal forecast corresponding to the DJF 2009/2010 precipitation from the FC allocated to them. They were then asked key questions to evaluate their ability to understand and interpret the forecast and its corresponding verification, and their visualisation preferences. Feedback was also sought on user perception of the main problems regarding the communication of forecast information, barriers to the wide-range dissemination and exchange of actionable climate information and possible solutions to these issues.

Exercise 4: Use of Forecasts in Decision-Making For the final exercise the participants of the CLIMRUN workshop had to evaluate a simplified decision-making case study that required the use of seasonal forecast information. They put themselves in the position of an energy manager in Northeast Brazil, who had to decide whether to use the hydropower water reserves now or save them for the coming months. To guide their decision, each group was given a probabilistic seasonal precipitation forecast from a different FC for DJF 2009/2010 and, if available, the corresponding verification information.

Note that the data collection was performed as part of a Masters degree project, that was submitted in April 2014. The information collected may not be representative of the actual or current information accessible online from each FC.

7.3 Results

The outcome of the comparison between the FCs was evaluated, followed by an assessment of the decision-making case study exercise.

7.3.1 Forecast Availability and Accessibility

It was found that all FCs, except APCC (needs login), BoM and MF have free access to their climate forecast data, although the BoM grants users access for research purposes. The web pages were not found to be entirely user-friendly: most of them were difficult to browse and not intuitive when seeking specific information. The number of clicks required to access the forecast data varied (see Table 7.1). This indicates that the forecast information at some FCs is more easily accessible, which can imply a preference for the user when choosing an FC, regardless of the quality of the information. The number of *clicks* from the initial website to the verification data was also assessed and varied considerably among FCs (see Table 7.1). The proximity of the forecast and verification information, and thus the ability for the end user to evaluate in which regions the forecast is useful, was also assessed. This varied strongly depending on the FC. Some display forecast and verification information together, and some have completely different paths to each source. The websites are in English with the exception of CPTEC, BCC (parts are translated to

Table 7.1 The availability and accessibility of online forecast information for each FC

Availability and accessibility criterion	FCs that meet the criterion	Exceptions and further explanation
Login needed to access forecast	APCC, BoM, MF	BoM grants users access for research purposes
Two "clicks" needed to access the forecast from homepage	BCC, MSC	
Three "clicks" needed to access the forecast from homepage	IRI, APCC, CPTEC, SAWS	
Several "clicks" needed to access the forecast from homepage	UKMO, ECMWF, CPC, NOAA, TCC, KMA,	BOM forecasts were provided upon request
		MF forecasts are not available online
Two "clicks" needed to access the verification from homepage	BCC	BOM verification was provided upon request
		MSC verification was found via a Google search
Three "clicks" needed to access the verification from homepage	ECMWF, TCC, NOAA	
Four "clicks" needed to access the verification from homepage	IRI, CPTEC, KMA	
Up to six "clicks" needed to access the verification from homepage	UKMO	
Website is not in English	CPTEC, BCC (only parts)	
	MF (in French)	
Incomplete scientific explanation or description of their forecasts	BCC (only some basic text)	
	SAWS (no explanation)	

English) and MF (website only in French). Most FCs include a scientific explanation or description of the forecast maps available on their website. BCC has some text but not a full scientific explanation, and SAWS has no explanation.

7.3.2 Forecast Visualisation and Description

It was found that the most commonly available forecast is displayed as "most likely tercile". It was not possible to obtain this kind of information from the MF website. Therefore, this FC was not included for further analysis. The UKMO was the only FC to show the forecast in terciles and quintiles (i.e. considering more than three categories). The most common lead time was 1 month (i.e. issued 1 November 2009 for a DJF 2009/2010 forecast) and 3 months lead (i.e. issued 1 September 2009 for a DJF 2009/2010 forecast). There is a range of climate variables available via different FC websites, although precipitation is the only one offered by all FCs. The geographical region of each FC forecast/verification was global, with the exception of BoM and KMA, which showed regional maps. The most common verification type is the ROC, although different aspects of ROC analysis are presented. It was not possible to open the verification display from SAWS or APCC due to technical problems. For the forecasts, the region, forecast period and variable of the forecast corresponded to the verification product in all FCs except BoM and KMA, where the forecast and/or verification were available for a different region, or only a subset of a region. It was often unclear if the lead times were the same, due to the lack of information in one or both graphics. When comparing the descriptive information provided alongside the graphics (i.e. title, labels, etc.), nearly all FCs indicate the forecast type, the period (with the exception of KMA) and the climate variable (precipitation). Other data, such as the indication of units, existence of a legend and longitude and latitude labels, are not shown in all forecast graphics (see Table 7.2). Some FCs do not display a unit or longitude/latitude labels or grid. The number of different colours used to represent different levels of skill also varied, depending on the FCs, from five to 30 colours. When comparing the information in the title of the verification, all centres indicated the verification type (except NOAA), the period (except NOAA and KMA) and the climate variable.

7.3.3 Forecast Evaluation by End Users

When shown the different forecast examples, all participants agreed that the graphics were difficult to understand because there was not enough explanatory information (e.g. a description or interpretation of the graphic). Overall, the meaning of the legend remained unclear, and units were not always included. A common observation was that the title did not explain the graphic, and the terminology used was considered ambiguous (e.g. tercile, ECMWF, DJF). Acronyms should be written

Table 7.2 The online forecast communication approaches followed by each FC

Forecast visual communication approach	FCs using this approach	Exceptions and further explanation
Forecast is displayed as "most likely tercile"	IRI, ECMWF, CPTEC, BCC, TCC, SAWS	MF, not possible to obtain information
Forecast is displayed as "most likely quintile"	UKMO	MF, not possible to obtain information
Forecast lead time is 1 month	ECMWF, BoM, BCC, NOAA, TCC, KMA, MSC	
Forecast lead time is 3 months	ECMWF, APCC, CPTEC, BoM, BCC, TCC	
Geographical region of forecast is global	All FCs	BoM, CPTEC and KMA, local regions shown
ROC score verification used	IRI, ECMWF, BCC, TCC, MSC, UKMO	ROC: UKMO, TCC Generalized ROC or GROC: IRI ROC skill score: ECMWF ROC area above-normal tercile: BCC ROC lower and upper tercile: MSC
Forecast units are not shown	UKMO, BoM, BCC, NOAA	
Longitude/latitude labels are not shown	MSC, SAWS	
Longitude/latitude grid is not shown	UKMO, IRI, ECMWF, APCFC, CPTEC, BoM, BCC, CPC, TCC, MF, SAWS	

out and jargon should be explained on first mention. Another critique was that longitude and latitude labels were missing or, where they were included, their meanings were not stated. The choice of colours was generally accepted, although it was mentioned that they do not depict exact numbers on the map. The meaning of the colour white or the percentages for each category were also not clear to some. The omission of units (e.g. percentages, metre per second, height, etc.) was criticised.

Interviewees suggested that the title and the descriptions should be modified to make the graphics more understandable with better information; they should be self-explanatory. At a glance, it should be clear what is shown and how the colours should be interpreted. Another suggestion was to include a zoom view.

The evaluation of the verification maps was similar to that of the forecasts. All agreed that the title was unclear, as well as the terminology (e.g. ensemble-mean correlation), which should be explained by specifying, for instance, what was being correlated. Some participants thought that the colour choice was reasonable, whereas others stated that the colours were misleading. A global map was mentioned to be too big to effectively illustrate this kind of information. Concerning the key question of how the communication of climate forecast information to decision-makers can be improved, individuals suggested the establishment of climate centres with advisors who know the subject and have connections to climate scientists/institutes/universities to help *translate* the information and to avoid incorrect statements.

It is important to create connecting links to enable the exchange of information for both forecast providers and users because often clarity and understanding in the communication are missing. Also, improving the quality of the forecasts can help to create better relationships among scientists and decisions-makers. Participants said that it would be helpful for climate scientists to synthesise their data further to meet their specific needs and improve the communication of uncertainty. Regarding the barriers that hinder a wide-range dissemination and exchange of actionable climate forecasts, the insufficient exchange of unfiltered and freely accessible data was highlighted, as well as that information is often not available free of charge. Another barrier was the difficulty in understanding or defining the meaning of *forecast uncertainty*.

7.3.4 Use of Forecasts in Decision-Making

For the decision-making exercise, each group was asked to interpret the forecast and verification graphics from one of the five preselected FCs in order to make decisions regarding the management of water reserves for hydropower generation in Northeast Brazil. The simplified decision scenario was whether the energy manager should use available water reserves to generate hydropower immediately or, depending upon forecast precipitation, to keep the reserves for the future season in case of forecast drought conditions (i.e. below-normal rainfall). A precipitation forecast for DJF 2009/2010 was provided, with one-month lead time. Participants were asked to integrate probabilistic forecasts and verification information in their decision-making process. Two groups were given forecast and verification graphics from the UKMO (see Fig. 7.1). One group decided that the forecast probability was not high enough for any of the three categories (below normal, near normal, above normal), and so they would not use the forecast information for this year. The other group stated that they were going to use the water now because the most probable tercile was above normal and the skill score, from the verification graphic, was between 0.7 and 0.8, which meant to them that the skill was relatively high.

The forecast is shown as most likely tercile categories, labelled as "below normal", "near normal", and "above normal", with probabilities included (although the unit (%) is not indicated). Each category is shown as an individual map. The probability range is from 20 to 80 % divided by 20 % intervals and is the same for each category, as are the colours which use blue to indicate low probability and orange for high probability. ROC score verification is used, where grey represents a low score and red a high score. However, this is only available for the near-normal and above-normal categories. The start date, ensemble size, 95 % confidence intervals or the baseline verification period are not shown. The next group analysed IRI forecast and verification maps (see Fig. 7.2). They concluded that the data was clear to understand. The forecast indicated a 40–45 % below-normal precipitation for the season. Although the verification does not show forecast skill, the group decided to use the hydropower water resources now.

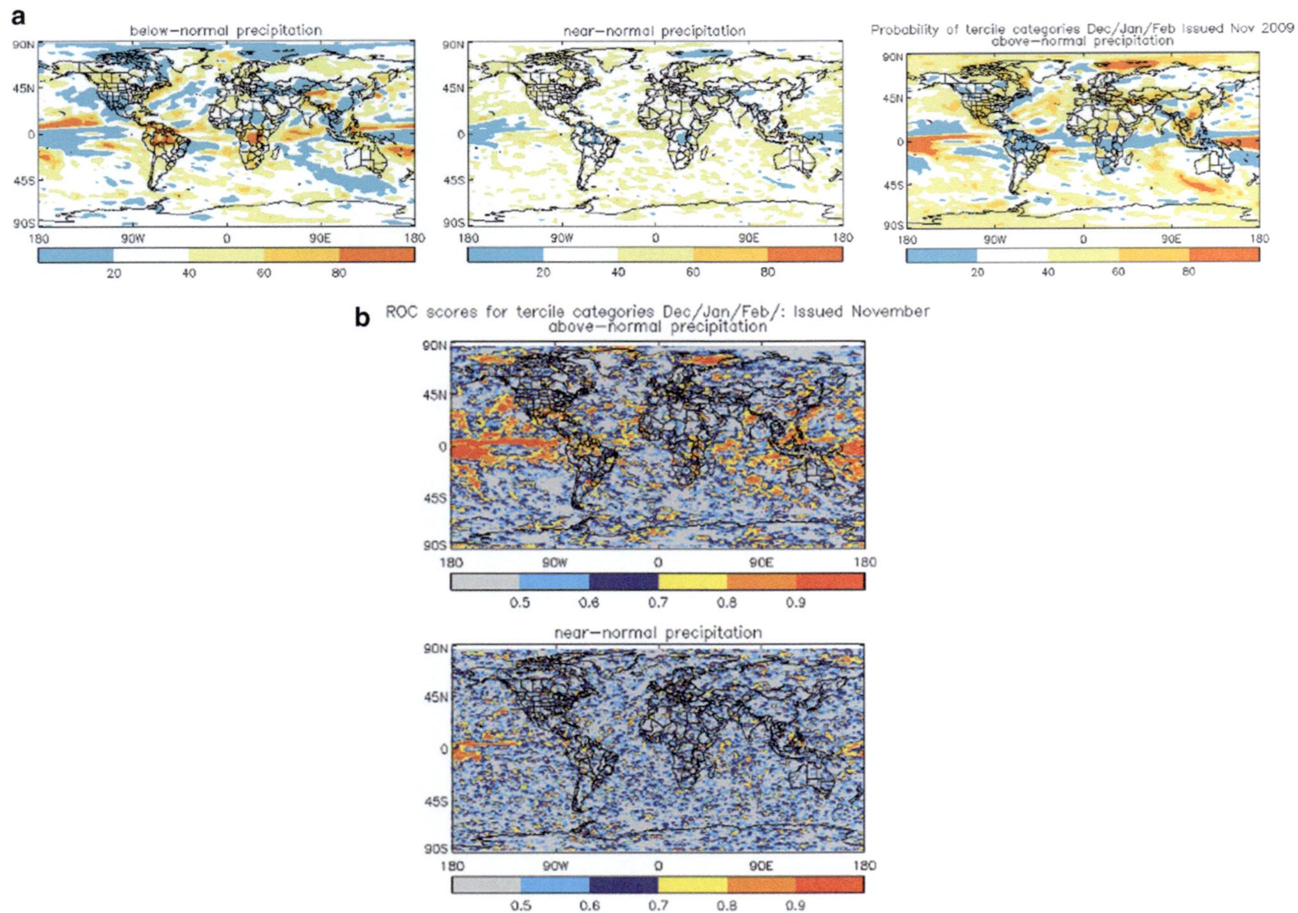

Fig. 7.1 UKMO (**a**) forecasts and (**b**) verification maps. The maps show global precipitation (**a**) forecast for probability of most likely category in DJF 2009/2010 and (**b**) the corresponding global forecast verification of the DJF forecast

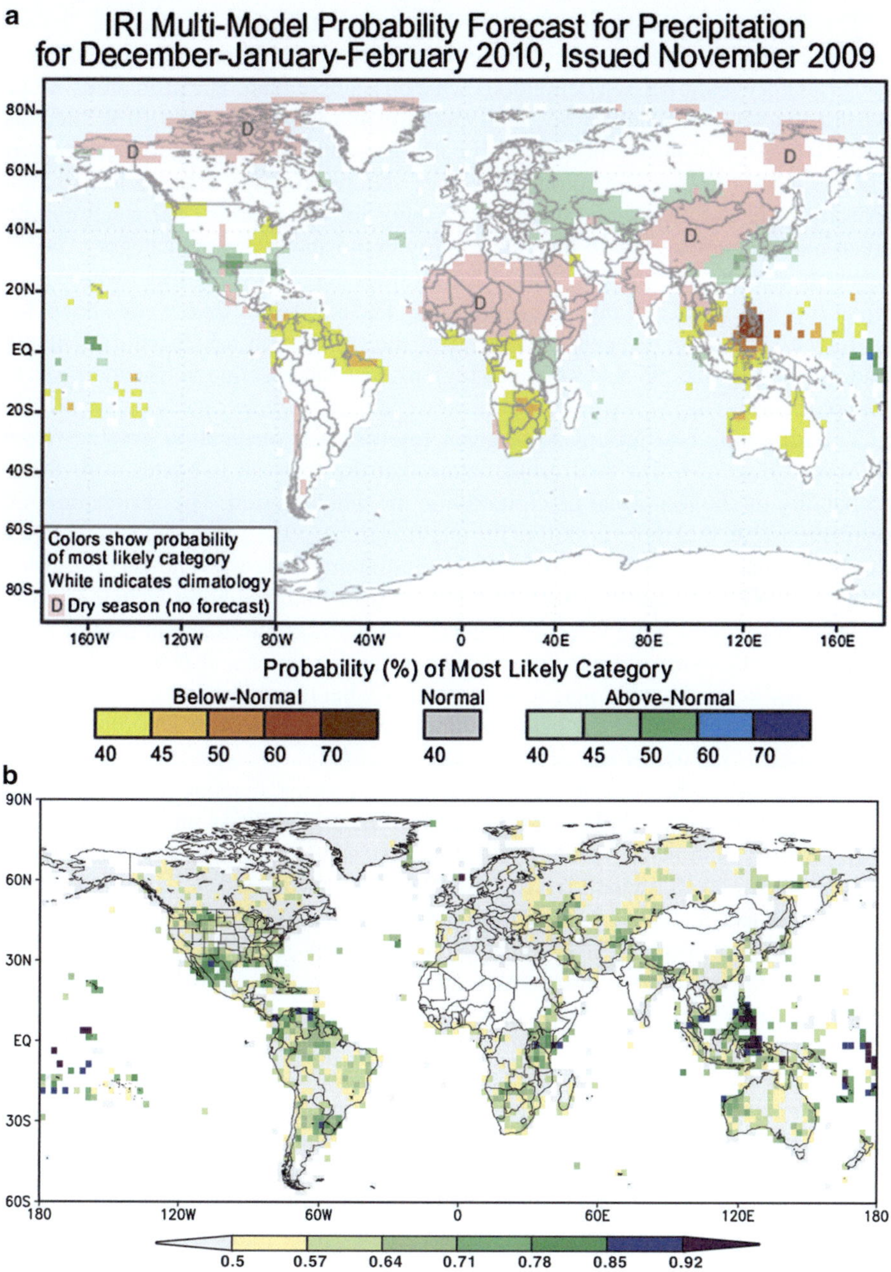

Fig. 7.2 *IRI (a) forecast and (b) verification maps*. The maps show global precipitation (**a**) forecast for probability of most likely category in DJF 2009/2010 and (**b**) the corresponding global forecast verification of the DJF forecast (Note, the following information was provided with the verification map: Title: Generalized ROC (GROC): Lead 0.5 months, precipitation forecast skill: DJF. Legend: *Gray* indicates no (or negative) skill. *Purple* indicates highest skill)

The forecast is shown as most likely tercile categories, labelled as “below normal”, “near normal”, and “above normal”, with probabilities and unit included. Each category is shown as a continuous scale on a single map. The probability range is from 40 to 70 % divided by 5 % intervals from 40 to 50 % and then 10 % intervals from 50 to 70 % in the lower and upper categories, although the normal category is a single value (40 %). The colour scale uses yellow-brown to indicate low to high probability in the below-normal category, grey for the near-normal category and green-blue to indicate low to high probability in the above-normal category. Single generalised ROC verification is used for all three forecast categories, where grey represents low score and purple a high score. The exact start date is not shown for the forecast, although it appears to differ from the verification, which has a 0.5 month lead. Ensemble size, 95 % confidence regions, and the baseline verification period are not shown.

The following two groups were given forecast and verification graphics from ECMWF (Fig. 7.3). The first group found that the forecast indicates a 40–80 % probability of above-normal precipitation in the whole region. The verification did not demonstrate that there is skill in the target region, so the group decided that they would not use the forecast in their decision-making process. The second group could not answer the questions. There were too many unfamiliar terms in the graphics in order to identify the forecast and verification information and to make a decision. The participants did not know what *tercile*, *System 3* and *DJF* meant. They also did not know what the percentages were or what the abbreviation *prob.* meant. This highlighted the need for better communication to end users starting with the basics, such as the use of full or well-defined terms.

The forecast is shown as most likely tercile categories, labelled as “below lower tercile”, “other”, and “above upper tercile”, with probabilities and unit included. Each category is shown as a continuous scale on a single map. The probability range is from 40 to 100 % divided by 10 % intervals from 40 to 70 % and then a 30 % interval from 70 to 100 % in the lower and upper categories, and the “other” category has no probability. The colour scale uses yellow-red to indicate low to high probability in the below lower-tercile category, white for the other category and turquoise-dark blue to indicate low to high probability in the upper-tercile category.

Anomaly correlation coefficient verification is used, which verifies only the forecast mean, not the probability categories. Purple indicates a strong negative correlation, white no correlation and red a strong positive correlation. A ROC score verification is also used, although only available for the below lower- and above upper-tercile categories, where purple-blue indicates a low score and yellow-red a high score. The exact start date, ensemble size, 95 % confidence regions and baseline verification period are all shown.

The next group analysed information from the APCC (see Fig. 7.4). At the time of data collection, the verification information was not accessible, due to technical issues. Therefore, only the forecast information was considered. They pointed out that the forecast shows a 60–70 % decrease in precipitation in DJF 2009/2010 (note: this is a misinterpretation of a high probability for below-normal precipitation). For

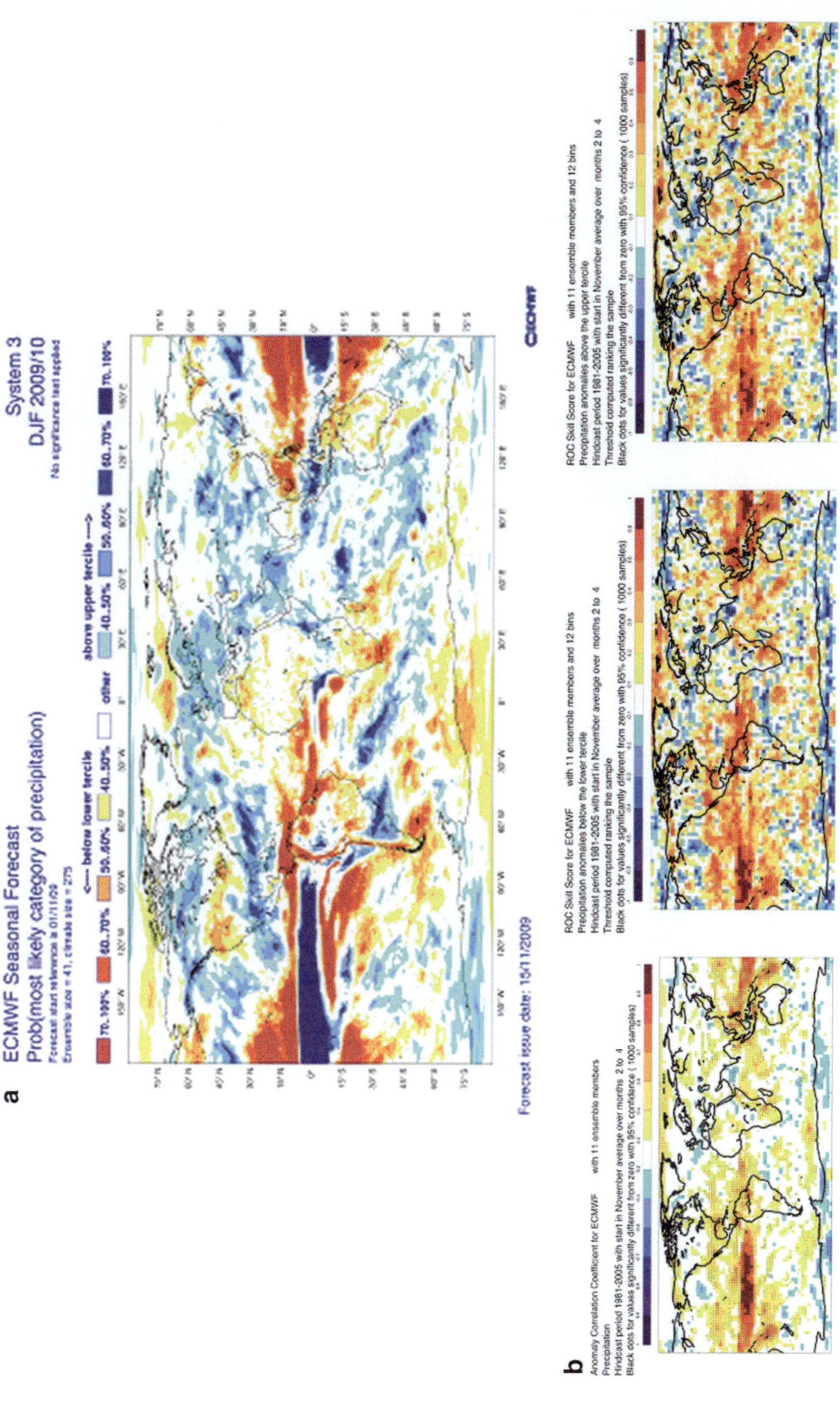

Fig. 7.3 ECMWF (**a**) forecast and (**b**) verification maps. Figures show global precipitation (**a**) forecast for probability of most likely category in DJF 2009/2010 and (**b**) the corresponding global forecast verification of the DJF forecast

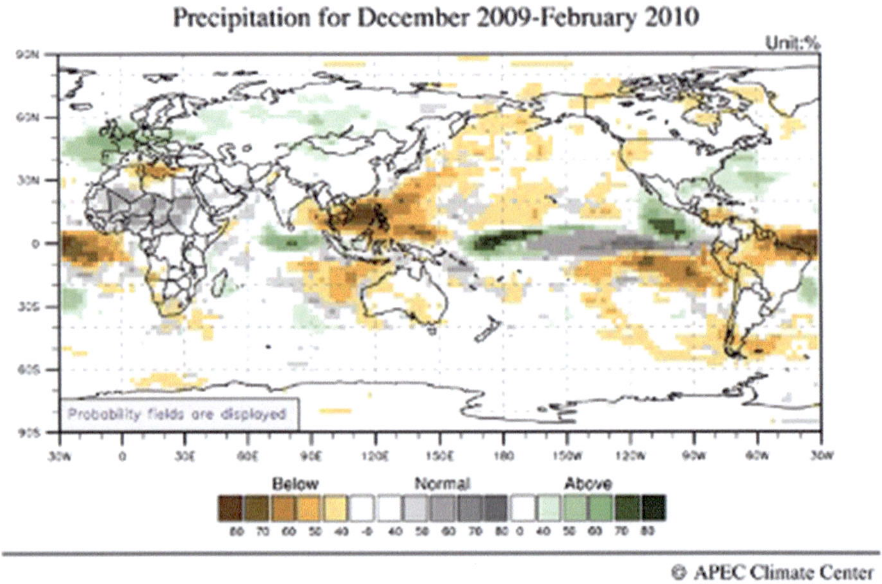

Fig. 7.4 APCC forecast map. Figure shows global precipitation forecast for probability of most likely category in DJF 2009/2010

hydropower planning, the available data, and especially the lack of verification, were not enough for decision-making. As the forecast indicated a possible decrease in rainfall, they decided that the reservoirs should be kept filled for the coming season for normal operation of the power plant, with reduced downstream flow, just in case the forthcoming precipitation season turned out to be below normal.

The forecast is shown as most likely tercile categories, labelled as "below", "normal", and "upper", with probabilities and unit included. Each category is shown as a continuous scale on a single map. The probability range is from 0 to 80 % divided by 40 % intervals from 0 to 40 % and 10 % intervals from 40 to 80 % in all three categories. The colour scale uses white-dark brown to indicate low to high probability in the below category, white-dark grey to indicate low to high probability in the normal category and white-dark green to indicate low to high probability in the upper category. No verification map was available from the APCC website. The exact start date, ensemble size, 95 % confidence regions and the baseline verification period are not shown.

The last two groups were given the CPTEC forecast and verification graphics (Fig. 7.5). The first group identified that the forecast indicated a severe decrease in precipitation (note: this is also a misinterpretation of a high probability for below-normal precipitation). However, from the verification graphic, there did not appear to be much skill for the DJF season in the area of interest. Therefore, this group decided not to use the forecast information for their decision-making process and chose to keep the reservoir water to ensure sufficient power for the forthcoming

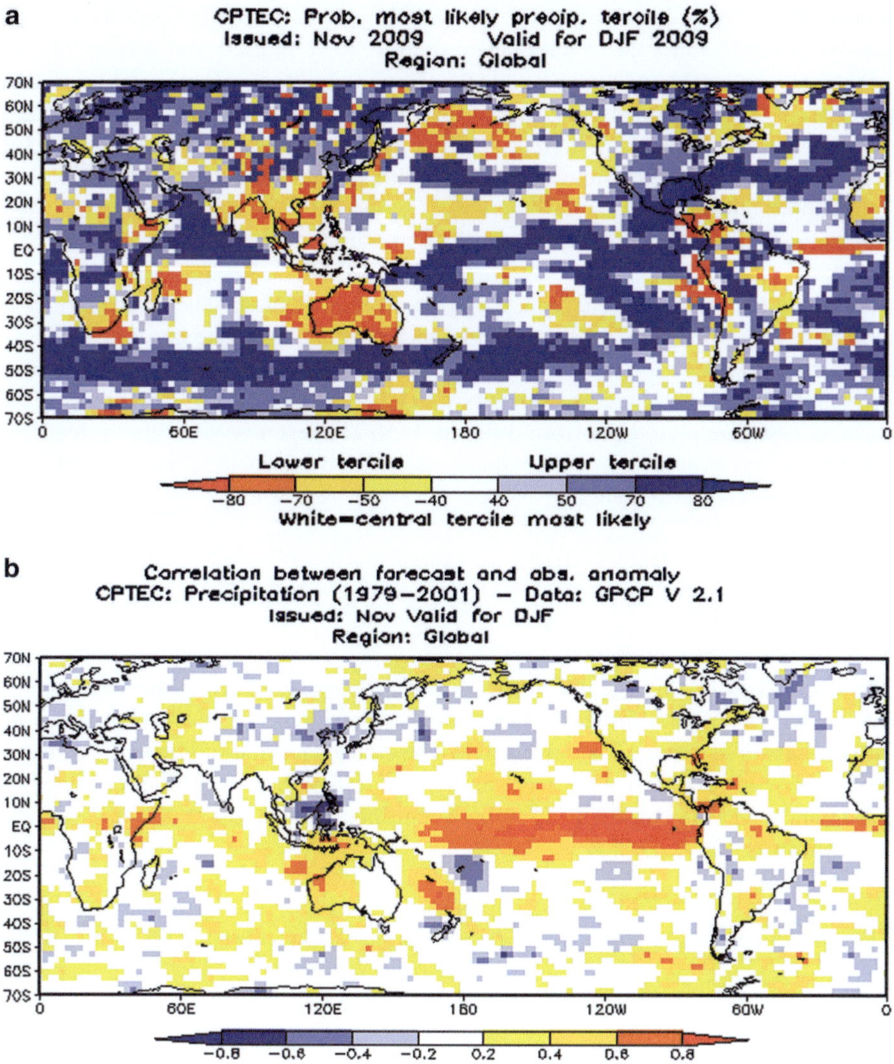

Fig. 7.5 CPTEC (**a**) forecast and (**b**) verification maps. Figures show (**a**) global precipitation forecast for probability of most likely category in DJF 2009/2010 and (**b**) the corresponding global forecast verification of the DJF forecast

season. The second group answered that the forecast predicted that there will be a likely below-normal rainfall and that the verification maps supported this. They concluded that it is best to use the hydropower in the next season and not at present.

The forecast is shown as most likely tercile categories, labelled as "lower tercile", "central tercile most likely", and "upper tercile", with probabilities and unit included. Each category is shown as a continuous scale on a single map. The probability range is from −40 to 80 % divided by 10 % intervals from 40 to 50 % and 70 to 80 %, and a 20 % interval from 50 to 70 % in the lower-tercile (shown with negative values) and upper-tercile (shown with positive values) categories, and the central-tercile category has no probability attached. The colour scale uses yellow-red to indicate low to high probability in the lower-tercile category, white for the central-tercile category and light to dark blue to indicate low to high probability in the upper-tercile category.

Anomaly correlation coefficient verification is used, which verifies only the forecast mean, not the probability categories. Blue indicates a strong negative correlation, white no correlation and red a strong positive correlation. The exact start date, ensemble size, and 95 % confidence regions are not shown. The baseline verification period is shown.

Note that a whole range of decisions was taken based on forecasts for the same variable and time period but from different FCs. This exercise also highlighted the human aspect of the use of probabilistic information versus the amount of risk an individual is willing to take.

7.4 Discussion

The main focus of this chapter concerns the challenges regarding the visual communication of climate forecast information to end users and decision-makers of climate-sensitive sectors. The results of this research show that fundamental changes have to be made to improve the accessibility and visualisation techniques of forecast information and the communication between scientists and users. A set of recommendations is outlined below.

Availability and accessibility:

- Users should be able to access information freely or be granted access for scientific reasons.
- Websites should be translated into English in order to improve the wide dissemination of the information.
- Forecast information should be easily accessible: No more than three clicks from the initial website.
- User-friendliness of the website could be improved to facilitate navigation, e.g. by introducing standardised web designs and search options at all FCs.

Visualisation and description:

- A standard forecast type and verification type should be used, as well as a common layout for forecasts and verifications that appear alongside each other.
- Explanations of the fundamental information regarding forecast terminologies should be consistent between FCs.
- Latitude and longitude labels and grids should be included in all forecasts.

- Titles should contain information detailing the variable, forecast period, issue date, and target region. A legend and an indication of the units displayed should be included. They must be self-explanatory.
- Every forecast should be accompanied by explanatory text, which clarifies terminologies and how the graphic and colours should be read/interpreted.
- Colour choice and selection should be standardised across all FCs, in order to know exactly which tercile is above normal, near normal and below normal. For example, the same colour should consistently be assigned to the above-normal and below-normal category, e.g. blue for above-normal (wetter) and red for below-normal (drier), when displaying probability categories for precipitation. A further study could be undertaken to clarify if these colours should remain consistent across all variables, i.e. is there any added value to switching the red and blue categories for temperature?

Decision-makers from various climate-sensitive sectors could benefit from improved accessibility, communication and understanding of climate forecast products if these recommendations are adopted across all FCs. This is of particular relevance during El Niño (or La Niña) events, when climate-related impacts in climate-sensitive sectors are more likely and also more predictable (Goddard and Dilley 2005; Halpert and Ropelewski 1992; Ropelewski and Halpert 1987). Following the exercises to evaluate the perception of the DJF 2009/2010 precipitation forecast, most answered that seasonal forecasts are helpful but not relevant to make decisions because they are too uncertain, based on the idea that climate is too chaotic to predict. There appeared to be a lack of understanding on the role of the verification data to support the forecasts, which needs particular attention in ongoing climate forecast communication. Nevertheless, there was a general agreement that it is useful to have these climate outlooks as an orientation, but that one has to be careful with their interpretation. All individuals agreed that, to overcome barriers in the use of seasonal forecasts, the degree of uncertainty is a key aspect to communicate effectively and consistently.

7.5 Conclusions

The general conclusion that arose from the analysis was that each FC uses different techniques to communicate climate information, which is confusing for the end user. Individuals also interpret information very differently and have diverging expectations, which highlight that decision-making is related to the amount of risk an individual is willing to take and individual tolerance towards risk.

The results of the participatory exercises have underlined the existing gap between the scientific data available and the information needed by decision-makers (Asrar et al. 2013). Some groups misinterpreted forecast probabilities as the magnitude of precipitation (e.g. a high probability of the below-normal category was interpreted as a severe decrease in precipitation). Therefore, training to interpret probabilistic information, tailored websites or information centres would be helpful to connect provider and user needs and understanding. It is thought that these initiatives should be sector tailored so that end users can relate to their specific application and, in turn, facilitate the integration of climate forecast information in decisions-making processes. Further, it is necessary to establish intercomparison exercises between forecasting centres to ensure that climate forecast products are communicated consistently.

Additional work is needed to implement the standardisation options at the FCs regarding accessibility and visual communication of forecasts towards the development of climate services. New communication approaches should be explored. In a more general context, a clearer distinction between climate projections (e.g. IPCC) and climate predictions (e.g. probabilistic seasonal forecasts) is urgently needed, as end users assume that these two products show the same information. Finally, the use of probabilistic forecast information also requires an evaluation of the corresponding risk and the economic cost for a range of possible decisions that are specific to a climate-sensitive sector (e.g. energy).

This is a preliminary study of the barriers to using climate information in decision-making, focusing on the visual and descriptive aspects of climate forecast communication. However, there are many other dimensions to the problem, such as tailoring climate information to user needs (Rasmussen et al. 2014), estimating their potential economic value (Sultan et al. 2010) and analysing risk management approach of end users (Crane et al. 2010). Research that brings together these themes could help to advance the use of climate forecasts in decision-making processes, to facilitate the adaptation of key sectors of society to climate variability and change and the corresponding societal impacts.

Acknowledgements The research leading to these results has received funding from the EUPORIAS project (grant agreement no. 308291) and SPECS project (grant agreement no. 308378) funded by the European Commission's Seventh Framework Research Programme.

References

Asrar GR, Hurrell JW, Busalacchi AJ (2013) A need for "actionable" climate science and information: summary of WCRP open science conference. Bull Am Meteorol Soc 94(2):ES8–ES12, 41

Crane TA, Roncoli C, Paz J, Breuer N, Broad K, Ingram KT, Hoogenboom G (2010) Forecast skill and farmers' skills: seasonal climate forecasts and agricultural risk management in the southeastern United States. Weather Clim Soc 2(1):44–59

Doblas-Reyes FJ, García-Serrano J, Lienert F, Biescas AP, Rodrigues LR (2013) Seasonal climate predictability and forecasting: status and prospects. Wiley Interdiscip Rev Clim Chang 4(4):245–268

Falloon P, Fereday D, Stringer N, Williams K, Gornall J, Wallace E et al (2013) Assessing skill for impacts in seasonal to decadal climate forecasts. J Geol Geosci 2:e111

Gámiz-Fortis S, Pozo-Vázquez D, Trigo RM, Castro-Díez Y (2008) Quantifying the predictability of winter river flow in Iberia. Part II: seasonal predictability. J Climate 21(11):2503–2518

García-Morales MB, Dubus L (2007) Forecasting precipitation for hydroelectric power management: how to exploit GCM's seasonal ensemble forecasts. Int J Climatol 27(12):1691–1705
Goddard L, Dilley M (2005) El Nino: catastrophe or opportunity. J Climate 18(5):651–665
Halpert MS, Ropelewski CF (1992) Surface temperature patterns associated with the Southern Oscillation. J Climate 5(6):577–593
Hewitt C, Mason S, Walland D (2012) The global framework for climate services. Nat Clim Chang 2(12):831–832
Hewitt C, Buontempo C, Newton P (2013) Using climate predictions to better serve society's needs. Eos Trans Am Geophys Union 94(11):105–107
IPCC (2013) Data Distribution Centre. http://www.ipcc-data.org/guidelines/pages/definitions.html. Accessed 1 Jul 2014
Jolliffe IT, Stephenson DB (eds) (2012) Forecast verification: a practitioner's guide in atmospheric science. John Wiley & Sons, Chichester
Jupp TE, Lowe R, Coelho CA, Stephenson DB (2012) On the visualization, verification and recalibration of ternary probabilistic forecasts. Philos Trans R Soc A Math Phys Eng Sci 370(1962):1100–1120
Lowe R, Barcellos C, Coelho CA, Bailey TC, Coelho GE, Graham R et al (2014) Dengue outlook for the World Cup in Brazil: an early warning model framework driven by real-time seasonal climate forecasts. Lancet Infect Dis 14(7):619–626
Rasmussen LV, Mertz O, Rasmussen K, Nieto H, Ali A, Maiga I (2014) Weather, climate, and resource information should meet the needs of Sahelian pastoralists. Weather Clim Soc 6(4):482–494
Rodrigues LRL, García-Serrano J, Doblas-Reyes F (2014) Seasonal forecast quality of the West African monsoon rainfall regimes by multiple forecast systems. J Geophys Res Atmos 119(13):7908–7930
Ropelewski CF, Halpert MS (1987) Global and regional scale precipitation patterns associated with the El Niño/Southern Oscillation. Mon Weather Rev 115(8):1606–1626
Ropelewski CF, Halpert MS (1989) Precipitation patterns associated with the high index phase of the Southern Oscillation. J Climate 2(3):268–284
Slingsby A, Lowe R, Dykes J, Stephenson D, Wood J, Jupp T (2009) A pilot study for the collaborative development of new ways of visualising seasonal climate forecasts. Proceedings of the 17th Annual Conference of GIS Research UK, Durham, 13 Apr 2008
Solomon S, Qin D, Manning M, Chen Z, Marquis M, Averyt KB, Tignor M, Miller HL (eds) (2007) International panel climate change: the physical science basis. Contribution of Working Group I to the fourth assessment. Report of the intergovernmental panel on climate change. Cambridge University Press, Cambridge/New York
Sultan B, Barbier B, Fortilus J, Mbaye SM, Leclerc G (2010) Estimating the potential economic value of seasonal forecasts in West Africa: a long-term ex-ante assessment in Senegal. Weather Clim Soc 2(1):69–87
Thomson MC, Doblas-Reyes FJ, Mason SJ, Hagedorn R, Connor SJ, Phindela T et al (2006) Malaria early warnings based on seasonal climate forecasts from multi-model ensembles. Nature 439(7076):576–579
Trenberth KE, Stepaniak DP (2001) Indices of El Niño evolution. J Climate 14(8):1697–1701
WMO (2013) Global producing centres for long range forecasts. http://www.wmo.int/pages/prog/wcp/wcasp/clips/producers_forecasts.html. Accessed 1 Jul 2014

Chapter 8
Modelling Climate-Sensitive Disease Risk: A Decision Support Tool for Public Health Services

Rachel Lowe and Xavier Rodó

Abstract In order to control the spread of diseases and prepare for epidemics, decision support systems are required that take into account the multifaceted array of factors that contribute to increased disease risk. Climate forecasts, which predict the average climate conditions for forthcoming months/seasons, provide an opportunity to incorporate precursory climate information into decision support systems for climate-sensitive diseases. This aids epidemic planning months in advance, for diseases such as dengue, cholera, West Nile virus, chikungunya and malaria, among others. Here, we present a versatile model framework, which quantifies the extent to which climate indicators can explain variations in disease risk, while at the same time taking into account their interplay with the intrinsic/internal features of disease dynamics, which ultimately shape their epidemic structure. The framework can be adapted to model any climate-sensitive disease at different spatial/temporal scales and geographical settings. We provide case studies, quantifying the impact of climate on dengue and malaria in South America, Southeast Asia and Africa.

Keywords Dengue • Malaria • Public health • Disease prediction model • Early warning system • Climate services

R. Lowe, Ph.D. (✉)
Institut Català de Ciències del Clima (IC3), Barcelona, Spain
e-mail: rachel.lowe@ic3.cat

X. Rodó, Ph.D.
Institut Català de Ciències del Clima (IC3), Barcelona, Spain

Institució Catalana de Recerca i Estudis Avançats (ICREA), Barcelona, Spain

J.L. Drake et al. (eds.), *Communicating Climate-Change and Natural Hazard Risk and Cultivating Resilience*, Advances in Natural and Technological Hazards Research 45, DOI 10.1007/978-3-319-20161-0_8

8.1 Introduction

In order to control the spread of diseases and prepare for epidemics, decision support systems are required that take into account the multifaceted array of factors that contribute to increased disease risk. Many of the most important diseases affecting health are mosquito-borne, including malaria and dengue (Gage et al. 2008). Climate has a potentially large impact on the incidence of mosquito-borne diseases either directly, by affecting the developmental rates and survival of both the mosquito and pathogen, or indirectly, through changes in land cover and land-surface characteristics, which impact the availability of mosquito oviposition sites (McMichael et al. 2003). Additionally, weather and climate interact with local conditions, affecting not only mosquito infestation, but also human susceptibility and the contact rate between mosquitoes and humans (Banu et al. 2011).

The many potential drivers of complex disease systems, both extrinsic (e.g. climate and socioeconomic factors) and intrinsic (e.g. population immunity, seroprevalence status and demography), are often difficult to disentangle. Therefore, decision support system tools are required to simultaneously consider the complex interaction of climate hazards, socioeconomic disparities and human vulnerability in predictive disease risk models. Policymakers are well aware of the effects of climate on the dynamics of many diseases. However, despite this understanding, climate information is rarely exploited as a means to help prevent and control such health risks (Jancloes et al. 2014; Rodó et al. 2013).

Seasonal climate forecasts, which predict the average climate conditions for forthcoming months/seasons, show predictive capacity for particular locations and seasons (Doblas-Reyes et al. 2013). This provides an opportunity to incorporate forecast climate information into decision support systems for climate-sensitive diseases, such as dengue and malaria, to aid epidemic planning months in advance (Connor and Mantilla 2008; Thomson et al. 2006). The El Niño-Southern Oscillation (ENSO) and in particular its two main phases (El Niño and La Niña) can provide some indication of the likely prevailing climate conditions in certain regions (Ropelewski and Halpert 1987). It is currently the main climate indicator used for anticipating impacts on sectorial domains of interest for human societies and economies (particularly health). Due to their predictability, these indicators can potentially extend predictive lead time in disease early warning systems (Cash et al. 2013; Stewart-Ibarra and Lowe 2013).

In this chapter, we present a versatile spatio-temporal Bayesian modelling framework to support public health decision-making. We outline the minimum data requirements to formulate models. Explanatory variables at various spatial and temporal resolutions can be incorporated into a hierarchical model in order to make spatio-temporal probabilistic predictions of disease relative risk. Potential disease risk factors include altitude, land cover, proximity to road/rail networks and water bodies, temperature and precipitation, oceanic indicators, intervention activities, air traffic volume, population movement, urbanisation and sanitation indicators. However, in many cases, data on important drivers of disease systems is not measured or available. This limitation traditionally detracts from adequate progress in developing useful prediction systems at the local scale of cities or small regions. In order to quantify unknown or unmeasured disease risk factors, we use spatio-temporal random effects in the model framework. Once unknown structures are accounted for, we can identify which of the available indicators could significantly contribute to an effective early warning system.

The model can be adapted to simulate any climate-sensitive disease at different spatial and temporal scales and geographical settings. Below, we present examples of applying the framework to model dengue and malaria in South America, Southeast Asia and Africa, to quantify the extent to which climate indicators can explain variations in disease risk and assess the potential utility of forecast climate information in health decision support systems, for a given disease and location. By taking advantage of lead times of several months provided by climate forecasts, public health officials may be more effective in allocating intervention measures, such as targeted vector control activities and medical provision, to those areas most at risk.

8.2 Methods

8.2.1 Data Requirements

The model response variable takes the form of counts of disease cases (per unit time: day/week/month) spanning several years across administrative areas (i.e. municipalities, districts, provinces, etc.). Sometimes this information is stratified by vulnerability group (e.g. age, sex). It is essential to obtain population data to reflect differences in the underlying population over time, between locations and for different vulnerability groups. Other demographic indicators such as education levels and housing conditions can be extracted from national census data. When available, vector population data can be incorporated, as well as seroprevalence to allow for the pool of asymptomatic individuals that can hide the disease but significantly contribute to its transmission.

Meteorological station data can be obtained or requested from meteorological services. Otherwise, gridded climate data sets are readily available via the Internet (e.g. KNMI Climate Explorer (climexp.knmi.nl/), IRI/LDEO Climate Data Library (http://iridl.ldeo.columbia.edu/) and NOAA/ESRL/PSD Gridded Climate Datasets (http://www.esrl.noaa.gov/psd/data/gridded/)). For research purposes, hindcast data (i.e. retrospective forecasts made for a historical period in pseudo-operational mode) from seasonal climate forecasting centres such as the Met Office and the European Centre for Medium-Range Weather Forecasts (ECMWF) can be requested. However, real-time seasonal climate forecast data are not always easily obtainable or freely available to public health decision-makers.

When spatial variations of disease are of interest, geographic information system (GIS) layers, such as land cover use and transport networks, can be collated and tested in the model framework. Gridded data (e.g. climate or topological) can be reconciled with spatial polygon data (e.g. disease counts and demographic characteristics) using interpolation methods (Bivand et al. 2008) or by assigning a grid point to each spatial polygon on the basis of the shortest Euclidean distance between the area centroid and neighbouring grid points (Lowe et al. 2011). Once all available explanatory data has been transformed to the same spatial and temporal resolution as the response variable, it can be incorporated into the model framework, to account for confounding factors and help correctly attribute variations in disease risk to variation in climatic factors.

In many cases, information regarding the complex factors that influence disease transmission (e.g. circulating serotypes, pool of vectors or even vector predators) is not measured or available. In order to quantify such unknown or unmeasured disease risk factors, spatio-temporal random effects can be used and included in a model framework. This helps to capture unknown structures and identify which available indicators are significant drivers of the disease system.

8.2.2 Model Framework

To model counts of disease cases for different vulnerability groups (e.g. stratified by age, j) in time, space or both time and space, a generalised linear mixed model (GLMM) can be formulated where for each spatial area ($s=1,\ldots, S$) and time step ($t=1,\ldots, T$), the count of disease cases, y_{jst}, follows a negative binomial distribution with an unknown scale parameter, κ, and mean, μ_{jst}. A negative binomial model formulation is adopted to allow for extra-Poisson variation (overdispersion) in the observed dengue counts caused by unknown confounding factors and possible correlations in both time and space. The expected number of cases, e_{jst} (i.e. the population for a given vulnerability group, j, area, s, and time, t, multiplied by the overall disease risk over all groups, space and time), is included as a model offset. The model can be rearranged so that the ratio of observed to expected cases (i.e. the relative risk) is modelled by a combination of explanatory variables.

Explanatory variables at various spatial and temporal resolutions (e.g. climate, socio-economic, health infrastructure, etc.) can be incorporated and tested in the model framework, to select a suitable combination of statistically significant variables, to account for confounding factors (see model later in this section). The disease relative risk in preceding time steps can sometimes act as a surrogate for spatio-temporal confounding factors (e.g. increased mosquito populations or intervention activities). Therefore, suitably lagged disease incidence can also be considered in the model framework, in a type of density-dependent approach (Ellner and Guckenheimer 2011), in the absence of more precise information about the disease system. This lag should be chosen as a compromise between the longest lag plausible to provide predictive skill and the shortest lag possible to allow enough time for public agencies to provide an early warning. Note that the inclusion of an autoregressive term causes the first few observations in each location to be lost, which shortens the length of the time series.

Unstructured random effects can be included, creating a hierarchical model framework, to help account for unknown or unobserved confounding factors in the disease system (e.g. mosquito abundance, population immunity, quality of health-care services and local health interventions). Such random effects introduce an extra source of variability (a latent effect) into the model, which can assist in modelling overdispersion, in addition to the single-scale parameter in the negative binomial model. To allow for correlated heterogeneity between locations or clustering, structured random effects can be included. One way to impose a spatial dependency structure is to assume a Gaussian intrinsic conditional autoregressive (CAR) model prior distribution (Besag et al. 1995) for the spatial effects, which takes the neighbourhood structure of the area under consideration into account.

When modelling disease risk over a large geographical area with distinct ecological biomes (e.g. arid desert areas or tropical rainforest), a spatial zone variable should be included in the model framework (see Lowe et al. 2014a). This variable can be treated as a categorical variable, or factor, with the first zone set as a reference level. Many climate-sensitive diseases have a marked annual cycle, which can vary and differentially affect the different regional biomes. To allow for this, a first-order autoregressive month effect (Gilks et al. 1996) can be interacted with the zone categorical variable. It is recommended that the month with the lowest average disease risk is set as the reference level.

Before moving to a Bayesian framework, preliminary exploratory analyses can be made using fixed effects in a simple generalised linear model (GLM) framework. An initial model might include the climatic, demographic, socio-economic and cartographic variables, with relevant lags and polynomial terms (to capture possible non-linear effects) and also categorical variables to account for the annual cycle, administrative regions and ecological zones and associated interaction terms. Model selection algorithms, such as Akaike information criterion (AIC) stepwise regression (Venables and Ripley 2002), can help to identify candidate predictor variables at suitable time lags, by removing non-significant interaction terms, quadratic terms and explanatory variables (Lowe et al. 2013b). Residual analyses can be performed to check model assumptions and the presence of outliers, influential observations or leverage points, as statistical associations may be distorted by influential values and may not be robust between all values (see Lowe 2011).

GLMs assume independent (or at least uncorrelated) observations. However, this assumption is usually violated when modelling disease risk. There could be strong temporal correlation effects within some areas, and there could also be spatial clustering effects between neighbouring areas. To allow for such latent effects and correlation structures, a GLM (including fixed effects) can be used as a starting model and further refined by including random effects in a Bayesian model framework, to better capture spatio-temporal variations in disease risk. The full hierarchical model framework comprises a combination of climate variables at suitable time lags, non-climate covariates and previous disease risk, as delay count variables, as well as spatially and temporally structured and unstructured random effects. The model is formulated as follows:

$$y_{jst} \sim \text{NegBin}\left(\mu_{jst}, \kappa\right)$$

$$\log\left(\mu_{jst}\right) = \log\left(e_{jst}\right) + \alpha + \alpha_{s'(s)} + \Sigma\beta_i x_{ist} + \Sigma\gamma_i w_{ist} + \delta z_{jst-n} + \varphi_s + \upsilon_s + \omega_{t'(t)} + \omega_{t'(t),s'(s)}$$

A negative binomial as opposed to a Poisson distribution is chosen for the disease case counts to allow for observed global overdispersion in these counts (i.e. variance of counts being substantially in excess of their mean values over time, in all areas). The spatially unstructured random effects in the model (see below) then additionally allow for local spatial variations in this overdispersion. The variables x_{ist} represent the selected climate influences (e.g. x_{1st} = precipitation, x_{2st} = temperature, etc.). The variables w_{ist} are non-climate influences (e.g. w_{1s} = altitude, w_{2t} = intervention investment, etc.). Variable z_{jst-n} is the log ratio of observed to expected cases, n time steps previously, $z_{jst} = \log\left(y_{jst-n} / e_{jst-n}\right)$. Spatial random effects are composed of unstructured φ_s and structured components υ_s. The spatially unstructured random effects, φ_s, are assigned independent diffuse Gaussian exchangeable prior distributions. The structured random effects, υ_s, can be assigned a CAR model prior (see Lowe et al. 2013a for details). A first-order autoregressive month effect $\omega_{t'(t),s'(s)}$ is included for each zone (when applied to a large geographical area with distinct ecological zones), with month 1 and zone 1 aliased to the model intercept, α, and subsequent months following a random walk or first difference prior in which each effect is derived from the immediately preceding effect. Note that the fixed effect for month 1 is denoted as $\alpha_{s'(s)}$. The random month effect for zone 1 is denoted as $\omega_{t'(t)}$.

Independent diffuse Gaussian priors (mean 0, precision 1×10^{-6}) are normally used for the fixed effects ($\alpha_{s'(s)}, \beta_i, \gamma_i, \delta$) and weakly informative independent gamma hyperpriors (values for shape parameter of 0.5 and inverse scale parameter of 0.0005) for the scale parameter κ and for the precisions of the spatial and temporal random effects. These represent relatively standard choices for spatio-temporal GLMMs (Wakefield et al. 2000). Parameters in a GLMM can be estimated using a Bayesian framework, where parameter uncertainty is accounted for by assigning prior distributions to the parameters (Best et al. 1999). Hierarchical models can be created by parameterising prior distributions with unknown *hyperparameters*, which have their own *hyperprior* distribution. Markov Chain Monte Carlo (MCMC) provides a practical method for the estimation of parameters in Bayesian models (Gilks et al. 1996). This is because associated MCMC sampling yields samples from full posterior predictive distributions, which automatically incorporate all components of variance at the different levels in the model. Therefore, a full assessment of prediction uncertainty is possible.

In the assessment of complex Bayesian models, posterior predictive distributions are often calculated by simulating new pseudo-observations using samples from the posterior distribution of the parameters in the model (Gelman et al. 1996). The distribution of simulated values can then be compared to observed values. This framework was first developed to model the spatial and temporal distribution of dengue in Brazil (Lowe et al. 2011). The model was then extended to address the viability of dengue epidemic early warnings in the Southeast of Brazil (Lowe et al. 2013a) and was recently adapted to incorporate real-time seasonal climate forecasts to provide a dengue early warning ahead of a major global event (Lowe et al. 2014a). The framework has since been developed to address dengue risk in Ecuador (Stewart-Ibarra and Lowe 2013) and Thailand (Lowe et al. 2014b). It has also been adapted to assess the potential use of climate information in a malaria decision support system for Malawi (Lowe et al. 2013b). The following section presents some examples to demonstrate the versatile nature of the model framework.

8.3 Results

8.3.1 The Robustness of El Niño as a Predictor for Dengue in Southern Coastal Ecuador

By adapting the framework outlined above to deal with a single province ($s=1$), Stewart-Ibarra and Lowe (2013) assessed, for the first time, the importance of climate and non-climate drivers of interannual dengue fever variability for the southern coastal province of El Oro in Ecuador. In this region, dengue fever transmission follows a distinct seasonal pattern, with cases peaking during the rainy season from February to May. On average, positive anomalies in Pacific sea surface temperatures (i.e. El Niño conditions) are associated with positive anomalies in rainfall and air temperature in southern coastal Ecuador (see Fig. 8.1). This implies that warmer and wetter conditions are more likely during El Niño events, which in turn is conducive to increased mosquito abundance.

Local climate data and Pacific sea surface temperatures (Oceanic Niño Index, ONI) were used to predict dengue incidence in the province of El Oro (1995–2010). Unobserved confounding factors were accounted for using non-structured yearly random effects. ONI, rainfall and minimum temperature were found to be positively associated with dengue, with more cases during El Niño events. However, this time period included the strong 1997–1998 El Niño event, which might have dramatically influenced the model inference. To assess the robustness of the ONI parameter as a predictor for dengue variations in El Oro, a comparison was made between the parameter estimates from models fitted to two time periods: 1995–2010 and 2001–2010 (excluding the 1997–1998 El Niño event), while accounting for observed (e.g. serotype circulation, mosquito abundance) and unobserved (temporal random effects) confounding factors. For both time periods, the ONI parameter had similar positive associations with dengue, although the uncertainty in the ONI parameter in the 2001–2010 model was greater (see Fig. 8.2). This suggested that the strong 1997–1998 El Niño event did not influence the 1995–2010 model and that ONI is a potentially robust predictor for dengue epidemics in southern coastal Ecuador.

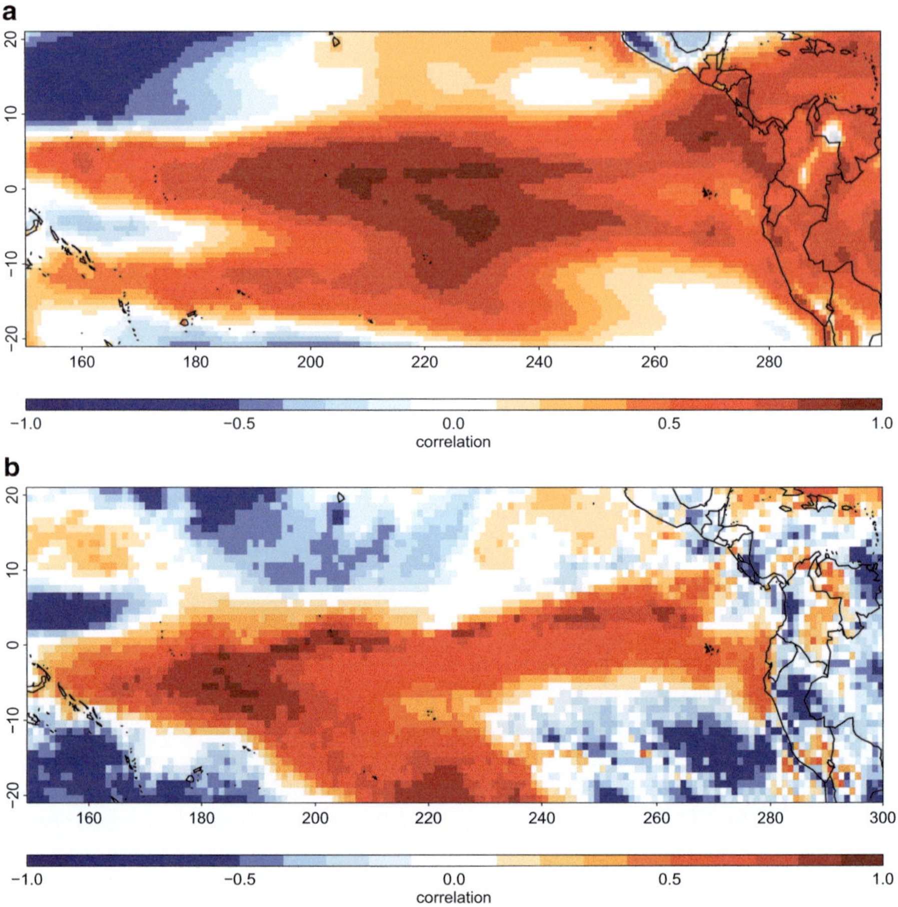

Fig. 8.1 The Oceanic Niño Index (ONI) (December–February) 2-month lagged correlations with (**a**) mean temperature (February–April 1980–2010) and (**b**) precipitation (February–April 1995–2009) (Sources: ERA-Interim Reanalysis daily, Global Precipitation Climatology Project (GPCP) daily 1.0 and the Climate Prediction Center of NOAA/National Weather Service)

Fig. 8.2 Kernel density estimates for the marginal posterior distribution of the Oceanic Niño Index (ONI) parameter in the dengue model for El Oro, Ecuador, fitted to data from 1995 to 2010 (*dashed curve*, note: includes 1997–1998 El Niño event) and 2001 to 2010 (*solid curve*) (Reproduced from Stewart-Ibarra and Lowe (2013))

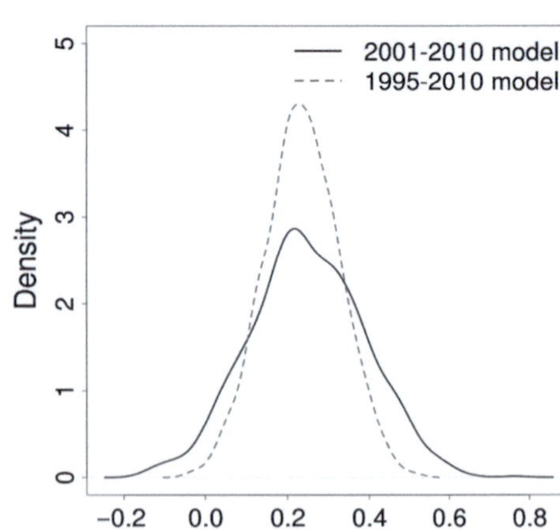

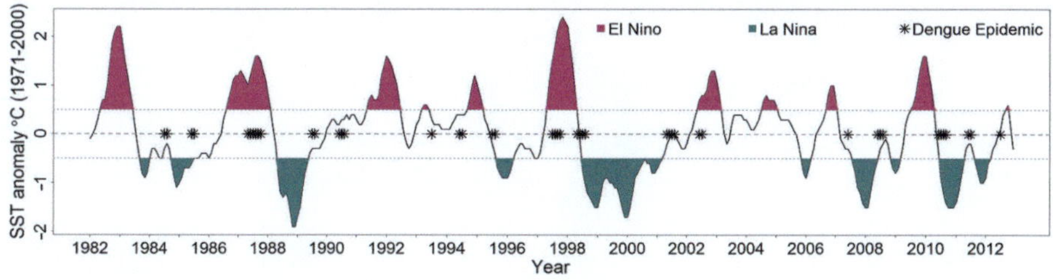

Fig. 8.3 El Niño (*purple*) and La Niña (*green*) conditions and dengue epidemics in Thailand (*) from 1982 to 2012

8.3.2 Accounting for Unknown Spatio-temporal Structures in Dengue Variation in Thailand

In recent years, the global burden of dengue disease has been rising dramatically (Guzman et al. 2010), and the need to both globally and regionally assess the role of climate in this expansion has become a major international priority. Among one of the most affected regions, Thailand provides a very detailed and exhaustive dengue surveillance and mosquito control system, with data sets reporting cases for more than three decades. ENSO has been identified as a potential driver of dengue in Thailand (Cazelles et al. 2005). However, dengue epidemics in Thailand do not always coincide with ENSO events (see Fig. 8.3). There may be other political or socio-economic changes that also influence outbreaks of the disease. To test if climate covariates (e.g. precipitation, temperature and ONI with appropriate time lags) were statistically significant predictors of dengue variation, monthly cases of dengue in the 76 provinces of Thailand for the period 1982–2012 were modelled using the above framework.

When analysing geographical data, it is important to account for the degree of spatial heterogeneity and correlation, in order to avoid underestimation of the credible intervals of model covariates. This was achieved by incorporating spatially unstructured (to account for heterogeneity) and structured (to account for correlation/clustering) random effects. To account for seasonality (the annual cycle), which could be attributed to climate and/or seasonal population movements, and dependencies of dengue incidence in 1 month to incidence the previous month, a random walk was introduced in the month effect. To allow for trends and unknown events that might have occurred over the three decades (e.g. changes in reporting or policy), exchangeable random year effects were included. The structure of the random spatial and temporal components of the model provides a combined measure of the various potential risk factors that might contribute to the underlying spatial and temporal variation in disease risk (see Fig. 8.4). By allowing for these structures, the impact of climate and non-climate variables on dengue relative risk can be more accurately assessed (Lowe et al. 2014b).

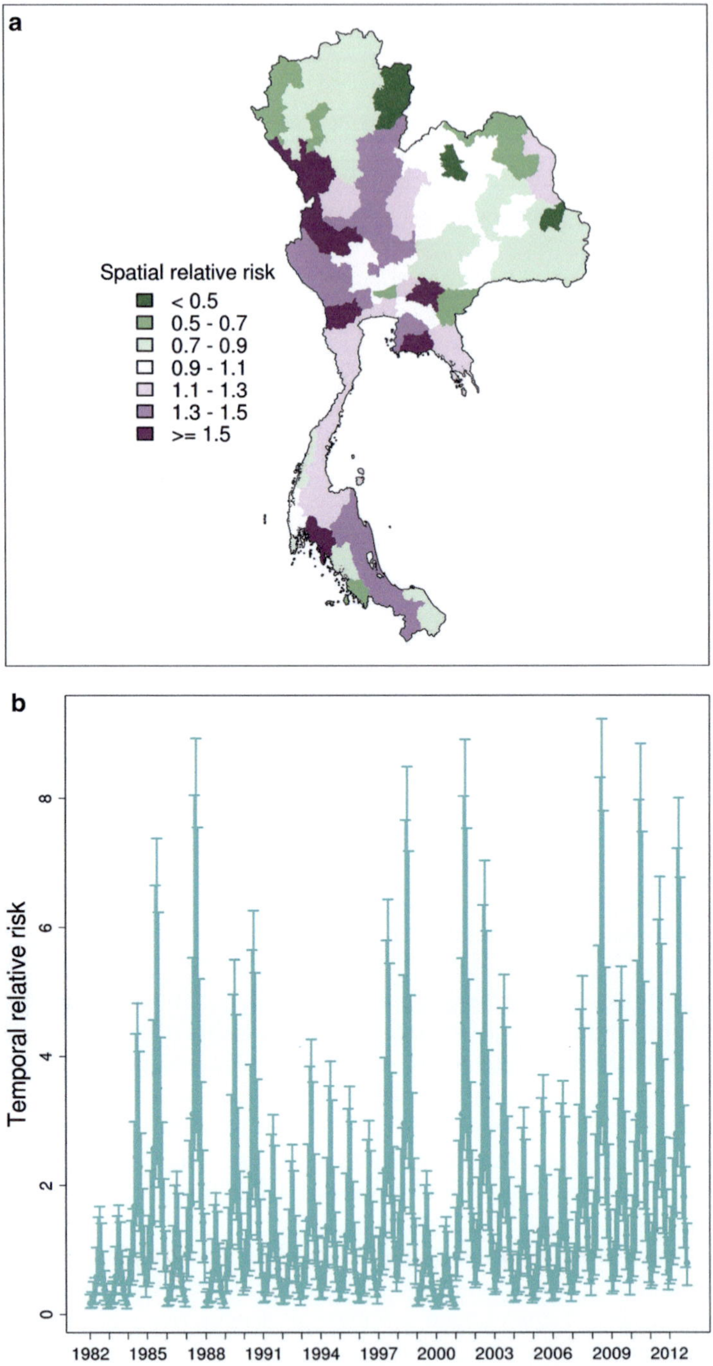

Fig. 8.4 (**a**) Spatial (structured and unstructured) and (**b**) temporal (monthly and yearly) random effect contribution to dengue relative risk for the 76 provinces of Thailand, 1982–2012

8.3.3 *Spatial Heterogeneity in the Usefulness of Climate Information for Malaria in Malawi*

Malaria imposes the biggest health burden in Malawi and is one of the leading causes of morbidity and mortality in children under 5 years of age. To investigate if climate information could help to explain the spatial and interannual variations in malaria morbidity in children under 5 years of age in Malawi, the model framework described above was adapted to simulate malaria risk at the district level. The model was formulated using an age-stratified spatio-temporal data set of malaria cases from July 2004 to June 2011 (Lowe et al. 2013b). Once intervention variations, unobserved confounding factors and spatial correlation were considered in the model framework, statistically significant climatic predictor variables were found to be average precipitation (in the form of a quadratic relation) and average temperature, during the 3 months previous to the month of interest.

To assess if the inclusion of climate information could improve the model estimation of malaria relative risk, over time, across space and for different age groups, the root mean square error (RMSE), a measure of the difference between model predicted and observed values, was calculated for a model excluding climate covariates. This was then subtracted from the RMSE of a model including climate covariates. Areas where the difference was less than zero, highlighted in grey (see Fig. 8.5), indicated that climate information improved the estimation of malaria relative risk, as the inclusion of these covariates results in a smaller difference between the model predicted values and the observations.

Climate information was found to improve the estimation of interannual variations in malaria morbidity rates in 41 % of the districts in Malawi that are mostly located in the north highland regions, subject to lower and intermittent malaria transmission intensity (with the exception of lakeside communities). In the southern regions, where malaria transmission is more intense, climate did not improve the model's capability to represent year-to-year variations present in the morbidity data, particularly for children under 5 years. By modelling age-stratified data, we can gauge an idea of how climate information may differentially impact groups with varying vulnerability.

8.3.4 *Probabilistic Dengue Early Warnings Driven by Seasonal Climate Forecasts in Brazil*

The efficacy of a climate-driven decision support system depends on the skill of the climate forecasting system. The EUROBRISA operational forecasting system (Coelho et al. 2006), a multi-model combined and calibrated system, produces 1-month lead climate forecasts for the following 3-month season. Seasonal climate forecasts used in this system have been reported to have the most skill in tropical regions of Brazil, with moderate skill in extratropical regions. In Lowe et al. (2014a), real-time seasonal climate forecasts from the EUROBRISA system and the

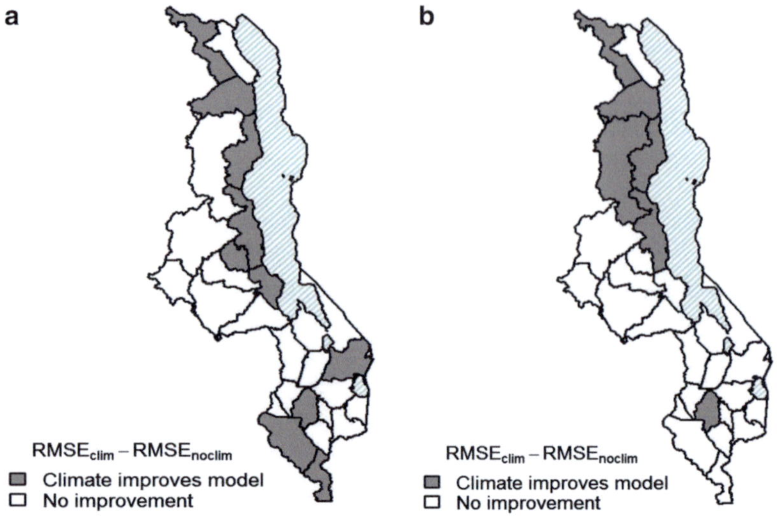

Fig. 8.5 Difference between RMSE for the model including climate information and RMSE for a model fit without climate information for (**a**) 5 years and over age group and (**b**) under 5-year age group. Districts with negative values (*grey*) suggest that climate information improves the model in these areas. Districts with positive values (*white*) suggest that climate information does not improve the model (Reproduced from Lowe et al. (2013b))

dengue epidemiological situation at the forecast issue date were integrated into the full model framework, defined above. The purpose was to provide a dengue forecast for Brazil, 3 months in advance of a major global event (the 2014 FIFA World Cup).

The probability of dengue incidence falling into predefined categories of low, medium and high risk was mapped using a visualisation technique that uses colour saturation to express forecast certainty (Jupp et al. 2012). As an indication of the trust a decision maker can place in the dengue predictions for a specific location, the forecast map was accompanied by a verification map, expressing the past performance of the model (see Fig. 8.6). This climate-driven dengue early warning was used to support the decisions of the National Dengue Control Programme several months ahead of the event, to help direct mitigation and control actions to those areas with a higher probability of dengue outbreaks. The provision of this tool, before a major global event, represents an unprecedented prototype of a climate-informed public health service.

Fig. 8.6 (continued) (**b**) Evaluation of past performance for each area based on out-of-sample retrospective dengue forecasts, June, 2000–2013. The skill score takes the value one for a perfect forecast and zero for the benchmark (long-term average) forecast. The darker the shade of *green*, the greater the skill of the forecasting system. Negative values (*white*) show areas where the model did worse than using the benchmark (Reproduced from Lowe et al. (2014a))

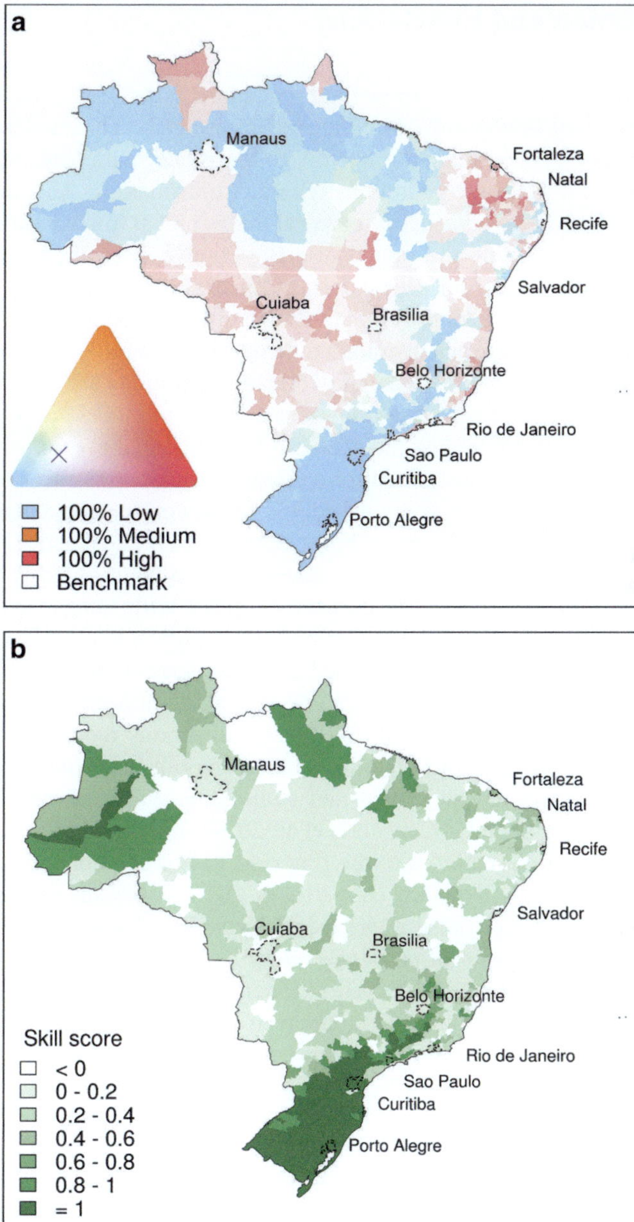

Fig. 8.6 (**a**) Probabilistic dengue forecast for Brazil, June 2014. The continuous colour palette (ternary phase diagram) conveys the probabilities assigned to low-risk, medium-risk and high-risk dengue categories. Category boundaries defined as 100 cases per 100,000 inhabitants and 300 cases per 100,000 inhabitants. The greater the colour saturation, the more certain is the forecast of a particular outcome. *Strong red* shows a high probability of high dengue risk. *Strong blue* indicates a high probability of low dengue risk. Colours close to *white* indicate a forecast similar to the benchmark (long-term average distribution of dengue incidence in Brazil), marked by a *cross*.

8.4 Discussion and Conclusions

This chapter presents a complex modelling procedure to determine the most important drivers of spatio-temporal variability in disease risk. The model framework combines climatic and non-climatic factors in the model parameterisation to correctly quantify variability captured by climate information. The methodology exploits recent advances in spatio-temporal hierarchical mixed modelling. An advantage of implementing the model in a Bayesian framework is the ability to address specific public health issues in terms of probabilities.

A major obstacle to developing climate-driven disease models is the lack of good quality climate and disease data over long time periods. Therefore, it can be difficult to empirically determine if slow-changing climate conditions are spuriously correlated with disease relative risk. It is important to perform model checks for robustness of explanatory variables and the potential for influential events, while keeping aside sets of data not previously used for calibration purposes.

Depending on the question being asked, the framework can be adapted to using a different probability distribution. For example, Barcellos and Lowe (2014) assessed the relative impact of climatic zones and population density on dengue permanence across Brazil, i.e. the number of years with dengue outbreaks. To address this question, a binomial distribution was employed to model the number of years in which dengue incidence exceeded a given epidemic threshold. The framework can also be extended to consider non-linear relationships using structured additive regression models. This allows the simultaneous modelling of possible non-linear effects of continuous covariates using smooth functions, while estimating usual fixed effects of categorical and continuous observed variables, spatial correlation and heterogeneity (Chirombo et al. 2014). Further work is under way to incorporate different indicators into the framework. For example, dynamical dengue (Favier et al. 2006) and malaria (Laneri et al. 2010; Tompkins and Ermert 2013) models can capture the physical relationships between climate hazards and other indicators, as well as non-linear processes.

The advantage of the framework presented here is that it allows the incorporation of empirical or physical relationships, between climate and vectors, with human vulnerability factors, in order to describe disease risk in space and time, for population groups with different exposures. The model framework provides a useful decision support system tool for national or regional public health authorities to control climate-sensitive disease risk. It is hoped that the tool could strengthen the capacity of surveillance and control teams across the globe, to more effectively plan targeted control interventions at the local level and manage scarce health resources, well ahead of an impending epidemic.

Acknowledgements The research leading to these results has received funding from the DENFREE project (grant agreement no. 282378) and EUPORIAS project (grant agreement no. 308291) funded by the European Commission's Seventh Framework Research Programme.

References

Banu S, Hu W, Hurst C, Tong S (2011) Dengue transmission in the Asia-Pacific region: impact of climate change and socio-environmental factors. Trop Med Int Health 16(5):598–607

Barcellos C, Lowe R (2014) Expansion of the dengue transmission area in Brazil: the role of climate and cities. Trop Med Int Health 19(2):159–168

Besag J, Green P, Higdon D, Mengersen K (1995) Bayesian computation and stochastic systems. Stat Sci 10(1):3–41

Best N, Arnold R, Thomas A, Waller L, Conlon E (1999) Bayesian models for spatially correlated disease and exposure data. Bayesian Stat 6:131–156

Bivand RS, Pebesma EJ, Gómez-Rubio V (2008) Applied spatial data analysis with R, vol 747248717. Springer, New York

Cash BA, Rodó X, Ballester J, Bouma MJ, Baeza A, Dhiman R, Pascual M (2013) Malaria epidemics and the influence of the tropical South Atlantic on the Indian monsoon. Nat Clim Chang 3(5):502–507

Cazelles B, Chavez M, McMichael AJ, Hales S (2005) Nonstationary influence of El Nino on the synchronous dengue epidemics in Thailand. PLoS Med 2(4):313–318

Chirombo J, Lowe R, Kazembe L (2014) Using structured additive regression models to estimate risk factors of malaria: analysis of 2010 Malawi malaria indicator survey data. PLoS ONE 9(7):e101116. doi:10.1371/journal.pone.0101116

Coelho CAS, Stephenson DB, Balmaseda M, Doblas-Reyes FJ, van Oldenborgh GJ (2006) Toward an integrated seasonal forecasting system for South America. J Clim 19(15):3704–3721

Connor SJ, Mantilla GC (2008) Integration of seasonal forecasts into early warning systems for climate-sensitive diseases such as malaria and dengue. In: Seasonal forecasts, climatic change and human health. Springer, Netherlands, pp 71–84

Doblas-Reyes FJ, García-Serrano J, Lienert F, Biescas AP, Rodrigues LR (2013) Seasonal climate predictability and forecasting: status and prospects. Wiley Interdiscip Rev Clim Chang 4(4):245–268

Ellner SP, Guckenheimer J (2011) Dynamic models in biology. Princeton University Press, Princeton

Favier C, Degallier N, Rosa-Freitas MG, Boulanger JP, Lima JRC, Luitgards-Moura JF, Menkes CE, Mondet B, Oliveira C, Weimann ETS, Tsouris P (2006) Early determination of the reproductive number for vector-borne diseases: the case of dengue in Brazil. Trop Med Int Health 11(3):332–340

Gage KL, Burkot TR, Eisen RJ, Hayes EB (2008) Climate and vectorborne diseases. Am J Prev Med 35(5):436–450

Gelman A, Meng X, Stern H (1996) Posterior predictive assessment of model fitness via realized discrepancies. Stat Sin 6:733–759

Gilks WR, Richardson S, Spiegelhalter DJ (1996) Markov chain Monte Carlo in practice. Chapman & Hall/CRC, Boca Raton

Guzman MG, Halstead SB, Artsob H, Buchy P, Farrar J, Gubler DJ, Peeling RW (2010) Dengue: a continuing global threat. Nat Rev Microbiol 8:S7–S16

Jancloes M, Thomson M, Costa MM, Hewitt C, Corvalan C, Dinku T, Hayden M (2014) Climate services to improve public health. Int J Environ Res Public Health 11(5):4555–4559

Jupp TE, Lowe R, Coelho CA, Stephenson DB (2012) On the visualization, verification and recalibration of ternary probabilistic forecasts. Philos Trans R Soc A Math Phys Eng Sci 370(1962):1100–1120

Laneri K, Bhadra A, Ionides EL, Bouma M, Dhiman RC, Yadav RS, Pascual M (2010) Forcing versus feedback: epidemic malaria and monsoon rains in northwest India. PLoS Comput Biol 6(9):e1000898

Lowe R (2011) Spatio-temporal modelling of climate-sensitive disease risk: towards an early warning system for dengue in Brazil. Dissertation, University of Exeter

Lowe R, Bailey TC, Stephenson DB, Graham RJ, Coelho CA, Sá Carvalho M, Barcellos C (2011) Spatio-temporal modelling of climate-sensitive disease risk: towards an early warning system for dengue in Brazil. Comput Geosci 37(3):371–381

Lowe R, Bailey TC, Stephenson DB, Jupp TE, Graham RJ, Barcellos C, Carvalho MS (2013a) The development of an early warning system for climate-sensitive disease risk with a focus on dengue epidemics in Southeast Brazil. Stat Med 32(5):864–883

Lowe R, Chirombo J, Tompkins AM (2013b) Relative importance of climatic, geographic and socio-economic determinants of malaria in Malawi. Malar J 12(1):416

Lowe R, Barcellos C, Coelho CA, Bailey TC, Coelho GE, Graham R, Rodó X (2014a) Dengue outlook for the World Cup in Brazil: an early warning model framework driven by real-time seasonal climate forecasts. Lancet Infect Dis 14(7):619–626

Lowe R, Cazelles B, Paul R, Rodó X (2014b) Towards a climate-driven dengue decision support system for Thailand. Paper presented at the EGU general assembly conference abstracts, Vienna, Austria, 27 April – 2 May 2014, id.5692. http://adsabs.harvard.edu/abs/2014EGUGA..16.5692L. Accessed 25 Jul 2014

McMichael AJ, Campbell-Lendrum DH, Ebi KL, Githeko AK, Scheraga JD, Woodward A (2003) Climate change and human health: risks and responses. World Health Organization, Geneva. http://apps.who.int/iris/handle/10665/42742. Accessed 25 Jul 2014

Rodó X, Pascual M, Doblas-Reyes FJ, Gershunov A, Stone DA, Giorgi F, Dobson AP (2013) Climate change and infectious diseases: can we meet the needs for better prediction? Clim Change 118(3–4):625–640

Ropelewski CF, Halpert MS (1987) Global and regional scale precipitation patterns associated with the El Niño/Southern Oscillation. Mon Weather Rev 115(8):1606–1626

Stewart-Ibarra AM, Lowe R (2013) Climate and non-climate drivers of dengue epidemics in southern coastal Ecuador. Am J Trop Med Hyg 88(5):971–981

Thomson MC, Doblas-Reyes FJ, Mason SJ, Hagedorn R, Connor SJ, Phindela T, Morse AP, Palmer TN (2006) Malaria early warnings based on seasonal climate forecasts from multi-model ensembles. Nature 439(7076):576–579

Tompkins AM, Ermert V (2013) A regional-scale, high resolution dynamical malaria model that accounts for population density, climate and surface hydrology. Malar J 12:65

Venables WN, Ripley BD (2002) Modern applied statistics with S. Springer, New York

Wakefield JC, Best NG, Waller L (2000) Bayesian approaches to disease mapping. In: Elliott P, Wakefield JC, Best NG, Briggs D (eds) Spatial epidemiology: methods and applications. Oxford University Press, Oxford, pp 104–127

Chapter 9
Shallow Landslide Hazard Mapping for Davao Oriental, Philippines, Using a Deterministic GIS Model

Ian Kaye Alejandrino, Alfredo Mahar Lagmay, and Rodrigo Narod Eco

Abstract Davao Oriental located at 7°30′N and 126°50′E is one of the many landslide-prone provinces in the Philippines experiencing severe rainfall throughout the year. With the increase in population and other infrastructural developments, it is necessary to map the landslide potential of the area, to assure the safety of the people and delineate suitable land for development. In order to produce rainfall-induced shallow landslide hazard maps, Stability Index Mapping (SINMAP) was used over a 5-m interferometric synthetic aperture radar (IFSAR)-derived digital terrain model (DTM). SINMAP is based on the infinite slope stability model. Topographic, soil geotechnical, and hydrologic parameters (cohesion, angle of friction, bulk density, infiltration rate, and hydraulic transmissivity) were assigned to each pixel of the DTM with the total area of 5,164.5 km^2 to compute for the corresponding factor of safety. The landslide hazard maps generated using SINMAP are found to be accurate when compared to the landslide inventory from 2003 to 2013. The landslide susceptibility classification was translated to zoning maps indicating areas that are safe from shallow landslides, areas that can be built upon with slope intervention and monitoring, and the no-build areas. These maps complement the structurally controlled landslide, debris flow, and other natural hazard maps that are being prepared to aid proper zoning for residential and infrastructural developments.

I.K. Alejandrino, B.S. (✉)
Institute of Civil Engineering, University of the Philippines,
Diliman, Quezon City, Philippines

Project NOAH, Deparment of Science and Technology, University of the Philippines,
Diliman, Quezon City, Philippines
e-mail: ian_931@yahoo.com

A.M. Lagmay, Ph.D. • R.N. Eco, B.S.
National Institute of Geological Sciences, University of the Philippines, Diliman,
Quezon City, Philippines

Project NOAH, Deparment of Science and Technology, University of the Philippines,
Diliman, Quezon City, Philippines

J.L. Drake et al. (eds.), *Communicating Climate-Change and Natural Hazard Risk and Cultivating Resilience*, Advances in Natural and Technological Hazards Research 45, DOI 10.1007/978-3-319-20161-0_9

Keywords Landslide • Hazard mapping • Deterministic model • Philippines • Davao Oriental • Natural hazards

9.1 Introduction

Landslides triggered by rainfall pose significant threat to human lives and property in the Philippines. As the country's population grows, landslides have become a major concern to the safety of citizens. Unlike flooding which causes damage to structures that more often can be fixed, landslides may leave irreparable damage. The province of Davao Oriental is classified as one of the areas with high susceptibility to landslides according to the Mines and Geosciences Bureau (MGB), the government agency tasked to produce landslide hazard maps for the country. While it is not advisable to locate communities or build infrastructure in areas identified by MGB as susceptible to landslides, displacement of people from their communities and consequent loss of their source of livelihood have great social implications. This calls for detailed landslide hazard maps that identify specific areas within an existing community (i.e., municipality or village) to be used as part basis for proper zoning of residential, industrial, and agricultural areas. By identifying safe zones from detailed hazard maps, unnecessary relocation of communities and their consequent ill effects may be avoided. Empirical and descriptive land models, because of their easy implementation, have been developed for hazard analysis and landslide monitoring and are applicable in a regional scale (Caine 1980; Cannon and Gartner 2005). Previously, MGB produced descriptive landslide hazard maps using a checklist of factors to be assessed in the field (see Table 9.1), which may contribute to the mapping of landslide susceptibility. Figure 9.1 shows the 1:50,000 landslide hazard map done by MGB for a part of Davao Oriental.

In recent years, physically based landslide models have been developed to assess landslide hazard using a range of topographic, geologic, and hydrologic parameters (Baum et al. 2008; Dietrich et al. 1995; Godt et al. 2009; Hong et al. 2007; Iverson 2000; Lu and Godt 2008; Ren et al. 2009; Wu and Sidle 1995). "Apart from the process of understanding, the quality of data and data availability, the degree of success of any particular research depends on the wise selection of statistical methods, guided by the knowledge of the limits and strength of each method" (Felicisimo et al. 2013, p. 175).

Statistical models require landslide inventory as an input. In creating regional maps for landslide susceptibility, the results of these models will incline toward the variables assigned to with more records of landslides which can lead to biased results, and the success of the model may vary for different locations due to the quality of data available per location.

Deterministic models, on the other hand, do not require landslide inventories as inputs but rather as a way to check and validate the acceptability of the model. The models rely on the spatial resolution of the terrain and the precision of the soil and geological parameters used to simulate processes that contribute to the change in the stability.

Table 9.1 Sample fact sheet, methodology, and rating system in the determination of landslide hazard susceptibility as used in the MGB geohazard mapping program (Open-source image from MGB-UNDP 2004)

Geohazard	Levels of susceptibility	
	Low to absent	Use inflection point for separation of levels
	Medium	
	High	

Weights of evidence method, expert driven

1. Landslide inventory, by previous reports, aerial photo interpretation/remote sensing images, topographic map interpretation, with or without actual field survey on budget and time
2. GIS

   Minimum requirements for thematic map inputs

   Geologic maps on (use lithology rather than formations) TMU, slope, faults, roads, gully heads

   Optional requirements (only if available) – landslide inventory, land use/land cover map, soil map, vegetation map

   For 1:50,000, expert driven, subject to field verification

   For 1:10,000, data driven, field mapping

   Map calculations

   Simple addition for thematic maps, uniform weights for all themes

   After field checking, the weights can be exchanged depending on the acquired field data

   Make histogram of the rated pixels and identify inflection points to get the different susceptibility levels

Buffer	Distance	Rating
1. Faults	0–50 km	3
	51–100	2
	>100	1
2. Roads	0–25	2
	>25	1
3. Gully heads	0–25	2
	>25	1
4. Slope	0–2 %	1
	3–7 %	2
	8–13 %	3
	14–20 %	4
	21–55 %	5
	56–140 %	6
	>140 %	7
5. Landslides	Old	5
	Active	7
6. Land cover	Classify accordingly	Rate accordingly
7. Lithology	Classify accordingly	Rate accordingly
8. TMU	Classify accordingly	Rate accordingly
9. Vegetation	Classify accordingly	Rate accordingly

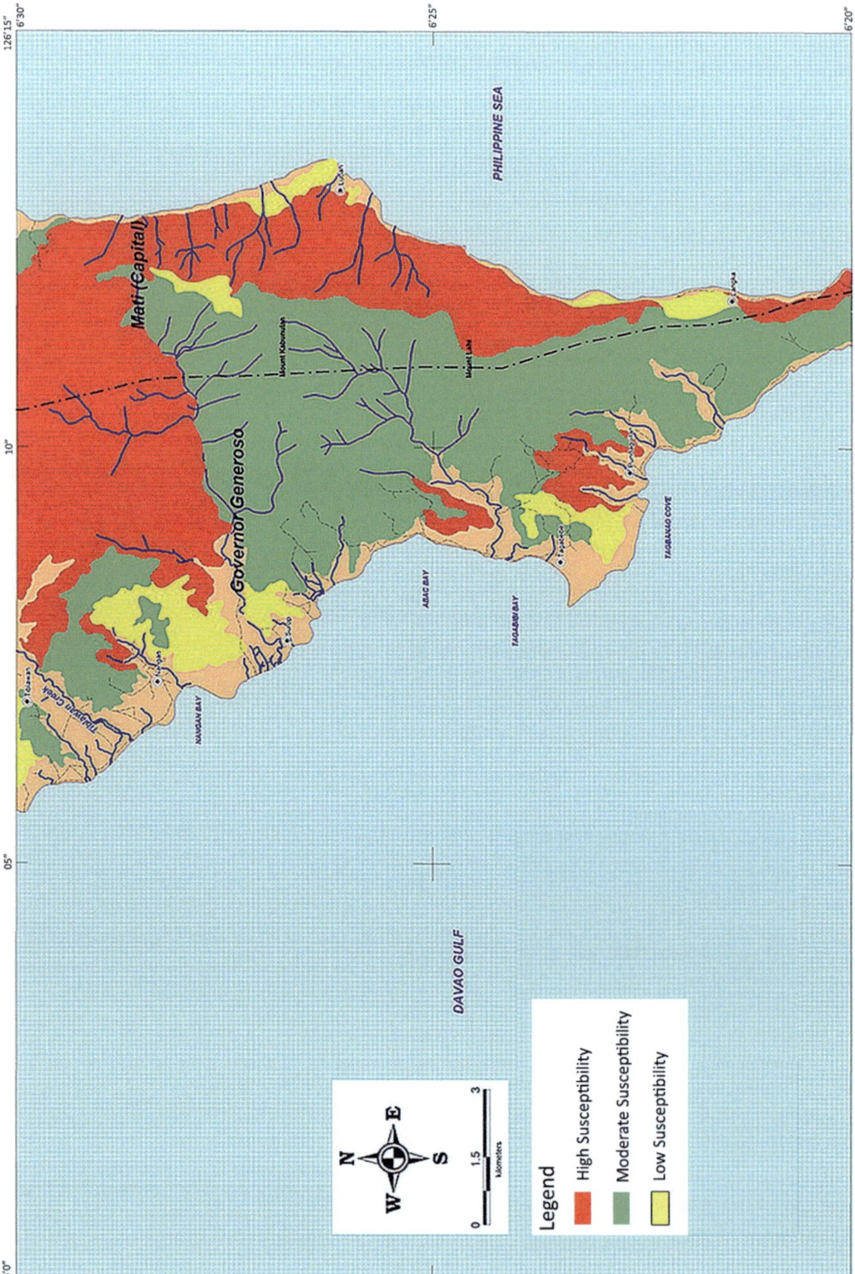

Fig. 9.1 Landslide hazard map from MGB (Open-source image from MGB Geoportal 2010)

A deterministic model can only address one type of landslide process. It is advisable to use a different model for each type (i.e., Debris-2D or Flow-R for debris flows and COLTOP-3D and Matterocking for deep-seated or structural failure). For the case of the Philippines, without a landslide inventory and a good coverage of satellite imageries to construct one, the option to create a landslide hazard map for the entire country using a statistical model cannot be completed. Having to resort on deterministic models, studies were initially conducted on provinces that have good quality of landslide inventories. Our study area Davao Oriental was one of those.

This study aims to generate a rainfall-induced shallow landslide hazard map for Davao Oriental using a deterministic GIS model called Stability Index Mapping (SINMAP). It is part of the flagship program of the Philippine government for disaster prevention and mitigation called Project Nationwide Operational Assessment of Hazards (NOAH), which seeks to use the best available tools and scientific methods for mitigating the impacts of natural hazards (Lagmay 2012). The shallow landslide hazard maps generated from SINMAP complement the deep-seated, structurally controlled landslide and debris flow hazard maps that are also being simulated for the province. These computer-simulated maps also complement the field data gathering effort by MGB using the empirical and descriptive methods of landslide mapping. By doing so, the landslide simulation output is verified and calibrated by MGB data for best results. It at the same time maximized utilization of newly available and high-resolution topographic data. Together with the storm surge and flood hazard maps, these will be freely accessible through the Project NOAH website to help in the disaster risk reduction and management efforts of the country.

9.2 Methodology

9.2.1 Stability Index Mapping (SINMAP)

SINMAP is an ArcView GIS extension and an objective terrain stability mapping tool that complements other types of terrain stability mapping methods. The analysis done by SINMAP is based upon the infinite slope stability model which balances the resisting components of friction and destabilizing components of gravity on a failure plane parallel to the ground surface. It is implemented for shallow landslide phenomena controlled by groundwater flow and convergence. It does not apply to deep-seated instability including rotational slumps and deep earthflows (Montgomery and Dietrich 1994).

Slope failures occur frequently during or following a continuous period of heavy rainfall. The mechanism for rainfall-induced landslides is mainly by the infiltration of rainwater from an initial unsaturated state, which then causes a decrease in negative pore pressure. This also leads to a decrease in effective normal stress acting through the failure plane and reduces the available shear strength to a point where equilibrium can no longer be sustained in the slope (Orense 2004). Thus, slope stability models that consider hydrologic wetness and shear strength-related properties of soil and topography of the region could possibly predict zones of mass movement in slopes.

The input data comprises the topographic slope, specific catchment, geotechnical and hydrologic soil parameters, and climate. Topographic variables such as catchment area and slope are computed from the digital terrain model (DTM). The DTM used was generated from an airborne interferometric synthetic aperture radar (IFSAR) survey of the Philippines and has 5-m horizontal resolution and 0.5-m vertical accuracy.

SINMAP does not require numerically precise input and accepts other input parameters in terms of upper and lower bounds on their range of values. The methods implemented rely on a grid-based data structure where geotechnical parameters are assigned to each cell in the grid over the study area. The accuracy of the output is therefore heavily reliant on the accuracy of the DTM used (Pack et al. 2005).

The primary output of the program is a stability index (SI) representing the possibility of landslide occurrence per cell. "Stability indices output by the analysis should not be interpreted as numerically precise and are most appropriately interpreted in terms of relative hazard" (Pack et al. 2005, p. 2). The SI is defined as the probability of the location as being stable assuming uniform distribution of parameters over the uncertainty ranges. It is not capable of predicting when landslides will occur but gives the location where it is most likely to take place. The most conservative of the SI values is defined as the factor of safety (ratio between stabilizing and destabilizing forces) in a given location. If the factor of safety is less than 1, it is then defined as the probability that the location is stable given the range of parameters used (Pack et al. 1998).

Table 9.2 represents the relative interpretation (upper threshold to stable) of the range of SI values based on the breakpoints given by the program. The program also suggests the relative degree of destabilizing factors required for stability and instability for each range of values. The term *defended slope* is used to classify regions where the slope is held in place by forces not considered by the model. These may

Table 9.2 Stability class definitions (Open-source image by Pack et al. 2005)

Condition	Class	Predicted state	Parameter range	Possible influence of factors not modeled
SI > 1.5	1	Stable slope zone	Range cannot model instability	Significant destabilizing factors are required for instability
1.5 > SI > 1.25	2	Moderately stable zone	Range cannot model instability	Moderate destabilizing factors are required for instability
1.25 > SI > 1.0	3	Quasi-stable slope zone	Range cannot model instability	Minor destabilizing factors could lead to instability
1.0 > SI > 0.5	4	Lower threshold slope zone	Pessimistic half of range required for instability	Destabilizing factors are not required for instability
0.5 > SI > 0.0	5	Upper threshold slope zone	Optimistic half of range required for stability	Stabilizing factors may be responsible for stability
0.0 > SI	6	Defended slope zone	Range cannot model stability	Stabilizing factors are required for stability

be areas with existing slope protection, exposed bedrock layer, and other factors that contribute to the stability of the slope.

For mapping purposes in the Philippines, the six classes are reduced to three major hazard ratings with corresponding interpretations. Class 3 refers to areas of low susceptibility where slope intervention is recommended, while Class 4 refers to areas of moderate susceptibility where slope intervention and engineering intervention are recommended. Lastly, Class 5 refers to areas of high susceptibility that are recommended as *no-build zones*. The defended region was not included in the reclassification due to its nature, and the possible trigger for these regions is most likely not governed by the analysis of the program.

9.2.2 *Stability Index*

SINMAP uses the following formula illustrated in Fig. 9.2 to calculate the stability index based on the infinite slope equation proposed by Hammond et al. (1992):

where:

FS = factor of safety
a = topographic catchment area
C = dimensionless cohesion
$= (C_r + C_s)/hpg$
h = soil thickness
p = soil density
g = gravity constant
r = water density to soil density ratio
T = soil transmissivity
Φ = soil internal angle of friction

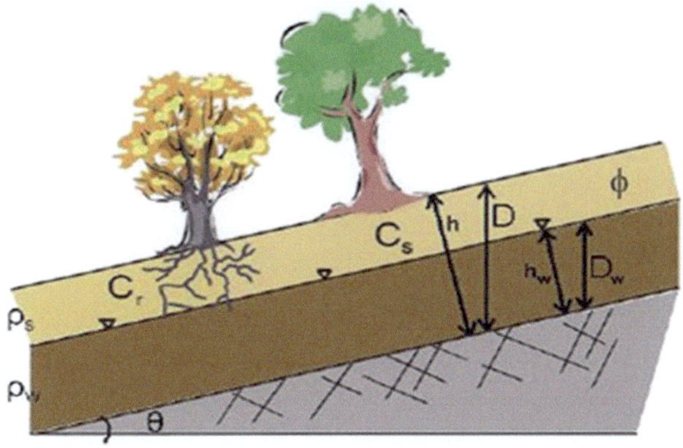

$$FS = \frac{C + \cos\theta\left[1 - \min\left(\frac{R}{T}\frac{a}{\sin\theta}, 1\right) r\tan\emptyset\right.}{\sin\theta}$$

Fig. 9.2 Infinite slope model (Adapted from open-source image (Hammond et al. 1992))

Variables a and θ are obtained from the DTM, while the rest of the geotechnical and hydrologic parameters are manual inputs to the model. The combination of the smallest C and angle of friction together with the largest R/T defines the most conservative analysis or worst-case scenario within the assumed variability in the input parameters (Pack et al. 1998).

9.2.3 Climate Type and Soil Cover

The province of Davao Oriental falls under type II of Philippine Atmospheric, Geophysical and Astronomical Services Administration (PAGASA) modified Coronas Climate Classification Scheme. The rainy season prevails the whole year-round with pronounced heavy rainfall during December.

Davao Oriental has soil cover which is mainly loam and sandy clay loam (see Fig. 9.3) with a section of rough mountainous land that has unidentified soil type. The other soil types cover the relatively flat sections and the coastline of the province.

The area of Davao Oriental was divided into two sets of input parameters: the upper portion's (Camasan sandy clay loam, mountain undifferentiated soil, Bolinao clay, and San Manuel silty clay loam) cells were assigned with parameter values from Sandy Clay Loam Classification, while the other part (Malalag loam, San Manuel silty clay loam, and a small part of Bolinao clay) was assigned with parameter values from loam. Other soil covers, which were smaller in area, were not reflected in the simulations due to their small contribution in the total land cover (Table 9.3).

9.2.4 Landslide Inventory

The landslide inventory was identified from high-resolution satellite imagery from 2003 to 2013. However, the type of landslide, trigger, or threshold was not available due to the lack of information on the exact dates when the landslides occurred. Figure 9.4 shows the DTM and the location of the landslide inventory for the province.

9.2.5 Process Flow

The outline for creating the landslide susceptibility map using SINMAP is shown as a schematic diagram in Fig. 9.5. The program extracts the topographic data from the DTM, and the soil map is used as calibration regions for the input parameters.

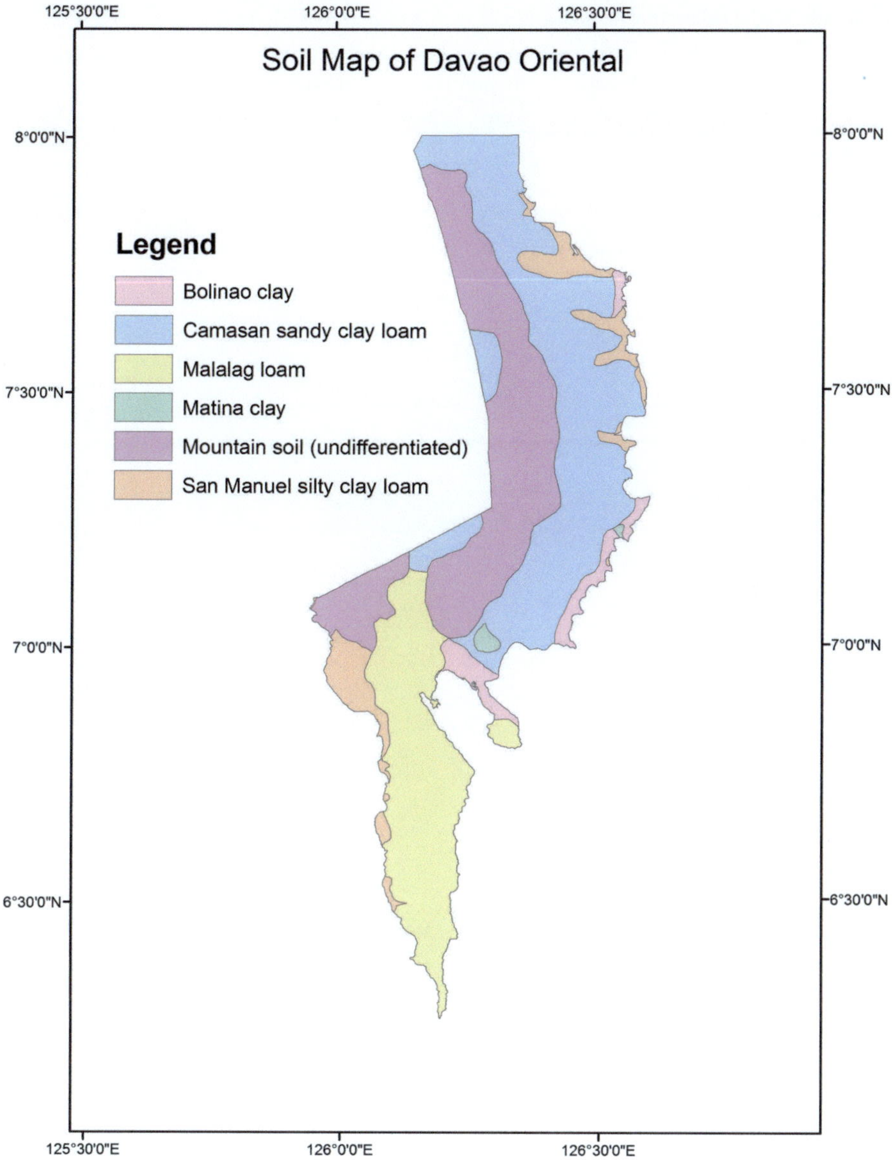

Fig. 9.3 Soil map of Davao Oriental, Philippines, from Bureau of Soils

Table 9.3 Model calibration parameters

	Soil density	Internal angle of friction		Cohesion		Approximate T/R	
	kg/m³	Min	Max	Min	Max	Min	Max
Loam	2114	28	32	0	0.642	0.97	40.29
Sandy clay loam	1681	31	34	0	0.579	3.75	119.25

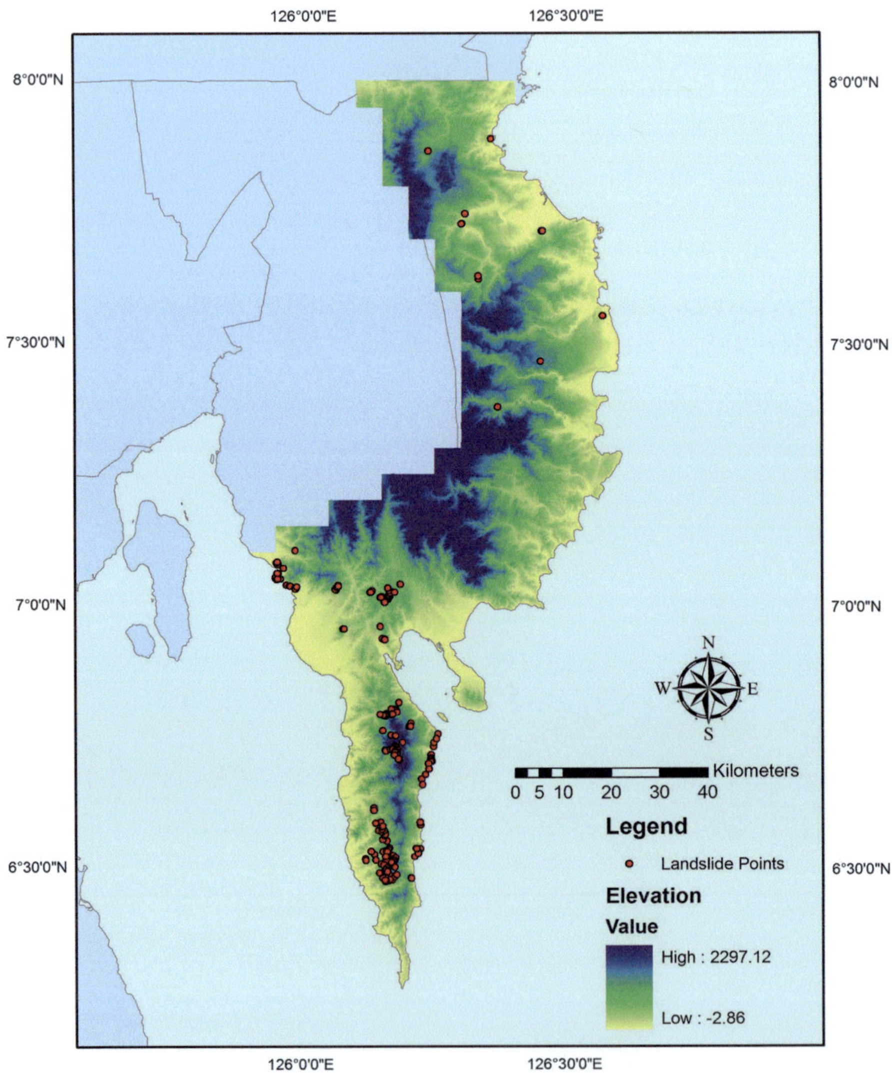

Fig. 9.4 IFSAR DTM and landslide inventory of Davao Oriental from various satellite imageries

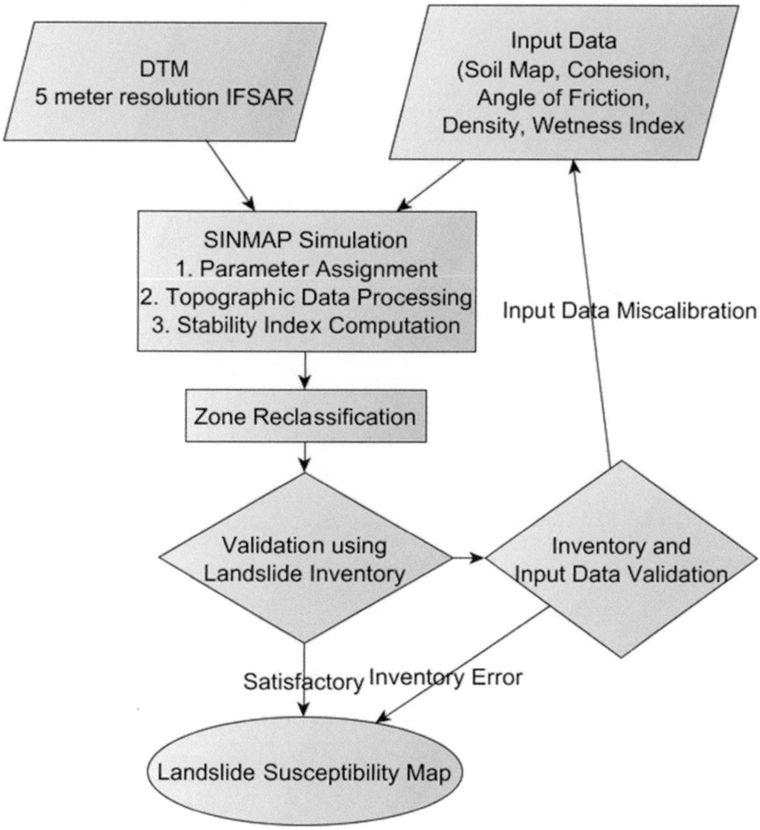

Fig. 9.5 Schematic diagram of the methodology

9.3 Results

The SINMAP simulations reveal that the mountainous regions of Davao Oriental are areas susceptible to medium to high hazards (see Fig. 9.6). Medium susceptibility is depicted as color orange in the hazard maps, while the high landslide hazard susceptibility is shown as red. Most of the high landslide-susceptible areas are confined to the very steep slopes and ravines. The flat areas especially near the coastlines are in general devoid of any indication of landslide susceptibility. Table 9.4 shows the area percentage with high, moderate, and low susceptibility compared to the total area of the DTM used in simulation. Landslide inventories were also used to evaluate the accuracy and acceptability of the generated hazard map.

Compared to the existing 1:50,000 scale landslide hazard maps generated by MGB, the maps produced from SINMAP simulations are more detailed and appear to delineate the degree of landslide hazard better (see Fig. 9.7). Areas depicted in the MGB map as high and moderate hazard are shown as very broad regions, while

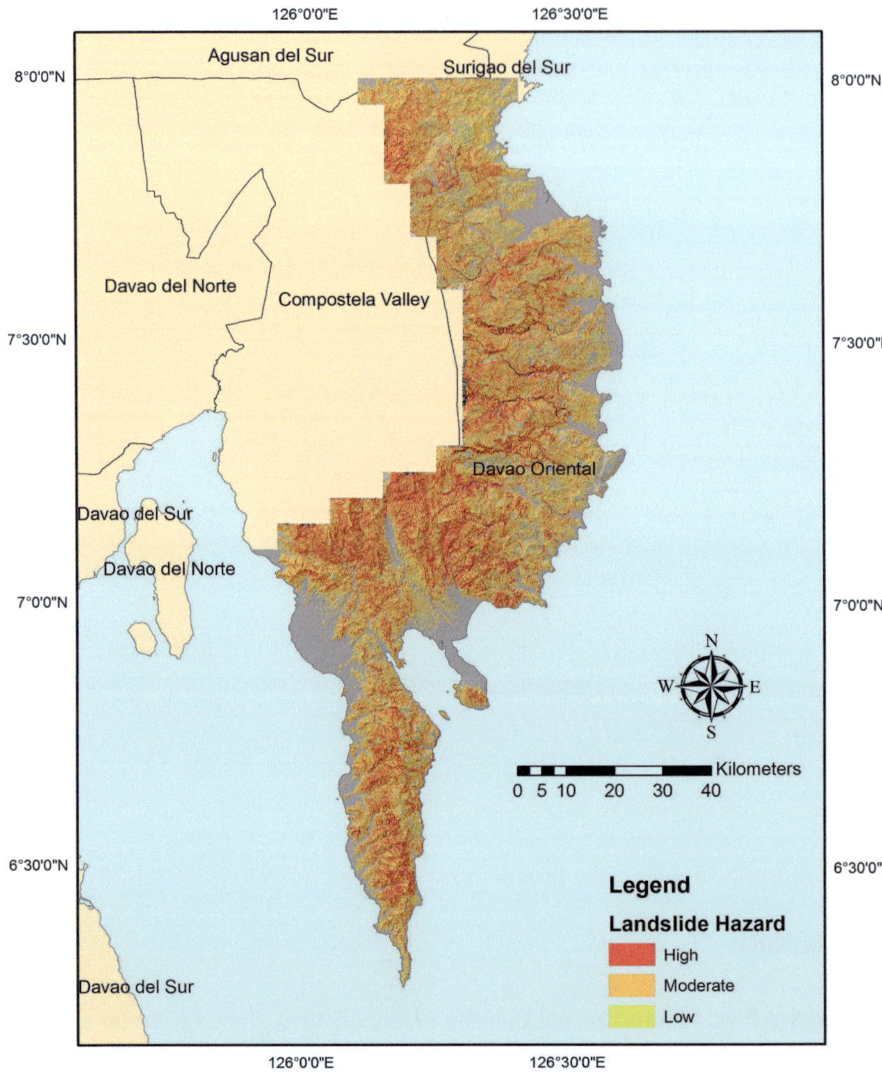

Fig. 9.6 Rainfall-induced shallow landslide hazard susceptibility map of Davao Oriental

those shown as high and moderate susceptibility in the SINMAP-generated hazard maps are site specific and localized. More areas are found to be stable and safe from landslide with the computer-simulated (SINMAP) hazard maps.

There are many deterministic models that were available to use such as SHALSTAB and TRIGRS, but we selected SINMAP because of its ease of use, and it accepts a range of input parameters, which is desirable for regional scale mapping because it allows room for uncertainty and nonuniform soil types. Though the inventory may be used as an input to improve the program's output, it was avoided

Table 9.4 Summary of rainfall-induced shallow landslide susceptibility and landslide inventory

	Stability classification				
	Stable	Low susceptibility	Moderate susceptibility	High susceptibility	Total area
Area (km^2)	1324.40	431.34	2512.26	863.88	5131.88
Percentage of the area	25.81	8.41	48.95	16.83	100.00
Number of landslides	4	3	77	153	237
Percentage of landslides	1.69	1.27	32.49	64.56	100

to see how the program would perform for areas with low number of available landslide inventory. Also there were parameter adjustments or back calculations that were done in other studies prior to ours to assign areas of landslide inventory to high susceptibility areas but were not applied to our study since they would defeat the purpose of the method being reproducible for other provinces of the country. Model results assigned moderate and high hazard areas to cover around 65 % of the province. In these areas fall 97.05 % of the total number of landslides observed from high-resolution imagery from 2003 to 2013 (see Fig. 9.8). The remaining 35 % of the province has little chance (around 3 %) of experiencing rainfall-induced shallow landslides. From this we can say that the program has very good accuracy (98.31 %) to determine possible unstable slopes.

Compared to the previous hazard map from MGB, the computer-simulated landslide hazard map was observed to be more detailed in terms of hazard delineation and characterization of the degree of hazard present in various areas (see Fig. 9.7). The classification in the MGB, due to its descriptive nature, is very subjective to the surveyor, and the assignment of the extent is greatly dependent on the coverage of the investigation conducted in the field. In comparison, SINMAP's accuracy and precision in identifying and delineating landslide hazards are greatly affected by the resolution of the DTM used in the program (Pack et. al 2005). Higher-resolution DTMs will likely yield more accurate results provided that the range of other input parameters are realistic. Furthermore, the effects of land cover to rainfall infiltration and root cohesion were not considered in this simulation. Thus, the results may be interpreted to be the worst-case scenario.

The applicability of the results as well as the classification of the relative degree of susceptibility depends on the intended use. In terms of identifying and detailed delineation of rainfall-induced shallow landslide hazards over a wide area, the program has produced very good results. The landslide inventory was found to be highly consistent with the areas that were assigned with high and moderate susceptibility (see Fig. 9.8). Even though almost all landslides in the inventory fall under areas classified as susceptible to landslides, the generated hazard maps are incomplete which require further analysis of landslides triggered by factors other than the effects of rainfall on shallow soil surfaces. Hence, the SINMAP results are best used

Fig. 9.7 Comparison between MGB landslide hazard map and SINMAP-simulated susceptibility map

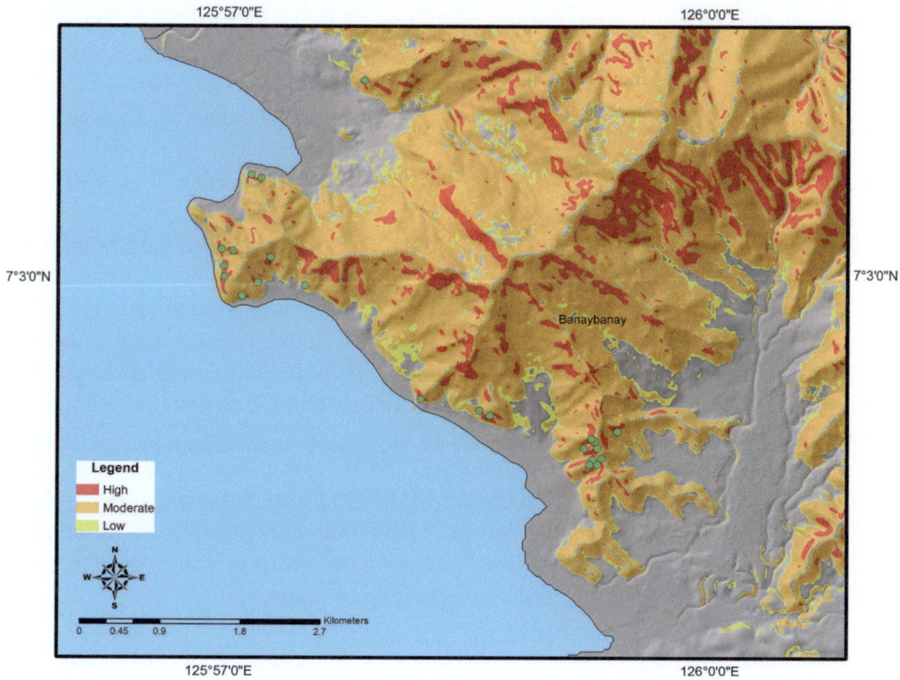

Fig. 9.8 Verification using landslide inventory

with other landslide susceptibility maps such as those that depict debris flow and deep-seated or structurally controlled landslide hazards.

9.4 Conclusions and Recommendations

The consideration of land cover can give a significant effect on the classification of shallow landslide hazards. Depending on the land cover, it will have varying influence on degrees of root cohesion and rainfall infiltration. A site investigation to verify the range of values of the input parameters will help improve the assessment for critical areas in a smaller scale such as inhabited areas and proposed areas for development. It was noted by Witt (2005): "Neither model [SinMap nor ShalStab] takes into account antecedent moisture nor the effect that geologic structure can have on concentrating groundwater flow" (p. 120).

Since we are mapping landslide hazard for regional scale, we opt not to address the effect of geologic structure on concentrating groundwater flow. A geological survey or validation may be done to incorporate its effects for critical areas (i.e., populated areas). Antecedent moisture or rainfall intensity and duration may be more important to consider with the goal of creating an early warning system that assesses stability in actual events. Though SINMAP has proven to be a good tool for detailed planning over a wide region, programs that consider transient dynamical response of the subsurface moisture to spatiotemporal variability of rainfall in complex terrains (e.g., TRIGRS -*Transient Rainfall Infiltration and Grid-Based Regional Slope-Stability Model* by Baum et al. 2008 and SEGMENT-landslide by Ren et al. 2008) are of great significance especially to communities currently residing on the identified hazard areas and for motorists traveling across roads that are constructed on steep slopes.

The MGB maps, if followed strictly, necessitate the relocation of communities or development of areas as not susceptible to landslides. Given the broad areas identified as landslide susceptible in the MGB maps, this can mean large-scale displacement of people from their communities and consequent loss of their source of livelihood which have great social implications. With the simulated landslide hazard maps that detail specific areas within an existing community (i.e., municipality or village), safe zones are identified and can be used as settlements, thereby avoiding unnecessary relocation of communities and their consequent ill effects. The computer-simulated maps complement the MGB field data gathering efforts, which use the empirical and descriptive methods of landslide mapping. The MGB data can be used to verify and calibrate the landslide simulations to produce more detailed and accurate maps for safe development planning of communities.

Acknowledgments We would like to thank the creators of SINMAP (Pack et al.) for making this program available to the research and development community and communicating with us in the early stages. We also thank the National Mapping and Resource Information Authority (NAMRIA) for the IFSAR DTM used in this simulation. Funding for the project titled *Enhancing Landslide Hazard Maps Through LIDAR and Other High Resolution Imageries* is from the Department of Science and Technology (DOST), government of the Philippines.

Other Notes DOST *Project NOAH* is a program implemented by the Philippine government to assess the different hazards present in the Philippines. Assessment of flood, landslide, and storm surge hazards is part of the program. Completed maps are to be added to the NOAH website (www.noah.dost.gov.ph) for free access to the general public to aid in the information dissemination to reduce effects of meteorological hazards in the country. The website, in partnership with *PAGASA*, also displays various weather sensors and visualizations to aid in the understanding of weather data.

References

Baum RL, Savage WZ, Godt JW (2008) TRIGRS—A Fortran program for transient rainfall infiltration and grid-based regional slope-stability analysis, version 2.0: U.S. Geological Survey Open-File Report, 2008–1159, p 75

Caine N (1980) The rainfall intensity: duration control of shallow landslides and debris flows. Geogr Ann Ser A Phys Geogr 62:23–27. doi:10.2307/520449

Cannon SH, Gartner J (2005) Wildfire related debris flow from a hazards perspective. In: Debris-flow hazards and related phenomena, Springer-Praxis books in geophysical sciences. Springer, Berlin, pp 321–344. doi:10.1007/3-540-27129-5_15

Dietrich WE, Reiss R, Hsu ML, Montgomery DR (1995) A process-based model for colluvial soil depth and shallow landsliding using digital elevation data. Hydrol Process 9:383–400. doi:10.1002/hyp.3360090311

Felicisimo M, Cuartero A, Remondo J, Quirós E (2013) Mapping landslide susceptibility with logistic regression, multiple adaptive regression splines, classification and regression trees, and maximum entropy methods: a comparative study. Landslides 10(2):175–189. doi:10.1007/s10346-012-0320-1

Godt JW, Baum RL, Lu N (2009) Landsliding in partially saturated materials. Geophys Res Lett 36:L02403. doi:10.1029/2008GL035996

Hammond C, Hall D, Miller S, Swetik P (1992) Level I Stability Analysis (LISA) Documentation for Version 2.0 (General technical Report INT-285). http://forest.moscowfsl.wsu.edu/cgi-bin/engr/library/searchpub.pl?pub=1992a. Accessed 16 Jun 2014

Hong Y, Adler RF, Huffman GJ (2007) Use of satellite remote sensing data in the mapping of global landslide susceptibility. Nat Hazards 43(2):245–256. doi:10.1007/s11069-006-9104-z

Iverson RM (2000) Landslide triggering by rain infiltration. Water Resour Res 36(7):1897–1910. doi:10.1029/2000WR900090

Lagmay AMF (2012) Disseminating near-real time hazards information and flood maps in the Philippines through web-GIS. Project NOAH Open File Reports. V1:21–36: ISSN-23627409

Lu N, Godt JW (2008) Infinite slope stability under steady unsaturated seepage conditions. Water Resour Res 44:W11404. doi:10.1029/2008WR006976

MGB Geoportal (2010) Landslide hazard map of Davao Oriental, Mines and Geosciences Bureau Geoportal. http://gdis.denr.gov.ph/mgbviewer/. Accessed 4 Jun 2014

MGB-UNDP (2004, March) Manual for standardization, geohazard mapping program. Mines and Geosciences Bureau project, funded by the United Nations Development Program, Montgomery DR, Dietrich WE (1994) A physically based model for the topographic control on shallow landsliding. Water Resour Res 30(4):1153–1171. doi:10.1029/93WR02979

Orense R (2004) Slope failures triggered by heavy rainfall. Philipp Eng J 25(2):73–90. http://journals.upd.edu.ph/index.php/pej/article/view/2343. Accessed 5 Apr 2014

Pack RT, Tarboton DG, Goodwin CN (1998) Terrain stability mapping with SINMAP, technical description and users guide for version 1.00, Report number 4114-0, Terratech Consulting, Salmon Arm, BC, Canada

Pack RT, Tarboton DG, Goodwin CN, Prasad A (2005) SINMAP 2- A stability index approach to terrain stability mapping. Utah State University. http://hydrology.usu.edu/sinmap2/. Accessed 5 May 2013

Ren D, Leslie LM, Karoly D (2008) Mudslide risk analysis using a new constitutive relationship for granular flow. Earth Interact 12:1–16. doi:http://dx.doi.org/10.1175/2007EI237.1

Ren D, Wang J, Fu R, Karoly D, Hong Y, Leslie LM, Fu C, Huang G (2009) Mudslide-caused ecosystem degradation following Wenchuan earthquake 2008. Geophys Res Lett 36:L05401. doi:10.1029/2008GL036702

Witt AC (2005) Using a GIS (Geographic Information System) to model slope instability and debris flow hazards in the French Board river watershed. Thesis, North Carolina State University. http://repository.lib.ncsu.edu/ir/bitstream/1840.16/552/1/etd.pdf. Accessed 6 Oct 2014

Wu W, Sidle RC (1995) A distributed slope stability model for steep forested basins. Water Resour Res 31(8):2097–2110. doi:10.1029/95WR01136

Part III
Mitigating Circumstances: Communicating Through Change, Uncertainty, and Disaster

Chapter 10
Comparative Analysis of Virtual Relief Networks and Communication Channels During Disaster Recovery After a Major Flood in Galena, Alaska, Spring 2013

Karen M. Taylor, Richard Hum, and Yekaterina Y. Kontar

Abstract In this case study, we examine disaster recovery in Galena, a small village in Interior Alaska. In May 2013, a massive ice jam on the Yukon River caused flooding that destroyed much of the community's infrastructure and forced the long-term evacuation of over 66 % of residents. In this work, we analyze the virtual relief networks that developed on the social media site Facebook immediately after the flooding in relation to more traditional media and agency press release activity. By contrasting communication channels, we establish the importance of a connection to outside populations during disaster recovery.

Keywords Disaster recovery • Flood • Communication channels • Virtual relief networks • Social media • Traditional media • Central Alaska • Ice jams • FEMA • Emergency management

10.1 Introduction

In May 2013, an ice jam caused major flooding in Galena, a remote village in Interior Alaska (see Fig. 10.1). Although the flood did not result in fatalities or major injuries, it still caused significant suffering to Galena's residents. Within 2 days, flooding destroyed nearly the entire region's infrastructure (see Fig. 10.2a, b) and displaced over 300 residents (Andrews and DeMarban 2013).

K.M. Taylor, Ph.D.
Department of Communication, University of Alaska Fairbanks, Fairbanks, AK, USA

R. Hum, M.A.
Alaska Native Knowledge Network, University of Alaska Fairbanks, Fairbanks, AK, USA

Y.Y. Kontar, M.B.A. (✉)
International Arctic Research Center, University of Alaska Fairbanks, Fairbanks, AK, USA
e-mail: ykontar@alaska.edu

J.L. Drake et al. (eds.), *Communicating Climate-Change and Natural Hazard Risk and Cultivating Resilience*, Advances in Natural and Technological Hazards Research 45, DOI 10.1007/978-3-319-20161-0_10

Fig. 10.1 Map of Alaska (Adapted from http://travelalaska.com)

Fig. 10.2 (**a**) Homes and other buildings are shown flooded in Galena, Alaska; (**b**) ice and water cover the roads in Galena, Alaska

Disaster response and recovery in Galena were challenged by logistical and cultural features of the far north (Salvation Army Alaska Division 2014). The village is located in central Alaska, just south of the Arctic Circle, approximately 270 air miles away from the nearest large town of Fairbanks. Galena has no connecting road systems; thus, the village can be reached only by air or river travel ("Galena City" 2014). All of the provisional and medical supplies must be delivered on aircraft or barges via the Yukon River.

The river, however, remains ice-free for only a few short summer months. During winter, it can be traveled solely via snow machines and dog sleds. Galena residents rely on subsistence hunting and fishing. Since the flood impacted nearly 90 % of Galena homes, it caused many residents to lose the opportunity to replenish all of their fish and game stocks (Salvation Army Alaska Division 2014). In addition, many Galena residents were affected by the loss of their dogs. In rural Alaska, dogs are more than "man's best friends." They assist locals in subsistence lifestyle by transporting people and heavy loads of supplies on snow in the winter. Dogs are also used for hunting and tracking, as well as early warning systems for bears and other forms of danger. Losing their dogs was both an emotionally and financially challenging experience for Galena residents.

Over the course of 5 days (May 26–30), the town was progressively inundated, first Galena's oldest and primary neighborhood—the Old Town, then parts of the New Town. Old Town contains homes and buildings including the town's post office, the Yukon Inn bar and restaurant, and municipal offices (Friedman 2013). It is situated within meters from the river, which makes it susceptible to annual spring flooding. Old Town is located between the Yukon River and a levee built in 1939 around Galena's former air force station, which is now a fully operational and state-owned airport ("Galena City" 2014).

Although the airport is situated in close proximity to the river, it is protected by levees. Flood assistance was possible mostly due to the fact that the airport runway remained operable. The majority of people evacuated within the first 3 days to Anchorage, Fairbanks, Ruby, and Tanana ("Alaska Prepares for More Spring Breakup Flooding" 2013). Some people left via private aircraft, while the majority used aircraft that were sponsored by the Tanana Chiefs Conference, a tribal consortium of 42 Interior Alaska villages (Andrews and DeMarban 2013). Approximately 30 people remained in Galena during the flood. Since they could not take their sled dogs on the airplanes, they chose to stay at the local school building and military dorms, which were converted into a temporary shelter (Hopkins 2013).

The shelters, however, did not have sufficient food, water, and sewage capabilities (Andrews and DeMarban 2013). The town lost both water and electricity, thus causing severe sanitation problems. Many residents criticized the delayed evacuation and response efforts. This delay can be explained by the limited communication capabilities from the village during the flood. Three weeks after the flood, a federal disaster was declared in Galena, which led to collaboration between local, state, tribal, federal, and nongovernmental mass care partners to provide residents with necessary life-sustaining services (Andrews 2013a; Salvation Army Alaska Division 2014).

10.2 Background

The flood, during the ice breakup along the Yukon, in Spring 2013 is the primary spur of the authors' research efforts. The flood has impacted several communities in Interior Alaska. Flooding on the Yukon River in Alaska occurred at Eagle on May 17, when a short-lived ice jam backed up water into several homes (Schwing 2013). Two days later, almost every building suffered flood damage in Circle, when an ice jam developed just below the town. Fortunately, the ice jam broke out quickly and the floodwater receded (Andrews 2013a).

In Far North, ice jams occur during the transitional periods of freeze-up and breakup and thus indicate the beginning and end of ice cover season (Beltaos 2008). In Alaska, spring is known as the ice jam season. For decades, locals in Nenana have been holding an annual ice pool contest of guessing the exact time the Tanana River ice will break up ("Nenana Ice Classic" 2014). As a rule, Alaskans are prepared for the annual ice jams but not for the severity of their consequences.

The extensive floods on the Yukon in Spring 2013 are attributed to abnormal weather patterns. April and early May 2013 were the coldest in decades in Interior Alaska (Andrews 2013b). As a result, the winter snowpack in the Yukon River basin remained in place weeks later than normal, and river ice remained solid (NWS 2014). The most stubborn ice jam formed on May 26 at Bishop Rock—a place notorious for ice jams formation, and caused a major flood 18 miles upriver in Galena (see Fig. 10.3) (NASA 2013).

Ice backed up more than 40 miles behind the jam before the jam released on May 29 (NWS 2014). As the flood inundated Galena, evacuation efforts began on May 26 ("Alaska Prepares for More Spring Breakup Flooding" 2013). Within 2 days, the majority of Galena residents were evacuated by air to neighboring communities. A few people, who chose to stay behind to look after their dogs, were stranded at the levee-protected temporary shelters at the Galena airport and a boarding school building without electricity or wastewater sanitation (Andrews 2013b; NWS 2014).

A federal disaster declaration was issued in late June, which led to collaboration between local, state, tribal, federal, and nongovernmental mass care partners to provide residents with necessary life-sustaining services (Andrews 2013a; Salvation Army Alaska Division 2014). However, rebuilding and recovery in Galena, which was estimated to exceed $80 million, was slowed by problems with transportation, supplies, and bureaucracy (Andrews 2013a; NWS 2014).

The remoteness of the community from infrastructure that Federal Emergency Management Agency (FEMA) typically relies on (e.g., international-capable airports and cellular towers) along with the seasonal changes of Alaska has meant that rebuilding is slower and yet simultaneously more time-constrained. The Yukon River is only reliably open for boat traffic from June through August and is the primary route for transportation in the region ("Galena City" 2014). Rebuilding efforts thus faced a de facto deadline of completion by 3 months after the flooding. A year later, reconstruction of Galena was still in progress, and nearly 10 % of residents remained in shelters in Fairbanks and other towns (Friedman 2014).

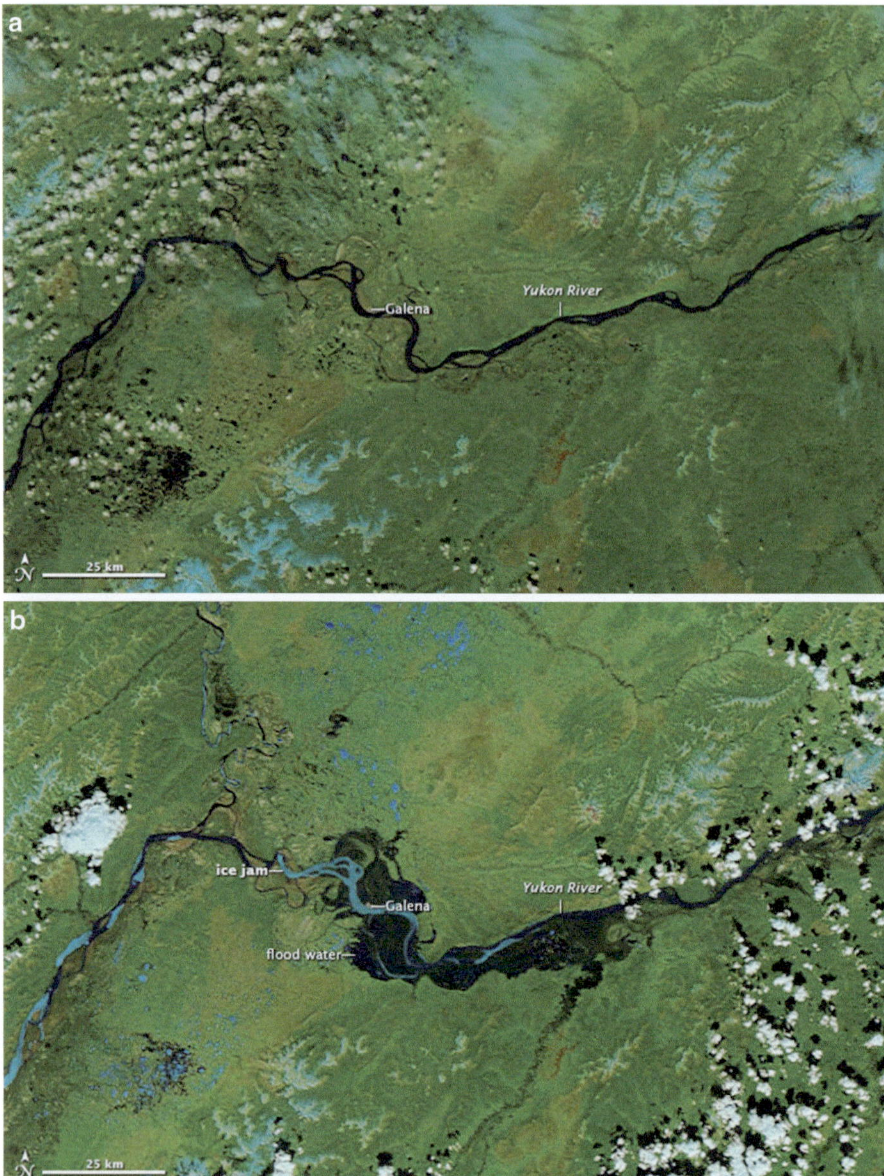

Fig. 10.3 (**a**) Galena, Alaska, and Yukon River in May 2012; (**b**) flood waters swamped Galena, Alaska, on May 28, 2013

10.3 Literature Review

Much has been written about risk, crisis communication, and the role of social networks. The recognition about the importance of social networks is sometimes theorized without extensive empirical data, but evidence exists to suggest that network dynamics can be understood as providing agency and opportunities to change and improve community resilience. Moore and Westley (2011), for example, highlight that while diffusion of innovation models is useful for understanding the change that can happen when social systems are forced to respond to novelty, only with the addition of understanding how individuals mobilize networks can we understand how innovations are able to cross scales (local to regional to national, e.g., in the Galena case).

Groscurth (2011) suggests four paradoxes organizations confront doing disaster relief, which show some of the lines of tension in institutional racism. One is referred to as the inclusion/exclusion paradox, which became particularly poignant for the American Red Cross in funneling minority volunteers in the recruitment of volunteers for responding to disasters that hit hardest in majority-minority communities—communities that often get hit hardest. The particularity/universality paradox and individual/institutional paradoxes are familiar in organizations other than disaster relief, but the speed of events can complicate each. The privilege/equality paradox is an inevitable complexity in crises, and critically recognizing the racial component is valuable in studying disaster response in Alaska. Race is relevant in Galena as well, and it may be necessary to reinvent language to get out of the racially fraught discourses that circulate. Though disagreeing with Groscurth, it makes more sense to think of these pairings not as paradoxes, because they are not contradictory but are mutually creating. For example, in the Galena flood, it is only possible to recognize the disparity by contrasting the recruitment efforts for help in Fairbanks vs. in Galena. The posts to coordinate laundry assistance in town entailed a series of exchanges in advance, whereas the assistance provided by a neighboring village, primarily Alaskan Native, involved primarily ex post facto responses in the form of threads expressing gratitude, with no advance planning organizing efforts having been visible.

McCosker (2013) highlights the ways that nonprofessional on-the-scene photography heightens disaster emotional impacts for an audience because we are now inside the disaster zone, rather than positioned to view it as if looking at a picture. But the de-framing argument is problematic in its assumption about passivity of audience and about *innocence* or authenticity of raw footage (i.e., professionals can create that too). The better argument here is for interactivity, not de-framing. McCosker (2013) provides a good reminder about the ontology of information as flow, something we experience as one piece of many ongoing elements. We expect the need to attend to things like medium of communication, because we are reminded about shared locus of attention leading to affectivity (pp. 386, 388). Spectacle is thought of as necessarily self-increasing; it reaches a state of perfection that is destruction of its own system.

Griffin et al. (2008) write about information sources and the divergent meanings that diverse audiences take from them. One point made is that agencies are well advised to create preemptive messages. An example observed in the Galena flood response was the use of a disaster expo forum that included organizing volunteer efforts (p. 308). The expo can help mitigate concerns of trust, efficacy, and attribution, which in turn correlate with anger. Anger as represented in headlines like "no money from FEMA" is going to play very differently across audiences. Among natives this will be read one way, given colonialist history and relations with government/Bureau of Indian Affairs (BIA), among urban Fairbanksans it will be read with different anger because of resentment about government intrusion in general, while outside of relatively urban areas the anger is aimed at bureaucracy, for contracting practices incompatible with the short building season and shipping.

Doerfel et al. (2010) used network analysis in studying the role of NGOs in disaster recovery (e.g., one study done in Croatia after war, another in New Orleans after Hurricane Katrina). One of the recommendations from this analysis is about the importance of indirect support, to form new social capital and network connections, as being equally important as any direct support. Certainly the importance of network connection formation is recognizable in the Galena data as well. However, organizations will not exist as part of the network, except as represented by key individuals.

Veil and Sellnow (2008) use a best-practice theoretical model adapted from the Department of Homeland Security (DHS). This model places emphasis on providing self-efficacy mechanisms, the role of pre-planning and strategy inclusion, and the importance of accepting uncertainty. The model warns against wasted effort as channels of communication multiply, instead arguing for a willingness to sacrifice transparency and some sense of additional info in return for faster decision-making, because as Weick and Sutcliffe (2007) pointed out, action first and then understanding later is one of the hallmarks of high reliability organizations. As with the after-action review of Galena responses conducted by the Government Accountability Office (GAO), one advantage also is that it reminds us to symmetrically study what works as well as what does not work.

Perrow (1999) puts forth that planning creates its own set of risks in relation to disasters. Since disasters by definition are events that are outside of the normal operating of any system, a system that invests more efforts unto preparing for disasters that are unpredictable is in danger of creating blind spots and overconfidence. Systems or organizations that are inelegant but robust are better than efficient and with redundancies only tacked on later; alternately, systems are sustainable if they are decoupled more and kept simple instead of complex. This represents a nice ideal, and granted that incremental growth can be risky, but at some point systems are sufficiently complex that they cannot be decoupled or robust. Planning does *NOT* mean centralized control necessarily, and one of the strengths of a community network model is that routine everyday interactions that strengthen the network are also simultaneously improving preparedness to deal with disasters.

The Social Amplification of Risk model by Renn et al. (1992) points out "The social experience of risk is not confined to the technical definition … Humans perceive risk as influenced by their values, attitudes, experiences, and cultural identity" (p. 137). Perception of risk thus varies by information sources, and also differences at the individual level, and is based deeply in cultures. We are able as observers to make some conjecture about risk perception based on risk behaviors, and if agencies learn more about a community's shared experiences of risks as well as their values, they will be better able to respond in ways that are deemed relevant by those impacted. One illustration that risk perception varies even within a population that is well informed about technical risk can be seen in the preflood variation in Galena. All members knew the risk of flooding during spring breakup, and many community members remembered the breakup floods of 2013 as a firsthand experience. Responses varied widely, from those who monitored information sources continuously or even preemptively evacuated to others who figured that what would happen was what would be and were willing to accept that flooding was outside of control by any methods worth taking. The risk perceptions of Galena residents may also be influenced by their longer cultural history, since the oral traditions for passing along native knowledge are still practiced and contain numerous flooding stories and reminders that the residents had originally only established semipermanent dwellings in part as a coping mechanism. Thus, the *risk* associated is as much part of the risk of having stayed in one place to build significant infrastructure, as well as the risk of ice jams on the river as it thaws.

Utz et al. (2013) used experimental crisis communication methods based on the Fukushima nuclear plant disaster to determine:

> Crisis communication via social media resulted in a higher reputation and less secondary crisis reactions such as boycotting than crisis communication in the newspaper. However, secondary crisis communication was higher in the newspaper condition because the traditional media is still considered more credible. (p. 45)

Traditional news is more likely to be forwarded. By contrast, the advantage of Facebook is that it is viewed as more positive. This finding seems consistent with observations in the Galena flood. Newspaper articles attracted almost universally negative commentary. While we see little evidence of any propensity to forward newspaper articles or other interaction effect with social media, the general expectation of greater positivity on Facebook appears to be a consistent norm.

Majors (1998) makes clear one final lesson. Agencies involved in disaster preparedness need to remember that there is no such thing as *The Public*, only a multitude of stakeholders that form a plethora of publics. This suggests that information, which can be interactive and tailored to provide personalized risk assessment, as social media are capable of doing, will likely be more effective than mass dissemination messages. There can be advantages to each of course. One reason to emphasize traditional mass media includes taking advantage of social influence and opinion leadership. This does emphasize the value of consistent messaging across channels, though that can be difficult given the uncertainty inherent in disaster situations and the complexity of trying to model disaster scenarios in advance.

10.4 Methods and Analysis

Our methods split along two functionally distinct yet structurally overlapping channels of communication. The first is a public Facebook group that formed in response to the flooding. The second is traditional news media and agency-derived press release activity. Functionally, these two channels are distinct. Communication through the Facebook group was established and sustained directly by Galena residents or those with close personal ties to the community. Traditional media and agency communication was established by outsiders and intended to serve a broader audience than those directly impacted. Structurally, however, they overlapped in that both traditional news media and agencies utilized Facebook as part of their communication efforts.

In collecting data for the Facebook group, NodeXL software was used exclusively for data retrieval, network analysis, word-pair analysis, and graph visualization. Longitudinal data was collected at both monthly and daily scales, May–December 2013 and May 28–July 31, 2013, respectively. Frequency curves were constructed for each time scale. Key points along this curve were identified for more detailed network and content analysis at both scales. NodeXL is designed to provide a simple and fast method to create explorative social network analysis studies ("NodeXL" 2014), as such it is an ideal tool to use when looking for dynamic network relationships because time-sliced networks can relatively easily be constructed.

The first step in our workflow was to retrieve the social media data. This was accomplished by using the NodeXL add-on Social Net Importer ("Social Network Importer for NodeXL" 2014) and taking advantage of their Facebook Group Network Import feature to pull time-sliced information from the public group Yukon River Rescue (YRR). YRR formed during the active evacuation of the community (as did a handful of other related groups) and quickly became the most active group, and hence the focus of our study. Two sets of data were collected. One aggregated group activity across monthly time intervals for a 7-month period following the flood. The second focused on the early phases of group development and aggregated group activity at a finer daily resolution.

Network analysis was conducted at the structural level and sought to characterize dynamics at a broad rather than individual level. As such, post/comment frequency, number of components, geodesic distance (diameter), and density measures were calculated using NodeXL's basic metric functions. Word-pair analysis was also conducted using NodeXL default language analysis functions. Again, graph visualizations were developed using NodeXL's visualization package. The networks in this study are small enough (< 7,000 edges for any one time slice) that filtering is not required and all networks were visualized using the same protocols—Fruchterman-Reingold layout, node color and size degree dependent, and edge thickness based on edge weight.

10.4.1 Facebook Data Analysis

The analysis of both the monthly and daily Facebook data began by examining the frequency of activity, both in terms of unique posts and comments on them, at both scales. Each scale shows a general power law form. This is interpreted to represent an initial broad public interest in the flood when the event was fresh and people's needs immediate, with a rapid tapering off of interest with time and a diminishing of immediate need. Looking at specific points along these curves, however, provides a more nuanced understanding of how the network changed over time.

First, in the monthly data (see Fig. 10.4), we see an almost immediate explosion in activity with very little buildup through May and June. This is likely a result of YRR mobilizing an already well-established network of users. The scope of this study makes it hard to know for sure, but it seems probable these users were linked through an assortment of unknown Facebook connections and/or any number of ties through different communication channels.

The networks in these early phases have a large number of components (63 and 73, respectively). A network component is a segment of the network with no ties to other parts of the network, but close examination of the early YRR networks shows that the vast majority of components are small in size (1–3 vertices) with most vertices belonging to a single large component. Within this large component, the geodesic distance, or diameter, is also fairly large. Individual node centrality in the networks is highly segmented, with a few vertices (nodes) being very central and the majority only weakly connected. We interpret these combinations of network properties to be indicative of a broadcast form of communication, where a few central players are driving a large portion of the communicative traffic (i.e., disseminating the majority of information).

As the summer turned to fall and then into winter, total activity in the group initially declined rather dramatically and then steadied and tailed off gradually through December. Network dynamics mimicked this pattern through both a decline in the number of components and the diameter of the largest component. Density increased within components as well, and centrality distributions became more even across the network. These patterns are interpreted to define a shift from large volume information dissemination and coordination early in the flood response to more intimate, bonded, and supportive communicative practices as time progressed.

Data aggregated at the daily scale gave us a more nuanced perspective on how this transition occurred. Our daily aggregates covered the time period from the initial formation of the group on May 28, 2013, until July 31, 2013. Like the monthly data, the daily frequency curve showed high volume activity instantly once the group was formed then a rapid tailing off. This, again, would seem to indicate that YRR tapped into well-established and well-evolved networks at its onset. At the daily scale, however, more variability can be seen in the frequency curve (see Fig. 10.5).

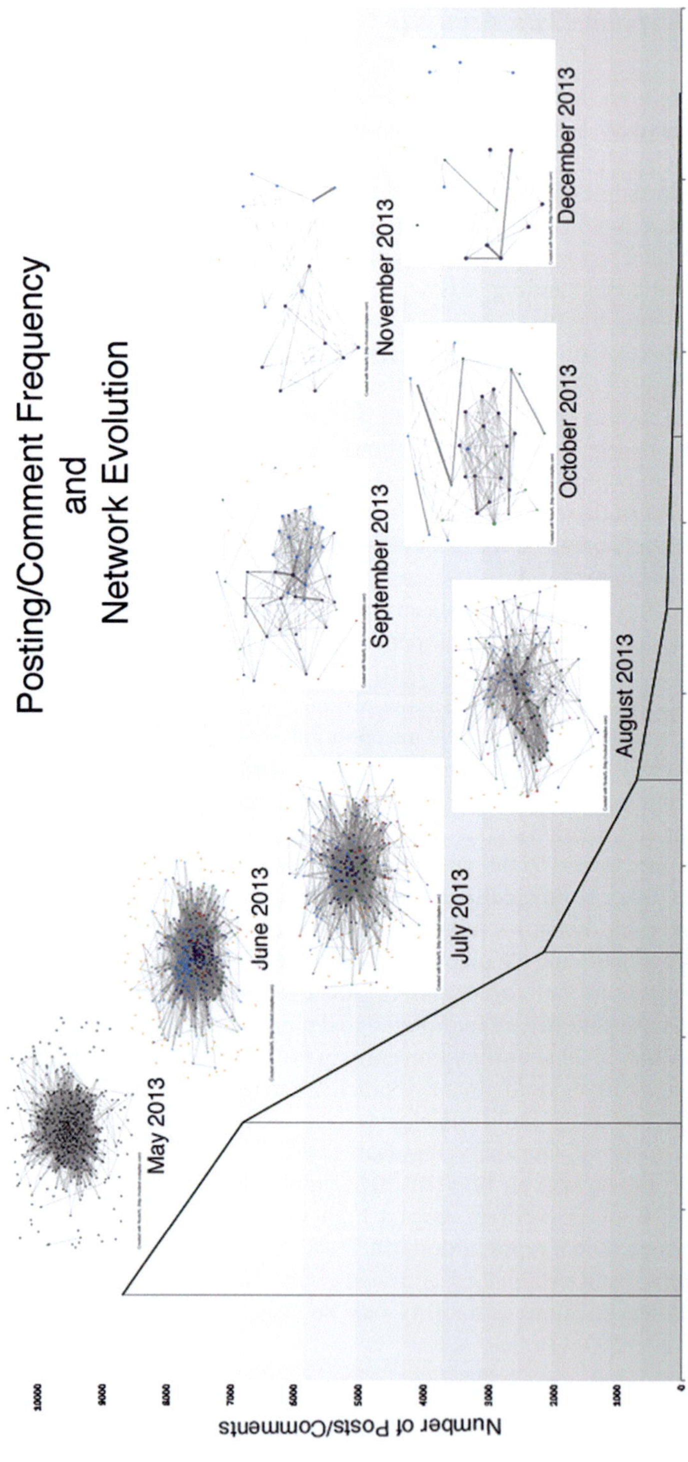

Fig. 10.4 Comment-posting and network activity aggregated at monthly intervals

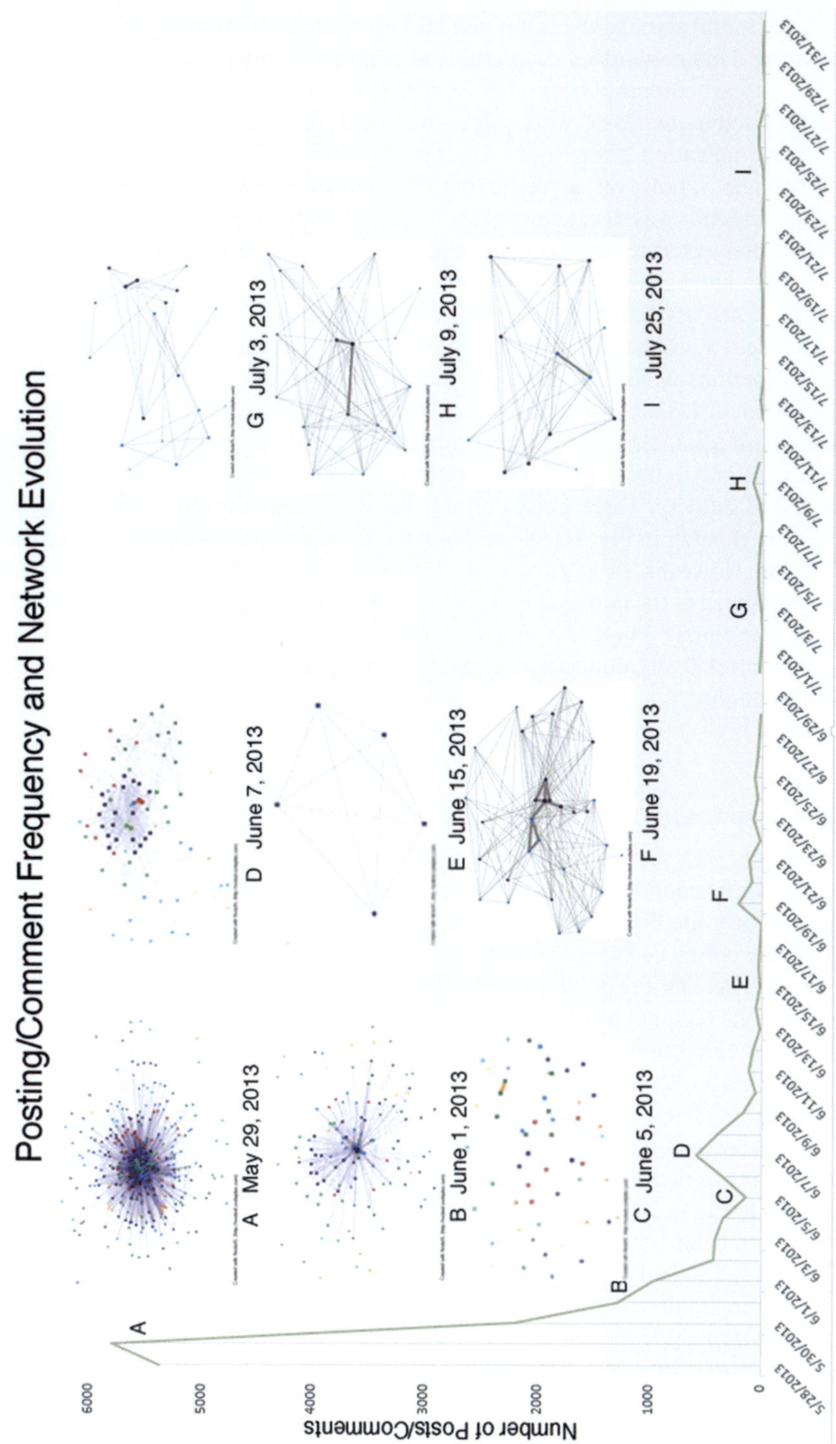

Fig. 10.5 Comment-posting and network activity aggregated at daily intervals

As stated, initial activity levels tapered off very quickly, within just a few days of the flood. The daily networks through this initial phase of group evolution look very much like the early monthly networks—a central component of large diameter and a few highly active members. This, again, was interpreted to represent a broadcast form of communication where a few key people were driving communication. We see that on June 5, however, as the activity began to decline, the strong broadcast form of the network was dissipating. The strong segmentation of a few highly central players connected to numerous less central group members was diminished, with centrality more evenly distributed across the network.

On June 7, activity flared back up. When this happened, the diameter of the network expanded, a number of small, disconnected components sprang up, and centrality distributions again differentiated before evening out again as the activity quickly diminished. Throughout the remainder of the time slices, examined centrality distributions, diameter, number of components, and density distributions remain fairly constant and indicative of a more tightly bonded community than seen in the initial burst of activity. Three pulses of activity were noted during this time frame (see points F, H, and I in Fig. 10.5), each of which was much smaller in size than the June 7 burst. However, during each of these flare-ups, broadcast network forms never materialized at the daily scale. From this, we concluded that it is at this stage, right after the June 7 burst, that YRR evolved from a predominately loose knit, interested, but relatively uninvested group to a more tight knit and deeply involved virtual community.

10.4.2 Newspaper Data Analysis

Traditional mass media that covered the flood included primarily regional newspapers, particularly the Fairbanks newspaper The Daily Newsminer and at the statewide level the online newspaper Alaska Dispatch. National coverage appeared only once via an Associate Press Wire (APWire) news article that appeared on May 30. Television at the regional level covered the story at least once (May 29, KTVF), and local radio provided coverage that is not included in the analysis here. Additionally, press releases from the local Department of Homeland Security were included in our mass media analysis, as these were disseminated widely. In all, a total of 20 news items were analyzed (see Fig. 10.6).

News coverage occurred in a series of clusters. The initial cluster began on May 13, which is 15 days earlier than our social media network coverage had begun. The spring breakup behavior of river ice is a news topic that can be predicted somewhat in advance to fall within a two- to three-week window, and local media begin reporting on it in preparation. The breakup of 2013 resulted in minor flooding in many communities along the Yukon River, progressing downstream, so photographers and reporters could arrange in advance to have coverage available. The dates of May 27 through 30 had one or more article per day.

A second cluster of news coverage occurred in September, around the time that the rivers' freeze-up is usually expected. This coverage included no sources other

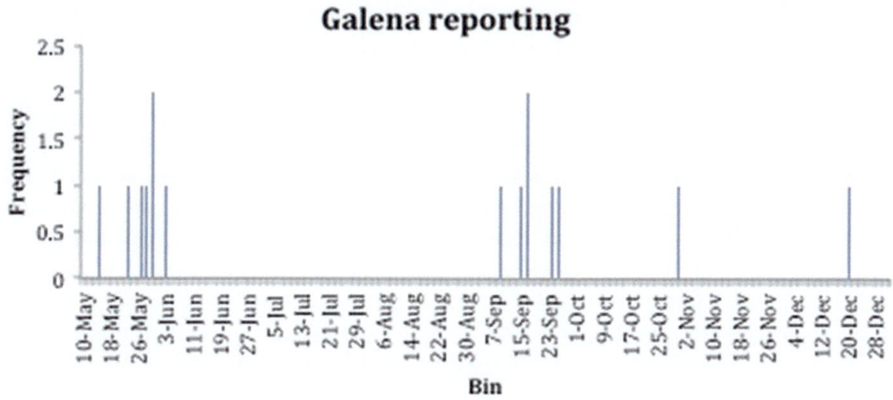

Fig. 10.6 A histogram illustrating the frequency of the data over time

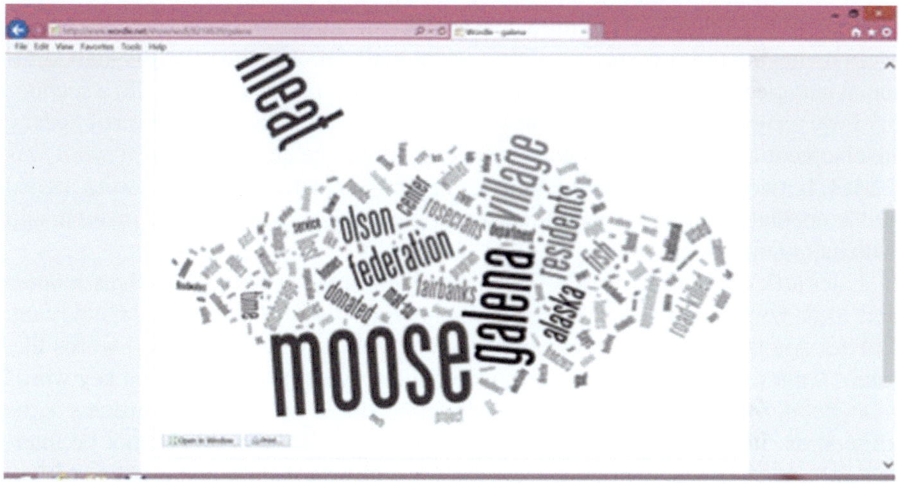

Fig. 10.7 Wordle of the final news article

than regional and indeed was all authored by a single Newsminer journalist. A final small cluster of two articles occurred in December, providing follow-up and serving as *human interest*, feel-good stories for the holiday season (see Fig. 10.7).

10.5 Discussion

Wordle provided a fast way to do a cursory content analysis and enabled a comparison of the content of the Facebook social media coverage and the traditional news media coverage ("Wordle" n.d.). The analysis focused on frequently used words and then as a secondary step looked specifically at issues of verb choice, since verbs are what carry the *action* in a sentence and a news story. A third step focused specifically on the individuals or organizations that were represented as playing a relatively more important role in the course of events. Names of individuals are left out from the presentation here, but in general terms it is still important to note how frequently single individuals were presented as the primary locus of action or agency.

In terms of overall frequency, it is unsurprising that in both traditional news and social media coverage, in the initial dates around May 27, the single most frequent word was *flood* and the second most frequent was *Galena*. Other shared high-frequency words in both newspapers and Facebook include *water*, *people*, and *ice*. Aside from these nouns, the word frequency shared little overlap. This suggests that there was relatively little initial exchange between the two channels. It contrasts to media theories that would tend to suggest a relatively high rate of interdependence. Agenda Setting theory, for example, would hypothesize a greater degree of overlap with journalists relying on information sources such as the town's residents who were doing the Facebook posting and in turn with news articles being posted on Facebook and stimulating further comments based on them. What we found instead was that in this initial cluster, little convergence occurred.

In reviewing the newspaper articles, additional high-frequency words that dominated included nouns such as *aircraft*, *residents*, *disaster*, and *Alaska*. On a secondary level were verbs such as *evacuating*, *flew*, and *going*. The first mention of agency involvement appeared in a May 17 article in which the single dominant word was *FEMA*. In two out of the seven articles in the initial cluster, individuals were named and immediately included an antonomasia identifying their role and membership with an organization.

Later articles focused on additional agency-oriented key words, such as *schools* and local government agencies in the metonymic form of Fairbanks, Circle, and Galena. Another September article discussing the recovery emphasized words like *funds*, *homes*, *federal*, and *state*. Even later, a December article included key words *food*, *flood*, *frozen*, and *program*. One characteristic to notice in later articles is the alliteration, increasing seemingly as time progresses. Though this cannot be interpreted as necessarily intentional, there may be significance, as the journalist responsible for these articles was a Fairbanks journalist writing for a holiday-focused audience. The poetic prosody elements may be part of creating a sense of festivity.

Facebook data showed a pattern of initial frequent postings dropping off exponentially, with spikes that also revealed a pattern of peaking and then rapidly decreasing usage. The spikes occurred at irregular intervals, presumably correlating with key events such as the first posting of photographs updating residents on post-flood conditions. The combination of visual and textual data that is characteristic of Facebook proved to have an important influence on shaping the crisis response dynamics among this community. We see an initial height of almost 6,000 messages at the beginning of the site's creation, May 28. This number needs to be understood in context, since initially the Facebook site was not the only one used by residents affected by the Yukon-Koyukuk flooding.

The messages included both initial posts to the wall of the site and replies to those postings. A week later, posts had dropped to a low on June 5 of only a couple hundred messages. Smaller peaks on June 7, June 19, and July 9 never rise above the one-thousand-message mark. None of these relative peaks in social media are reflected in traditional news media coverage, and none of the news media coverage appears to correlate with even small relative peaks in Facebook activity.

Content of the initial burst of messages tended to focus on individuals and village-level concerns. As in traditional media, high-frequency descriptors included *flood*, *Galena*, and *water*. Words that did not overlap but were seen with high frequency on Facebook included the verbs *know*, *get*, *thank*, and *help*. Indeed, throughout the initial steep power curve portion of the timeline and consistently for the first 2 weeks, *thank* remains constantly among the highest-frequency words per day. The first mention of any organization occurs on June 1, referencing the Tanana Chiefs Tribal Council (TCC on Facebook). The June 7 spike, reaching a height of about 500 messages, centered on conversations thanking the village of Fort Yukon, upriver from Galena, which had organized a fundraiser and a cleanup party to help. No reference to FEMA appeared until July 9, at which point it becomes the single most frequent word. In looking at some of the low-frequency posting dates, words such as *kids*, *back*, *awesome*, and *delicious* provided some sense of a steady, always ongoing conversation about the mundane, routinely positive small talk typical of Facebook anywhere.

A burst frequency analysis allowed us to examine the Facebook data using an algorithm that focused on *stop* words and selected terms based on frequency of use relative to frequency in the overall expected use based on the language as a whole, a statistical technique useful for *big data* sets. When we explored this burst frequency data, we found that personal names, place names, and even personal phone numbers are among the terms that come up throughout the summer months. Verbs used include *assist*, *answer*, ***hate***, *hire*, *join*, *loan*, *limit*, *pass*, ***pile***, ***pitch***, ***play***, *receive*, *realize*, *remind*, ***sell***, *sign*, *ship*, ***spread***, *suggest*, *support*, and *work*. The boldface terms are those that were most statistically significant. Notice that all of these represent actions by a single person, at least in this context.

The word choices make visible initial actions necessary in first getting out of danger and establishing a less-precarious temporary dwelling place. For example, the verb *sell* was so prominent because it was being used for two important purposes. The first is asking questions about where vital items (e.g., glasses and baby clothes) could be acquired. The second reason for this frequency is that one of the first things done was establish a way for individuals to bring in goods and get those goods out to those who might need them; in particular, a church basement was used to gather any used clothing donations from any surrounding communities (in this case extending all the way to Anchorage, a 7-hour drive from the nearest urban hub where the items were gathered) and both gave out clothing items that were essential and began selling the items cheaply that were not immediately necessary but also could not be stored by a group rendered reluctantly nomadic. Soon, items beyond strictly clothing were being exchanged in this venue, such as pots/pans, beads/beadwork, and cleaning/pet supplies.

The initial peak for posting dropped to 400 posts per day by the second day after the site was created (bear in mind that is still a lot of traffic given that the overall population of the Yukon-Kuskokwim river valley is only about 3,000). As of the end of the month (June 2), daily posting was down below 100 and then spiked upward again on June 7 to the level of 400 again. The rate of drop-off after the June 4 spike was slower. On June 11, a low point was reached (no posts with that date) and then returned to a short-term plateau of approximately 100 postings per day for almost a week. After that a rapid decline occurred. Posting continued at a low but fairly steady rate averaging 40 per day for the remainder of the time during which the site was observed, which was about 6 months.

During the second month, burst frequency analysis revealed a shift that included a higher representation on the Facebook site of various disaster response organizations. During the first month, the only organization mentioned on the site was the Small Business Administration (SBA), with only a low frequency. During the second month, we saw the SBA, the FEMA, the Catholic Diocese, the Yukon tribal conference, the *school* (in reference to a regional boarding school located in Galena), and the softball leagues all occurring with varying degrees of frequency. Of those, the Catholic Diocese was the highest burst frequency. The verbs that were represented on the burst frequency analysis are ***adopt***, *contact*, ***contribute***, *distribute*, *enjoy*, ***include***, *learn*, *manage*, ***pick***, ***plan***, *provide*, *rebuild*, and ***went***. Again, the verbs occurring with highest frequency are in boldface. It might be interesting to note that these verbs are on average longer, only four single-syllable examples instead of the half-dozen in the first month's burst frequency analysis.

Most of these are words we associate with actions that are coordinated across multiple individuals. For example, the verb *adopt* has occurred with such frequency because it conveys a care-taking relationship and those responsibilities were being shifted and re-negotiated (at least temporarily, with potential for longer-term implications). One of the most pressing issues for the Galena community was to figure out temporary dwelling spaces for their dogs. In the forced transition to an urban hub, finding suitable placement for not just one or two dogs but for entire teams of dogs, in some cases, proved a difficult challenge. The verb *adopt* also appeared frequently because one of the patterns that helped during this upheaval would be that an urban resident (member of the host community) would look for ways to coordinate with an entire family rather than trying to help individual by individual. Thus, efficiencies of aid were established, and the phrase *adopt a family* was used to represent that relationship.

During the third month, verbs included *continue*, *cover*, *donate*, *follow*, *gave*, *hope*, *inform*, *need*, *send*, and *touch*. During the final months of our analysis, few words were added that had not been part of previous conversations. Of those that did appear in the burst frequency analysis, most marked a shift to a strongly positive tone. *Elder*, *glad*, *good*, *happy*, *love*, and *prize* were the most notable. It would be nice to believe this represents that by this time period, the Galena story was wrapping up toward a happy ending. Of course, it could also represent a return/reinforcement to the Facebook norm that encouraged primarily positive posting.

An interesting characteristic to notice about these verbs is the initial clustering. Notice that each month tends to have clusters of initial letters. In terms of the verbs, the first month clustered around *s* (six out of 21, or 29 %, or 50 % of the highest frequency terms), and the second clustered around *p* (three out of 13, or 24 %, or 33 % of the highest frequency terms). Instances like this in conversation are referred to as examples of *convergence*, that is, of styles shifting to reflect each other's speech patterns (like accommodation, but where there is no clear answer as to who is in power) (see Soliz and Giles 2014). High levels of convergence usually would be taken as a sign of emerging community, or already existing community.

10.6 Conclusions: Lessons Learned

Consistent with this study's analysis, FEMA and the State of Alaska have stated the agencies were slow to respond and were hampered by their own internal procedure (*red tape* to use the colorful analogy). The OIG (Office of Inspector General) report that came out the following summer suggested several remedies, including greater reliance on local authorities and local practice, including local contracting practice. Our analysis here is of use as an indirect confirmation of those findings, along with a partial suggestion for one way to help achieve the worthwhile goal of greater local adaptation. Social network analysis demonstrates clearly that the local response coheres independently and rapidly and simultaneously gives outsiders a useful mechanism for efficiently tapping into that local knowledge base and identifying key strategic partners in that network. Network analysis tools can allow the masses of interaction happening on Facebook to be turned into a resource for identifying who is able to get various types of work done, including who is effective as a local channel for information flow in both directions (because recovery is necessarily a two-way interaction at all times).

We might expect, given the organizational constraints of the media cycle, that we should see a fairly close relation between spikes in communication of outsiders and spikes of communication by residents. It should be difficult, without a more fine-tuned unit of analysis than date of publication, to make any cause-effect relationship claims. We might like to hypothesize that newspaper and press release coverage would then in turn generate greater activity on social media, or we might like to hypothesize that increased social media activity should be picked up by external reporters and responded to with a corresponding spike in external communication.

In point of fact, however, it appears that the social media generated by residents and the written record generated by outsiders are entirely uncorrelated and that increases in frequency or intensity of message producing does not influence in any recognizable or synergistic way cross-channels. If this is atypical of what we know about the interdependency of media channels, then we should be exploring why this context is an exception to that usual convergence. It is tempting to explore here whether the racial lens that Groscurth (2011) notes often occurring in crisis communication might not partially explain the disjuncture. A less critical explanation could be offered that focuses instead on the ways in which physical distances, for all that digital communication tempts us to ignore them, still matter.

References

Alaska prepares for more spring breakup flooding (2013) http://www.redcross.org/news/article/Alaska-Prepares-for-More-Spring-Breakup-Flooding. Accessed 2 Mar 2015

Andrews L (2013a) Hard lessons for FEMA: Galena rebuilds with federal help, in spite of federal help, Alaska Dispatch News, 20 Sept 2013. http://www.adn.com/article/20130920/hard-lessons-fema-galena-rebuilds-federal-help-spite-federal-help. Accessed 2 Mar 2015

Andrews L (2013b) 2013: a year of Alaska weather extremes, from Barrow to Juneau, Alaska Dispatch News, 31 Dec 2013. http://www.adn.com/article/20131231/2013-year-alaska-weather-extremes-barrow-juneau. Accessed 2 Mar 2015

Andrews L, DeMarban A (2013) Rescue aircraft evacuating more Galena residents as flooding worsens, Alaska Dispatch News, 28 May 2013. http://www.adn.com/article/20130528/rescue-aircraft-evacuating-more-galena-residents-flooding-worsens. Accessed 2 Mar 2015

APWire (2013) Floodwaters drop after Alaska Yukon River ice jam breaks, Associate Press Wire. Retrieved 2 Mar 2015 from http://www.newsminer.com/news/local_news/floodwaters-drop-after-alaska-yukon-river-ice-jam-breaks/article_7014ea94-c944-11e2-9ddf-001a4bcf6878.html

Beltaos S (2008) Progress in the study and management of river ice jams. Cold Reg Sci Technol 51(1):2–19. doi:10.1016/j.coldregions.2007.09.001

Doerfel ML, Lai C-H, Chewning LV (2010) The evolutionary role of interorganizational communication: modeling social capital in disaster contexts. Hum Commun Res 36(2):125–162. doi:10.1111/j.1468-2958.2010.01371.x

Friedman S (2013) Galena won't receive federal funds for rebuilding in flood-prone area. Newsminer, 18 Sept 2013. http://m.newsminer.com/news/local_news/galena-won-t-receive-federal-funds-for-rebuilding-in-flood/article_40c73f36-1f6f-11e3-af14-001a4bcf6878.html?mode=jqm. Accessed 2 Mar 2015

Friedman S (2014) Galena making progress in flood recovery. Newsminer, 13 Jun 20014. http://www.newsminer.com/news/local_news/galena-making-progress-in-flood-recovery/article_e6a7aa46-f2d0-11e3-9f6f-001a4bcf6878.html. Accessed 2 Mar 2015

Galena City (2014) http://www.ci.galena.ak.us. Accessed 2 Mar 2015

Griffin RJ, Zheng Y, ter Huurne E, Boerner F, Ortiz S, Dunwoody S (2008) After the flood: anger, attribution, and the seeking of information. Sci Commun 29(3):285–315. doi:10.1177/1075547007312309

Groscurth CR (2011) Paradoxes of privilege and participation: the case of the American Red Cross. Commun Q 59(3):296–314. doi:10.1080/01463373.2011.583498

Hopkins K (2013) With Galena a ghost town, Koyukuk braces for 'flash flood'. Alaska Dispatch News, 29 May 2013. http://www.adn.com/article/20130529/galena-ghost-town-koyukuk-braces-flash-flood. Accessed 2 Mar 2015

KTVF (2013) Flooding in Galena, Alaska, KTVF News. https://www.youtube.com/watch?v=jY1mhxF_3xA. Accessed on 2 Mar 2015

Major AM (1998) The utility of situational theory of publics for assessing public response to a disaster prediction. Public Relat Rev 24(4):489–508. doi:10.1016/S0363-8111(99)80113-1

McCosker A (2013) De-framing disaster: affective encounters with raw and autonomous media. Continuum J Media Cult Stud 27(3):382–396. doi:10.1080/10304312.2013.772109

Moore M, Westley F (2011) Surmountable chasms: networks and social innovation for resilient systems. Ecol Soc 16(1):1–13

NASA (2013) Ice jam on the Yukon River floods Galena, Alaska: image of the day, 28 May 2013. http://earthobservatory.nasa.gov/IOTD/view.php?id=81227. Accessed 2 Mar 2015

National Weather Service (NWS) Alaska Flood Information (2014) http://www.floodsafety.noaa.gov/states/ak-flood.shtml. Accessed 2 Mar 2015

Nenana Ice Classic (2014) http://www.nenanaakiceclassic.com/. Accessed 2 Mar 2015

NodeXL: Network overview: discovery and exploration for excel (4 Dec 2014). http://nodexl.codeplex.com/. Accessed 23 Mar 2015

Perrow C (1999) Organizing to reduce the vulnerabilities of complexity. J Conting Crisis Manag 7(3):150–155. doi:10.1111/1468-5973.00108

Renn O, Burns WJ, Kasperson JX, Kasperson RE, Slovic P (1992) The social amplification of risk: theoretical foundations and empirical applications. J Soc Issues 48(4):137–160. doi:10.1111/j.1540-4560.1992.tb01949.x

Salvation Army Alaska Division – Alaska's EDS director recognized nationally for Galena work (31 May 2014). http://www.salvationarmyalaska.org/alaska/news/VOAD_Recognition. Accessed 2 Mar 2015

Schwing E, Alaska Public Media (19 May 2013) Second largest flood on record hits Eagle as Yukon breaks up. http://www.alaskapublic.org/2013/05/19/second-largest-flood-on-record-hits-eagle-as-yukon-breaks-up/. Accessed 2 Mar 2015

Social Network Importer for NodeXL (4 Dec 2014). http://socialnetimporter.codeplex.com/. Accessed 23 Mar 2015

Soliz J, Giles H (2014) Relational and identity processes in communication: a contextual and meta-analytical review of communication accommodation theory. In: Cohen E (ed) Comm yearbook, vol 38. Taylor & Francis, Hoboken, pp 107–143

Utz S, Schultz F, Glocka S (2013) Crisis communication online: how medium, crisis type and emotions affected public reactions in the Fukushima Daiichi nuclear disaster. Public Relat Rev 39(1):40–46. doi:10.1016/j.pubrev.2012.09.010

Veil SR, Sellnow TL (2008) Organizational learning in a high-risk environment: responding to an anthrax outbreak. J Appl Commun 92(1&2):75–93

Weick K, Sutcliffe K (2007) Managing the unexpected: resilient performance in an age of uncertainty, 2nd edn. John Wiley & Sons, New York

Wordle – Beautiful Word Clouds (n.d.) http://www.wordle.net/. Accessed 23 Mar 2015

Chapter 11
Development of the Stakeholder Engagement Plan as a Mining Social Responsibility Practice

Alexandra Masaitis and Glenn C. Miller

Abstract This chapter addresses the development of social responsibility practices for Newmont Mining Corporation's Emigrant Rain Project in the State of Nevada, USA. It results from Newmont's concerns related to conflicts on permitting the Emigrant Mine over the past 12 years, which brought the University of Nevada, Reno, and the Newmont Mining Corporation together to investigate the concept of a *Good Neighbor Agreement*. A Good Neighbor Agreement can be used to understand the relationships between various stakeholders, to improve communication, and to help resolve disagreements. For this purpose, we identified groups of relevant stakeholders based on location, which include the Carlin group, the Twin Creeks Group, and the Long Canyon Group. For the first group, which is geographically close to the examined site, we established levels of influence and their interest in the Emigrant Rain Project, composed stakeholders' matrixes and maps, and identified the stakeholders' engagement risks and stakeholders' engagement methods. For all groups together we conducted a survey to identify the opinions and concerns related to the mining industry in Nevada. The results of this research will be used to develop Good Neighbor Agreements as a general tool for conflict resolution.

Keywords Sustainable mining development • Good Neighbor Agreement • Stakeholder engagement

A. Masaitis, M.S. (✉) • G.C. Miller, Ph.D.
University of Nevada, Reno/0199, Reno, NV 89557, USA
e-mail: Alexandramsts@gmail.com

J.L. Drake et al. (eds.), *Communicating Climate-Change and Natural Hazard Risk and Cultivating Resilience*, Advances in Natural and Technological Hazards Research 45, DOI 10.1007/978-3-319-20161-0_11

11.1 Introduction

Recognition of local community needs and problems is important for reducing conflicts during permitting and operation of mines. The purpose of this paper is to develop and analyze a stakeholder engagement plan as an important step for evaluating the concept of a Good Neighbor Agreement (GNA), which is currently being investigated at the University of Nevada, Reno, for Newmont's Emigrant Rain Mine site, located in Nevada, USA. The goal of this project is to create an open dialog between mining companies and all interested and potentially impacted parties. The goal of a GNA is to minimize the possible conflicts and disagreements and create negotiation tools that can promote open communication between the mining company and stakeholders.

This study utilized a combination of literature and regulatory resources, as well as the primary source of new information, from people living in communities directly affected by mining. This includes employees of mining companies, Native Americans, direct neighbors, local and regional community leaders, regulators, and people in the NGO community who have interests in mining. This study employed cross-cultural research for inquiry across different groups of stakeholders and involved survey, direct observation, and participant evaluation. One of the primary difficulties revealed was the requirement that opposing interests agree to negotiation, which requires recognition that the other participants have a right to participate, and that participation requires giving up some level of control. This initial recognition is expected to be one of the largest barriers to overcome in order for GNAs to be effective.

11.2 Newmont Mining: Introduction of the Company

Newmont Mining Company was founded in 1921 in the USA as a "holding company for private acquisitions in oil and gas, mining and minerals enterprises" (Newmont Mining Company n.d.). Newmont is currently one of the world leaders in gold production with operations in five continents including the USA, Australia, Peru, Indonesia, Ghana, New Zealand, and Mexico. During operations at their mines, Newmont implements a spectrum of mining social responsibility projects. For example, the Newmont contributions to mining sustainable development associated with mining include: S&P 500 Index and Fortune 500, Sourcing Strategy, and Code of Business Ethics and Conduct. Newmont participated in the development of several initiatives: the International Cyanide Management Code, the United Nations Global Compact, the United Nations Declaration of Human Rights and the Voluntary Principles on Security and Human Rights, and the Initiative for Responsible Mining Assurance (IRMA). These projects show the importance of sustainability practices for conducting successful mining operations in both national and international settings. However, establishing critically important relationships with affected stakeholders often requires additional steps to promote good communication.

11.3 Description of the Emigrant Rain Project, Nevada, USA

The Emigrant Rain Project is an open-pit gold mine located approximately ten miles south of Carlin, Nevada, USA. This project has been controversial over the past decade, primarily due to potential impacts on surface and groundwater. The deposit is very near a now-closed acid-generating mine site, the Rain Mine, which remains a source of acid mine drainage that will require treatment for the foreseeable future. The US EPA commented extensively on the project, as did the NGO community, and the associated regulatory community, particularly the US Bureau of Land Management, and these concerns required extensive additional data on the mine. Ultimately the mine was permitted, and from the data developed since it became operational, the likelihood of acid generation is low. However, during those discussions, Newmont and NGOs began discussions on whether this mine could serve as an example for development of a generic GNA. The communities being considered in the present mine site include the Elko, Battle Mountain, Eureka communities, environmental NGOs, and the Western Shoshone tribal communities. While the Environmental Impact Statement for the Emigrant Rain Project showed that nearby communities will have mostly positive economic impacts from the Emigrant Project, many local residents still had concerns about the project.

11.4 Stakeholder List Development for the Emigrant Rain Project, USA

Identification of stakeholders is important for the development of mine social responsibility efforts. One of the difficulties for a project such as the Emigrant Rain Mine is the residents' physical distance from the mine. For the Emigrant Rain Project, the local community is spread over 100 miles from the mine site, and it often is not possible to hold a meeting and involve all residents. Nevada has a strong mining industry, particularly in the northeastern part of the state where the Emigrant Rain Mine is located, and numerous permitting and regulatory actions are taken each year by both federal and state agencies. Thus, there is a general acceptance of mining in the rural communities, and involvement of stakeholders generally reflects specific concerns of specific mines.

Additionally, environmental NGOs that often express the most interest in the mine are located in urban centers far from the mine. Most state government regulators are located in Carson City, which is nearly 300 miles from Emigrant Rain Project. The local ranchers, who potentially are recipients of a larger impact from the mine, are also spread across wide areas. Additionally, many stakeholders simply are uninterested in participating in the communication process, except when direct impacts occur. Involvement of Native American representatives is sometimes particularly difficult, perhaps due to cultural issues of not wanting things to change, but a feeling of powerlessness to affect the complex permitting and technical complexity of mining.

A survey was conducted to assess the views of a wide variety of stakeholders. The two local groups that responded strongly to the survey and can be easily contacted are Newmont employees and people from the surrounding services sector. This is not unexpected, since these groups have a direct economic interest in the mine. The goals for stakeholder engagement should be set before beginning the process. The International Association for Public Participation (IAP2 2007) provided the public participation spectrum for these goals:

1. *Inform* To provide balanced, objective, accurate, and consistent information to assist stakeholders to understand the problem, alternatives, opportunities, and/or solutions.
2. *Consult* To obtain feedback from stakeholders on analysis, alternatives, and/or outcomes.
3. *Involve* To work directly with stakeholders throughout the process to ensure that their concerns and needs are consistently understood and considered.
4. *Collaborate* To partner with stakeholders including the development of alternatives, making decisions, and the identification of preferred solutions.
5. *Empower* To place final decision-making in the hands of the stakeholder. Stakeholders are enabled/equipped to actively contribute to the achievement of outcomes. The methods of engagement will depend on stakeholder engagement goals and usually include surveys, public meetings, workshops, dialog with the government, public comments, and open houses.

The benefits from stakeholder engagement include the development of a dialog between the mining company and community representatives, and good communication is critical to reduce potential conflicts around the mining operation. Other benefits, according to the engagement framework (Stakeholder Engagement Framework 2011, p. 2), are shown in Table 11.1. The general list of stakeholders should include persons and organizations interested in the project that can be involved in the decision-making process regarding the mining operation. The Stakeholder Engagement Framework identified the next outcomes and outputs for engagement (The Stakeholder Engagement Framework 2011, p. 10): "What do you want to achieve at the end of the process; what tangible products do you want to produce from the stakeholder engagement process?"

For the Emigrant Rain Project the lists of local stakeholders were separated into the three different groups and include the Carlin group, the Long Canyon Group, and the Twin Creeks Group based on their distance from the mine and their general location. Stakeholders from Carlin group were located closest to the Emigrant Mine and were separated in the following categories: (1) academic organization, (2) chamber of commerce, (3) Farmer-rancher, (4) federal government, (5) healthcare, (6) industry association, (7) industry partner, (8) local government, (9) media group, (10) nongovernmental organization (NGO) (advocacy on mining issues), (11) nonprofit (nonadvocacy on mining issues), (12) regulator, (13) special interest group, (14) state government, (15) tribal, and (16) utility. Representatives from different categories have a different level of influence on the Emigrant Project. The stakeholder matrix will discuss this in more detail below.

Table 11.1 The benefits from stakeholder engagement

Benefits for the project	Benefits for stakeholders
Higher-quality decision-making	Greater opportunities to contribute directly to policy and program development
Increased efficiency in and effectiveness of service delivery	More open and transparent lines of communication
Improved risk management practices, allowing risks to be identified and considered earlier, thereby reducing future costs	Increasing the accountability of government and driving innovation
Streamlined policy and program development processes	Improved access to decision-making processes, resulting in the delivery of more efficient and responsive services
Greater engagement with stakeholder interests	Early identification of synergies between stakeholder and government work, encouraging integrated and comprehensive solutions to complex policy issues
Ensuring services are delivered in collaboration with stakeholders and provide outcomes which meet community needs	
Enhanced community confidence in projects undertaken	
Enhanced capacity to innovate	

11.4.1 Emigrant Rain Project Stakeholder Map and Matrix

The stakeholder matrix and maps were based on the several characteristics, which were adopted from stakeholder classifications discussed in the ICMM report Breaking New Ground (Breaking New Ground 2002). The stakeholder matrix shows the level of influence and interest in the Emigrant Rain Project. These characterizations are based on common experiences of interest groups involved with mining. Characteristics for the matrix include stakeholders with veto (veto), stakeholders with a right to be compensated (right/compensation), stakeholders with a right to consultation (right/consult), and stakeholders who should be informed (inform). Interest categories represent commitment to status quo vs. openness to change (gradation from "–5" to "5"), where 1 is a low rank and 5 is a high rank.

This gradation was made by the research group, which has professional contacts with stakeholders and ability to measure the level of interest and level of influence to the project for each stakeholder. Influences are based on the number of factors: "strength of their interest in the outcome, their legal rights, their access to external support, or their ability to block the outcome" (Breaking New Ground 2002, p. 354). This numerical characteristic for stakeholders, together with the stakeholder maps, will show the priority of stakeholder groups during the negotiation process, as a step of the Good Neighbor Agreement. For example, stakeholders with level of influence 5, will have more impact on the process than stakeholders with the level of influence 2. An example for the stakeholder matrix with the level of interest and influence for the Carlin group, Nevada, USA, is shown in Table 11.2.

Table 11.2 Stakeholder matrix with level of interest and influence for Carlin group, USA

Group category	Stakeholder name	Characteristic	Interest	Influence
Academic organization	Elko County School District; Great Basin College; University of Nevada, Reno; chamber of commerce: Elko Area Chamber of Commerce	Right/consult Inform	2	1
Farmer-rancher		Right/compensation Right/consult Inform	2	2
Federal government	Forest Service, US Congress	Veto Right/consult Inform	3	5
Healthcare	Elko Clinic, Golden Health Family Med Clinic, Golden Health Family Medical Center, Northeastern Nevada Regional Hospital	Right/consult Inform Veto	−3	1
Industry association	CAT Logistics Shared Services Distribution Center, National Mining Association, Nevada Mining Association	Right/consult Inform	3	1
Industry partner	Barrick, General Moly	Right/compensation Right/consult Inform	2	−1
Local government	Carlin City Government, Elko City Government, Elko County Government; Elko County Sheriff's Office, Elko Fire Department, Elko Police Department, Eureka County Government, Eureka County Sheriff's Office	Right/consult Inform Veto	4	4
Media group	Elko Broadcasting Company, KENV News, KRJC Radio, Ruby Radio Corporation	Inform	2	−1

Nongovernmental organization (NGO) (advocacy on mining issues)	Great Basin Resource Watch, Northeastern Nevada Stewardship Group, Western Shoshone Defense Project	Right/compensation Right/consult Inform Veto	4	3
Nonprofit (nonadvocacy on mining issues)	Elko Bighorns Unlimited, Elko Grammar School #2 PTA, Elko Senior Olympic Games, Elko Velo Cycling Club, Friends of the Elko County Library, Great Basin College Continuing Education, Horizon Hospice, Idaho State University Foundation, National Wild Turkey Federation – Elko, Northeastern Nevada Museum, PACE Coalition, POW/MIA, Spring Creek Christian Academy	Right/consult Inform	1	−2
Regulator	Bureau of Land Management, Nevada Division of Environmental Protection	Right/compensation Right/consult Inform Veto	4	5
Special interest group	Downtown Business Association, Elko Convention and Visitors Authority, Local Union #3, Northern Nevada Regional Development Authority, Spring Creek Association	Right/consult Inform	0	1
State government	Governor's office, Nevada Department of Transportation, Nevada Division of Wildlife Tribal: Elko Band Council, South Fork Band, Te-Moak Tribe of the Western Shoshone	Right/compensation Right/consult Inform Veto	4	5

After evaluation of the level of influence and interest for all stakeholder groups, the results can be presented in stakeholder maps. Mapping is helpful for the visual determination of the priority for the future negotiation process between the public and mining company (see Fig. 11.1).

Following stakeholder mapping, the next step in the process of mine social responsibility development is to analyze the linkage among stakeholders. Finally, the stakeholder engagement methods must be determined. For the Emigrant Rain Project the linkage among stakeholders is not particularly strong, and most communication occurs through public meetings with the local government and mining representatives. The mining social responsibility project should manage these relationships by creating open lines for communication and establishing a dialog among stakeholders, related to their characteristics and level of interest in the mine project.

The stakeholder engagement plan should be based on stakeholders who will want to contribute their resources, time, and availability. Similarly, the mining company needs to understand the goals for the engagement before the developing the stakeholder engagement plan. A certain risk should be recognized for this process and should be included in the stakeholder engagement plan. Some of these *risks* have been identified by the International Association for Public Participation and include the following: (1) stakeholders having a different understanding of the engagement objectives and different expectations about the outcomes of the engagement process; (2) stakeholders feeling excluded from the process, for example, if they are unable to attend engagement activities due to their geographic location; and (3) stakeholders having insufficient time to contribute fully or to raise concerns, for example, due to short project timelines (Stakeholder Engagement Framework 2011, p. 16). For the Emigrant Rain Project we chose the following goal and methods of engagement:

1. *Goal* To keep stakeholders informed about the project and to provide balanced objectives, accurate and consistent information is required to ensure that stakeholders understand the problem, alternative opportunities, and solutions regarding the mine site.

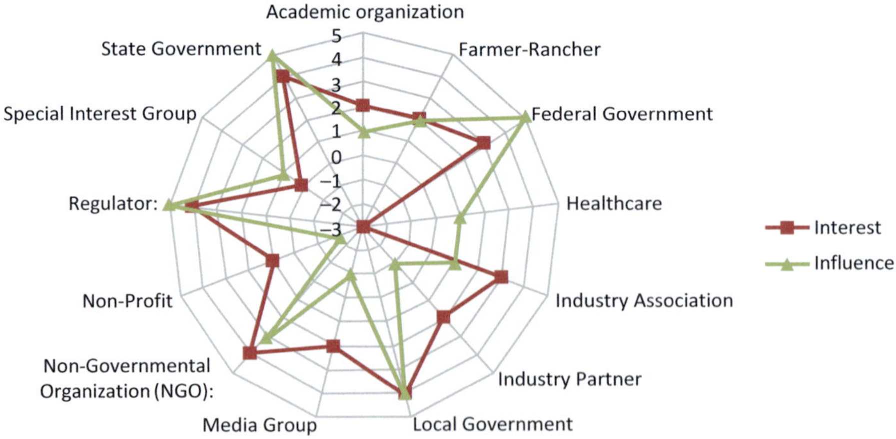

Fig. 11.1 Carlin group stakeholder map

2. *Methods of Engagement* The first method is collection of data related to public opinion about the Emigrant Rain Mine site, as well as major concerns and issues related to the impacts on the environment and community. The next method of engagement is media involvement, which was implemented through local newspapers to keep stakeholders informed. The last method is a series of public meetings to educate stakeholders about the mining operation and create a dialog to discuss the possible alternatives.
3. *Stakeholder Expectations* Managing risk strategy includes a discussion for all stakeholders that engagement may not involve major changes in the mining operation, but is an exchange of information and opinions, which can be used for the future purposes. Expectations for substantial changes in the mine by the stakeholders should not be created, although an openness to hear and consider mine plan changes is appropriate.

In conclusion, stakeholder engagement is a volunteer process and should provide a clear understanding of the benefits for all involved parties, and people should be motivated to participate in the process. The exchange of information regarding the Emigrant Rain Mine between all stakeholders needs to be beneficial for each party, and this information should be used for future decisions between Newmont and stakeholders.

11.4.2 Stakeholder Engagement: Determination of Local Issues Related to Mine Operations

In 2014 the University of Nevada, Reno, conducted a survey among various stakeholder groups who mostly live in Northern Nevada where Newmont has most of its Nevada mines and other surrounding communities regarding the Emigrant Rain Project and mining industry in Nevada. This survey is an initial step for the development of stakeholder engagement. The purpose of the survey is (1) to investigate perceptions of mining in the various communities and the mining industry in Northern Nevada, (2) to understand the level of communication between mining companies and the various publics, and (3) to examine potential methods for improving communication and resolving potential conflicts between neighbors and mining companies, their economies, and the environment.

Participants in the survey include: the employees from Newmont Mining Co.; residents from the Elko, Battle Mountain, and Reno communities; geologists who are working in the different mining sites in Nevada; and Nevada scientific and academic sphere representatives. This is an ongoing survey and the present participants include the following: 59.38 % indicate "I or someone in my family is working for the mining industry," 20.31 % of participants shows that "I am primarily concerned about the social and environmental impacts of mining," and 25.78 % indicate "I am not working in an occupation that is directly associated with the mining industry." This is a continuing survey of groups and individuals who have, or may have, substantially different opinions on mining, and the results should be evaluated in that context. The total number of responses which we received on the current period is 137.

We separated the responses for the periods when surveys were received electronically. A general public opinion survey from Elko, Battle Mountain, and Battle Mountain communities was distributed prior to June 16, 2014, and surveys of the Nevada Geological Society network were distributed on June 30, 2014, and responses from each survey were most received within 5 days of distribution. When this time separation was used to primarily isolate the mining-related responses, the differences were generally minor between the two surveys. While this lack of difference was initially unexpected, the general public surveys were related primarily to the communities in Northern Nevada, where mining is the economic driver of the economies.

The Northern Nevada communities of Elko, Battle Mountain, and Winnemucca exist in a mature mining area, where gold mining has been the dominant economic driver for over 30 years and where both the employment and economy are dependent on mining. Thus, the lack of substantial differences in the survey results is not surprising. Following the review of collected data, the following issues and concerns were established:

1. *Information Availability* Lack of availability of information/knowledge about the mining operation. Residents appear to know in general where they can find information related to a specific mine and the regulations governing mining in Nevada. The question about availability of information relevant to specific mining operations shows that updates of the operational status of the mine and national laws and mining permitting processes are priorities for many of the participants. The question from the survey is: "Would you find it useful if a website containing information relevant to a specific mining operation was available and, if so, what information would you want it to contain?" See results in Table 11.3. The numbers in the table represent the percentage of responses from total responses and the total numbers of answers.

Table 11.3 Interest in the potential website topics

	Extremely useful (%)	Useful (%)	Moderately useful (%)	Low useful (%)	Extremely low useful (%)	Total
Permits	21	41	23	9	4	
						133
Question-answer section	19	50	22	7	1	
						134
State and national law section	13	43	28	11	4	
related to the project						135
Update of the mining status	38	50	8	2	2	
						135

2. *Information Reliability* The level of trust regarding reliability of information regarding mining operations varies for the different sources. Not surprisingly that mining companies scored higher than the other sources, and environmental organizations scored the lowest. The question from the survey is: "How much trust do you have in the information regarding a mining operation received from the following sources?" Complete results are presented in Table 11.4.
3. *Health and Safety Program Awareness* Residents do not appear to have information about health and safety programs which Newmont Mining Company implements in NV, even though Newmont has extensive information available. The information about general background and issues associated with mines is also apparently important for the public. The question from the survey is: "If you were provided an opportunity to learn more about the mining industry, would you be interested in attending an informational event? If so, which of the following topics would you find interesting?" Results are shown in Table 11.5.
4. *Air and Water Quality* The survey also shows that air and water quality are large concerns for respondents. However, they are generally unaware where they can review the permits for operations and how the permits are designed to minimize the impacts on the surrounding community. Permits are complicated and, in general, community members can potentially utilize independent consultants who can educate and create trust between the community and the mining com-

Table 11.4 Level of trust of the various stakeholders

	Extremely high (%)	High (%)	Moderate (%)	Low (%)	Extremely low (%)	Total
Mining company related to their operation	14	46	28	9	3.6	137
Academia/educational sources	11	32	43	11	3	136
Environmental organizations	0.7	9	24	27	40	136
Government, regulatory agency	7	35	34	18	6	136

Table 11.5 Public interest in the potential topics related to the mining industry

	Extremely high (%)	High (%)	Moderate (%)	low (%)	Extremely low (%)	Total
Global mining market	**13**	**31**	**32**	**19**	**5**	131
Understanding of the mining operations and techniques	**20**	**46**	**21**	**11**	**0.8**	132
Understanding of the mining economy	**20**	**41**	**29**	**8**	**2**	133
Environmental impact from mining industry	**25**	**36**	**23**	**12**	**4**	132
Health concerns related to the mining operations	**19**	**38**	**26**	**14**	**4**	132
Mining sustainability practices	**17**	**45**	**27**	**8**	**3**	132
Health and safety on the mine site	**17**	**28**	**37**	**13**	**5**	131

pany related to the water/air quality issues. The opportunity to participate in an on-site inspection of a nearby mining operation was considered useful by half of the participants.

5. *Employment and Educational Opportunities* The large concerns for most participants are the employment opportunities and educational opportunity for the community. The question from the survey was: “If you had an opportunity to participate in a direct dialog with the nearby mining company, which questions and issues you would find most important?” See results in Table 11.6. The numbers in the table represent the percentage of responses from total responses and the total numbers of answers.

33 % of the general public answered that “employment opportunities” are important; from mining relatives: 40 % find it important. The answer “Provide educational opportunity for the community” received equal responses from general and mining participants.

6. *Economic Benefits* People have a clear understanding of the economic benefits, which they accrue from mining operations, where the job availability from a mining company received strong responses. The question from survey is: “In your opinion which of the following is the most important economic benefit of the mining company?” See results in Table 11.7.
7. *Public Meetings* Slightly more than half of the residents have not participated in public meetings related to the mining development. The question from the survey was: “Have you ever participated in a decision-making process regarding the development of the nearby mining operation?” The responses from this question show 49 % for “yes” answer and 51 % for “no” answer.

Table 11.6 Questions and issues related to the mining operation

	Extremely important (%)	Important (%)	Moderate (%)	Low (%)	Extremely low (%)	Total
Employment opportunities	**32**	**38**	**21**	**7**	**2**	130
Provide educational opportunity for the community	**14**	**43**	**37**	**4**	**2**	129
Hire independent experts for consultation about issues related to the mining operation	**9**	**26**	**36**	**22**	**7**	129
Create an open Internet forum for the communication between the mining company and the community	**7**	**29**	**4**	**18**	**6**	131
Negotiate the postclosure plans with mining company	**13**	**29**	**30**	**20**	**9**	127
Permit the regular citizen site inspections and sampling	**4**	**15**	**30**	**29**	**24**	129

Table 11.7 Ranking of perception of the benefits from the mine

Answer choices	Responses (%)	
Development of supply services	**7.75**	10
Professional job opportunities	**55.81**	72
Development of the recreational areas	**1.55**	2
Education support	**8.53**	11
Infrastructure and transportation	**10.08**	13
Others (please specify)	**16.28**	21
Total	**129**	

In summary, the survey shows the need for continued stakeholder engagement. Concerns about mining impacts exist, although community involvement is not high, except for a small number of individuals. The question related to the opportunity for participating in a dialog with mining company representatives regarding a specific mine shows the moderate interest from participants. The question from the survey is: "If given an opportunity, would you be interested in participating in a dialog with mining company representatives regarding a specific mine, recognizing that it takes a commitment of time?" (See Fig. 11.2.)

The results of this survey also reinforce the requirement of defining the list of "stakeholders" in a manner that is complete. In Northern Nevada communities, mining has strong support and forms the basis of the economy. In part due to an arid climate, the impacts are largely localized, and conflicts between agriculture and other land uses are minimized. This lower level of conflicts is likely not the case in areas where existing communities will be affected by water use, water contamination, and associated impacts on fisheries, recreation resources, and agriculture. We have observed that once mining becomes a dominant part of the economy, local support for mining is likely to increase, although not always. Many of the controversial mining projects also involve environmental NGOs that can nationalize the issues and affect how projects are perceived. Indeed, permitting of mines in the USA and Canada involve public participation processes where issues arise from larger concerns. For the Emigrant Mine, in particular, conflicts arose during the early permitting process that involved environmental NGOs, the Bureau of Land Management, and the US Environmental Protection Agency. Local community concerns were not common.

11.5 Good Neighbor Agreement as a Newmont Company Strategy for the Social Responsibility Development

The overall goal of utilizing a Good Neighbor Agreement is to create an open company-community dialog that will help identify and address social and environmental concerns associated with mining, as well as to find methods for communication and conflict resolution. The Emigrant Rain Mine, located in the State of Nevada, USA, is an example for this investigation. A stakeholder engagement matrix, maps,

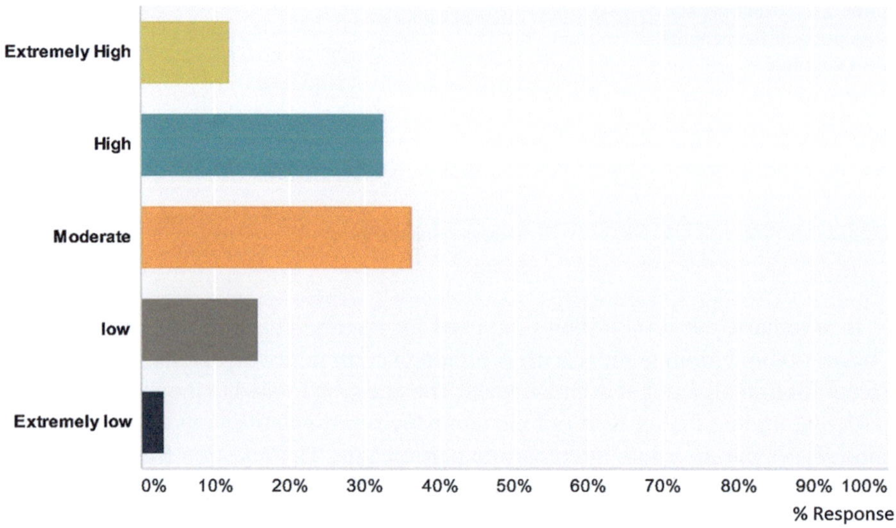

Fig. 11.2 Public interests in a dialog with mining company

and plans are the first step in the Good Neighbor Agreement implementation, while the determination of local issues and concerns is the second step. Interested individuals and groups, as well as residents living near the mine, will have a chance to participate in audit and plant inspections that include, but are not limited to: accident prevention, monitoring programs, emergency response, and regulatory compliance.

Ultimately a GNA will involve discussions and agreements on the variety of issues that the stakeholders feel are important, including such topics as hiring independent experts to interpret data produced by the mine, noise, transportation routes, protection of wildlife, land disturbance, and special aspects of reclamation and closure that may not be covered under existing regulations. However, the first step is to understand who the stakeholders are, whether they are interested in participating and the concerns that they may have regarding a specific mine.

11.6 Conclusion

Conflicts in the development of mining projects are now common between the mining proponents, NGOs, and communities. Newmont Mining Corporation, one of the world's largest producers of gold, recognizes the diverse communities that exist near their mines and recognizes the importance of minimizing conflicts with those communities. However, the highly variable social and economic conditions increase the possibility of conflicts and demonstrate the need to develop responsible mining strategies.

One of the methods that can formalize this process is to establish Good Neighbor Agreements, which deal specifically with challenges in relationships between mining operations and the local communities. This relatively new practice offers a process, which, while it will not be a substitute for regulatory oversight, can be used to address social needs and concerns of local communities that arise during the normal operation of a mine. These agreements will ultimately require compromises from all parties, including the mining companies, but can be integrated into a sustainable community where communication is critical for maintaining trust between the company and the community.

Ultimately, however, the first step of developing the GNA or any practice related to the mining-public relationship is to create a stakeholder engagement plan, establishing a list of stakeholders, which should include all of those interested or affected by the mine project. In the present case, the stakeholders were selected primarily due to proximity and were primarily supportive of the mine. Less emphasized were the more distant stakeholders, particularly from the environmental community, who are likely to have a different opinion on mining in general.

The next step is to analyze characteristics of all these parties and determine the level of interest and level of their influence on the project. This step includes the stakeholder matrix and mapping development. Finally, methods for engagement of the stakeholders should be chosen. The methods usually will depend on the goals of engagement and available resources as time, people, and budget. The goal of this study was to develop a process for engagement, and, while initially controversial, this mine ultimately has not been a major concern for any of the affected public. It is also important to recognize the possible risks from the engagement and include the risk evaluation in the development of the stakeholder engagement plan.

References

IAP2 (2007) International association for public participation. LAP'S2 Public Participation Spectrum. http://www.iap2.org.au/documents/item/84. Accessed 10 Jun 2014

Mining, Minerals, and Sustainable Development Project (2002) Breaking new ground: mining, minerals, and sustainable development: the report of the MMSD project. Earthscan Publications, London

Newmont Mining Company (n.d.) http://www.newmont.com/. Accessed 14 Jun 2014

Stakeholder Engagement Framework (2011) USA department of education. Melbourne, Australia. https://www.eduweb.vic.gov.au/…/oct2011stakeholderengagement.pdf. Accessed 8 Jun 2014

Chapter 12
Controlling Environmental Crisis Messages in Uncontrollable Media Environments: The 2011 Case of Blue-Green Algae on Grand Lake O' the Cherokees, OK

Alicia M. Mason and James R. Triplett

Abstract This chapter documents a content analysis of 62 media reports related to the 2011 blue-green algae (BGA) outbreak on Grand Lake O' the Cherokees, Oklahoma. A three-stage crisis model is used to understand the media framing and crisis communication related to the event. Media reports were categorized according to modality. The data set included: traditional media reports ($n=21$, 33 %), online blogs ($n=7$, 11 %), and online press releases ($n=34$, 54 %). These units of analysis represent both controlled and uncontrolled media representations of the crisis event. The objectives of this analysis are to understand how risk and crisis communication strategies were utilized before, during, and after the BGA outbreak. Five strategies and techniques for improving crisis communication effectiveness are detailed. Limitations and implications are provided.

Keywords Environmental communication • Crisis communication • BGA • Water quality • Media framing

12.1 Introduction

It has been over half of a decade since Cox (2007) framed environmental communication as a *crisis* discipline in the flagship issue of *Environmental Communication: A Journal of Nature and Culture*. He argued that "like perturbations in biological systems, distortions, ineptitudes, and system pathologies occur in our communication[s] about the environment" (p. 10). Today we see flourishing areas

---

A.M. Mason, Ph.D. (✉) • J.R. Triplett, Ph.D.
Pittsburg State University, Pittsburg, KS, USA
e-mail: amason@pittstate.edu

J.L. Drake et al. (eds.), *Communicating Climate-Change and Natural Hazard Risk and Cultivating Resilience*, Advances in Natural and Technological Hazards Research 45, DOI 10.1007/978-3-319-20161-0_12

of environmental communication scholarship relating to environmental rhetoric and discourse, media and environmental journalism, environmental conflict resolution, and crisis communication, among others. This case analysis utilizes the summer 2011 blue-green algae (BGA) outbreak in the waters of Grand Lake O' the Cherokees, to narrowly focus on key aspects of environmental journalism and crisis communication.

Scientists argue that climate change will have many effects on freshwater and marine environments. In some cases and for some communities, harmful algal blooms (HABs) can create crises. According to the EPA (2014), warmer water temperature due to climate change may favor harmful algal blooms in a number of ways. These blooms are made up of blue-green algae (BGA), which have higher temperature tolerances and float on the surface. This allows them to displace the beneficial green algae. Increased water temperatures prevent mixing and allow the water to stratify, which further favors the blue-green algae.

This case analysis employs a three-stage crisis model in order to understand the media reporting and crisis communication related to this BGA outbreak. Richardson (1994) elaborates on the differences in the model components: (1) *precrisis phase* (warning signs are present and attempts are made to eliminate risks), (2) *crisis impact phase* (the crisis strikes and communicative and operational support is provided), and (3) *postcrisis or recovery phase* (stakeholder confidence is restored). Following a procession through this crisis cycle, organizations can return to precrisis planning and preparation efforts.

12.1.1 Overview of the Case and Threat

The Fourth of July is one of the most heavily trafficked and profitable holidays in Northeast Oklahoma's Grand Lake area. Following a harsh winter and tornadic spring, the stage was set for an outbreak of blue-green algae. The Grand River Dam Authority's (GRDA 2011) July 1 announcement "discouraging bodily contact" (para. 1) with the waters of Grand Lake resulted in significant public outrage.

Established in 1935, the GRDA is a quasi-governmental organization, overseen by a board of directors, and designated to represent the people of Oklahoma. The GRDA is the managing authority of Grand Lake, which controls access and investigates water quality issues. The lake-based contagion that was present in the water is known scientifically as *cyanobacteria.* These blue-green algae are considered highly undesirable because they are potentially dangerous to human and animal populations (Turner 2011). While appearing harmless, some cyanobacteria develop into harmful algal blooms (HABs), which produce *neurotoxins* (toxins that affect the nervous system), *hepatotoxins* (toxins that affect the liver), and *dermatoxins* (toxins that affect the skin).

Four key environmental conditions must be present to cause harmful algal blooms: sunlight, warm temperatures, phosphorus, and nitrogen. The HABs resemble green scum, foam, or even paint on the surface of the water. Although HABs are not always toxic, scientists have yet to determine what causes the same species of algae to produce toxins during one bloom and not produce toxins during another (Casaletto 2011). Human poisoning and illness from exposure to BGA is rare, but can be severe.

The most widely documented case of human fatalities resulting from BGA occurred in Brazil, where 69 dialysis patients died after direct exposure to toxins in their dialysis fluid (Goering 1997). In the current case, the GRDA estimated the toxicity of the Grand Lake BGA 18 times greater than the World Health Organization's (WHO) acceptable level. The rarity of the event and uncertainty regarding the contagion posed a challenge for GRDA's crisis communication team.

12.1.2 Comparative Case Scenarios

The Grand Lake community is not the first community to be affected by cyanobacterial HABs, but this was the first time they were impacted. In an effort to understand the regional framework concerning water-based environmental threats, two incidents, the 2003 BGA outbreak on the Marion Reservoir and the 2009 *Escherichia coli (E. coli)* outbreak on Lake of the Ozarks, were selected for a comparative analysis. The rationale for selecting these events was based on: (a) the geographic proximity to the Grand Lake region, (b) the similarity in the nature of the threat (water-based bacteria), and (c) the differences related to the crisis communication responses.

The first case, the Marion Reservoir, is a 200 square-mile section of the Grand Lake watershed located near the headwaters of the Cottonwood River in Marion County, Kansas. Completed in 1968, this US Army Corps reservoir is the public water source for three cities Hillsboro, Marion, and Peabody. In June of 2003, Marion Reservoir experienced the first occurrence of heavy concentrations of blue-green algae (BGA). The US Army Corps notified the community water treatment facilities, who then notified state and federal entities managing public health matters. The corps immediately posted warning signs on the beaches, restricted public access, and provided educational fliers regarding the health risks of exposure to blue-green algae. In addition to this response, the water treatment plant added activated carbon to the treatment process to help remove the toxin from filtering into the public water supplies.

Following the outbreak $1.84 million was invested in the Marion Reservoir water treatment plant. In 2003, the Watershed Restoration and Protection Strategy (WRAPS) program received funding from the US Department of Agriculture EQUIP program and the Kansas Department of Health and Environment Non-point Source program to implement Best Management Practices (BMP) to help local communities improve and maintain high-quality water. Since the initial BGA outbreak over $1 million in federal, state, and local taxes and donations has been allocated to decrease the amount of nutrient-bearing sediments entering the reservoir, which are known contributors to the development of BGA. In 2007, Marion

Reservoir received additional funding to continue studying the watershed in an effort to prevent future BGA outbreaks (Blackman 2009).

In contrast to this approach, during the summer of 2009 the Lake of the Ozarks in Missouri faced a dangerous *E. coli* outbreak. At two of the lake's beaches, 60 water samples were taken on May 26, 2009. Half of samples showed abnormally high levels of the bacteria, *E. coli*, with one sample registering at eight times the maximum allowable level (988.3 *E. coli*). Similar to BGA there are many types of *E. coli* and most are harmless, but exposure to toxic forms can result in several adverse health effects including urinary tract infections, kidney failure, nausea, vomiting, and even death. The Missouri Department of Natural Resources disclosed the high levels of *E. coli* on June 26, one month after initially discovering the contamination (Miller 2010).

Unlike the Marion Reservoir case, DNR officials from Lake of the Ozarks (LOO) did not restrict public access to the water. They instead allowed the public to unknowingly swim in the contaminated water. This strategy is known as *the mum effect*, acting to block the flow of negative or unpleasant information (Tesser and Rosen 1972). It was later determined that the DNR failed to report the high levels of *E. coli* to the public for fear of hurting the lake's economy during the busiest time of the year (Miller 2010). This choice led to a severe public backlash against the DNR. The postcrisis phase included a suspension of the director of the Department of Natural Resources (DNR) for two weeks, a restructuring of the DNR department, and termination of several DNR officials.

These two comparative scenarios gauged the severity, hazard, and potential damage of these water-based environmental toxins differently and resulted in notably different crisis response strategies. The objectives of this Grand Lake case analysis are to (1) understand how crisis communication strategies were operationalized before, during, and after the BGA outbreak, (2) analyze the media framing of the BGA event, and (3) discuss five key communication strategies practitioners and crisis communicators may consider in this context.

12.2 Review of Literature

The belief that general audiences receive most of their information about science from the mass media has been well established (Nelkin 1995; Wilson 1995). The media often play a key role in interpreting scientific findings for the public, in addition to providing key information, selective summaries, and overall assessments of the quality and relevance of the information (Gregory 1989, p. 2–3). There is a general consensus among practitioners that the media provide vital links between individuals, groups, and organizations.

Media gatekeepers determine the newsworthiness of a story based on factors such as the topic's *timeliness*, *relevance*, *proximity*, *prominence*, *rarity*, *trendiness*, and *human interest.* In the context of environmental journalism, several studies have concluded that the economic drivers of the news industry commonly result in event-driven, reactionary coverage of environmental issues. Furthermore, a considerable amount of research indicates that the balanced reporting of issues such as global warming creates informational biases. Traditional news reports that include both sides of the issue, *balance*, have been found to create more uncertainty among audiences regarding climate-related issues. We should note that these are not one-sided communication failures only on the part of the media to relay environmental information. Extant literature has previously identified failures on behalf of scientists to transmit their high-level expertise to journalists (McComas and Shanahan 1999; Ungar 1992; Zehr 2000). As Schwarze (2007) argues, there are risks involved when conveying highly technical, scientific information to lay audiences.

Media are heavily utilized as a channel for transmitting crisis-related messages to groups who may be or are being impacted by environmental crises. Framing theory often serves as a useful theoretical framework for investigating how environmental crises are communicated to audiences. Audience framing involves invoking a "schemata of interpretation" which enables individuals to "locate, perceive, identify, and label" information attended to in the environment (Goffman 1974). The original definition posited by Entman (1993) holds that to frame is "to select some aspect of a perceived reality and make them more salient in a communicating text, in such a way as to promote a particular problem definition, causal interpretation, moral evaluation, and/or treatment recommendation" for the item described (p. 52) (for a review of framing research, see Levin et al. 2002). In this case we use framing theory to understand how the severity of the BGA outbreak and public susceptibility to the threat was communicated to general audiences.

The realm of uncertainty, which shrouds naturally occurring environmental risks that pose public health threats, is a challenging crisis communication context. In an effort to understand how lay audiences may have encountered information relating to the BGA outbreak from the media, the first research question is offered:

RQ1: How did the media frame (a) the severity of BGA's public health threat and (b) the potential public susceptibility?

In order to further understand how the case was presented to audiences, we also aimed to understand how media sources, who served as subject matter experts (SME) or opinion leaders, viewed BGA's threat to the community. A key crisis communication tenet is the need for the development of shared perspectives among stakeholders particularly toward the problem definition, the crisis itself. One guideline for evaluating the risks associated with crisis events is through the application of the precautionary principle.

The precautionary principle was formulated to prevent harm to both the environment and human health. The principle originated from within West German environmental law, *Vorsorgeprinzip*, meaning *foresight* (Freestone and Hey 1996; O'Riordan and Jordan 1995). The precautionary principle posits that when a threat, or *risk*, to human health and/or the environment occurs, precautionary measures must be in place to ethically ensure the safety of the affected publics and regions. Just as some academics have argued the field of environmental communication emerged out of need, the precautionary principle resulted from ongoing systematic and systemic failures of environmental regulations to prevent and protect human health and the environment. Although the principle has been criticized for its subjectivity (Maguire and Ellis 2009), it remains a method for evaluating the decision-making processes when matters of public health and the environment are at stake.

By sharing a perspective, including the evaluation of known risks, crisis communicators, political leaders, and public information officers are more aptly able to speak in one voice. This consistency helps to provide clarity for stakeholders and lay audiences. Stakeholders in this case were defined as groups or individuals with vested interests in the outcomes of the crisis and included residents, visitors, victims, policymakers, business owners, and the GRDA. We wanted to understand if key opinion leaders were able to speak in one voice; therefore, the second research question was posed:

RQ2: Did key opinion leaders (e.g., policymakers, business owners, and the GRDA) speak in one voice regarding the BGA outbreak?

Heath et al. (2007) argued that in the precrisis stage "as people become aware of risks, [a] social amplification of the risk (Kasperson 1992) occurs, during which media coverage and public persuadability [will be] sufficient to motivate mitigation" (p. 39). Providing the public with instructional and psychologically adjusting information within the amplification period of a crisis is crucial in achieving behavioral outcomes which empower stakeholders to cultivate a sense of agency, self-efficacy, for their personal protection. Coombs (2008) distinguished two key components of crisis messages by defining instructing information as that which "informs stakeholders of what to do to protect themselves [both physically and financially] from the crisis" (p. 108), while "adjusting information helps stakeholders cope psychologically with the effects of a crisis" (p. 109). Based on these message content considerations, the third research question is posed:

RQ3: Were there differences in the instructional and adjusting message content provided to general audiences over the course of the BGA outbreak?

12.3 Method

In order to answer the research questions, an exploratory content analysis of 62 media reports was conducted using a grounded theory approach. Grounded theory is a qualitative research methodology often used by scholars to analyze communication processes and content. The approach allows for emergent themes to develop from the data, opposed to using a predetermined frame to categorize data (Scott 2009).

The dates of these reports preceded the first confirmed illness resulting from exposure and continued through August 8, three weeks past the July 13, 2011 date when the public health warning was removed. Media reports surrounding the event were extracted from the LexisNexis database using the search terms *Grand Lake*,

BGA, and *outbreak*. The reports were categorized according to modality. The data set included traditional media reports ($n=21$, 33 %), online blogs ($n=7$, 11 %), and online press releases ($n=34$, 54 %). Five reports were omitted due to duplication. These units of analysis include both controlled and uncontrolled media representations of the crisis event.

12.3.1 Coding Procedures

Reports were coded separately based on their headline and text content. The headline content was coded separately because even if audiences did not read the news article content, they would more than likely be exposed to crisis-related information in the headline (Mason and Wright 2011). Media content was examined for distinct attributes including *quality of instructional/adjusting information*, *media sources*, and *quote comments.* Because BGA has been found to be highly toxic in several animal populations, including feline and canine (Peter 2011), and many residents and visitors maintain relationships with their pets, the inclusion of *animal welfare* information was additionally coded. These units were analyzed by groups of coders, with two to three persons per variable. Reliability for these measures was calculated using Scott's Pi, which discounts the level of *observed agreement* by the level of *expected agreement* due to chance and is commonly used for nominal data in communication studies (Potter and Levine-Donnerstein 1999). Categorical descriptions and calculated reliability scores are provided below.

Headline Content Two categories were constructed for analyzing the headlines (e.g., instructional and generalized headlines). Instructional headlines were coded based on the directional information delivered to the general public: *stay out of water*, *avoid contact*, and *use caution.* Generalized headlines held no specific instructional information for the audience and were coded as *warning/threat announced/issued*, *economic impact*, *warning/threat removed*, or *others.* Reliability was established at .86.

Instructional/Adjusting Information Based on a grounded theory approach, the emerging themes coded were *generalization of the advisory*, *minimization of the threat*, *specific public safety measures*, *acknowledgment of threat*, *possible health symptoms*, *vigilance*, and *warning removal.* Reliability was established at .92.

Media Sources Media sources were classified as *business owners*, *policymakers*, *local residents*, *victims*, *holiday visitors*, *GRDA spokesperson(s)*, *subject matter experts (SMEs)*, and *others (e.g., eyewitnesses).* Reliability was established at .94.

Quote Content The content of the quotes was coded as *economic impact*, *diminishment of threat*, *fear of threat*, *confirmation of threat*, *specific threat location*, and *warning removal.* Reliability was established at .87.

Additional variables related to *animal welfare* (caution and safety information related to animal safety) and *response measures* (i.e., fliers posted, orange tape around impacted areas of the lake, and warning signs), which were coded separately, if present, in each media source. Reliability ranged from .83 to .88.

12.4 Results and Findings

In an effort to answer *RQ1* and *RQ2* and better understand the societal implications of the BGA outbreak, opposed to the ecological, researchers sought to understand how sources were reportedly on-record responding to the BGA outbreak and whether key subject matter experts (SMEs) were able to speak in one voice; therefore, a series of Pearson's Chi squares was computed.

We found significant differences $\chi^2=(35, 51)=71.2, p<.001$ between the sources and their threat response characteristics (see Table 12.1). For example, policymakers and local business owners accounted for 77 % of the responses concerning the economic impact of the BGA outbreak, while policymakers and residents accounted for approximately 90 % of the public comments, which diminished the BGA threat. The GRDA accounted for 72 % of the publicly available comments concerning the confirmation of the threat in the crisis phase, with a marginal amount (4.9 %) additionally resulting from both the residents and others (usually eyewitnesses). Outside of the initial report of human illness, no voices from victims of the BGA outbreak were noted within the data set. The GRDA accounted for a majority (93.4 %) of the threat removal announcements that gave the all clear and demarcated the transition into the postcrisis phase of this event.

RQ3 sought to understand the instructional and psychological adjusting information that was delivered to high-risk populations potentially impacted by this environmental threat. To look at how this information was conveyed over time, a Pearson's chi-square was conducted. Results found significant differences $\chi^2=(15, 29)=46.5$, $p<.001$ between the report date and the form of instructional and psy-

Table 12.1 Crosstabulations for source variations across threat response characteristics

Source type	Economic impact	Diminishment of threat	Confirmation of threat	Fear of threat	Removal of threat	Specification of threat
Business owner	7(2.2)	0(1.3)	0(2.8)	0(.7)	1(.8)	0(.3)
Policymaker	7(2.5)	2(1.5)	0(3.1)	0(.7)	0(.9)	0(.3)
Resident	0(.8)	4(1.6)	3(3.4)	3(.8)	0(1.0)	0(.3)
Victim	0(.3)	0(.2)	0(.3)	0(.1)	0(.1)	0(.0)
Visitor	1(1.7)	1(1.0)	2(2.1)	1(.5)	1(.6)	0(.2)
GRDA	1(5.6)	2(3.3)	11(6.9)	1(1.6)	4(2.0)	1(.7)
SME	0(.6)	0(.3)	0(.2)	1(.7)	0(.2)	1(.1)
Others	1(1.4)	1(.8)	3(1.7)	0(1.4)	0(.5)	0(.2)

Note: expected frequencies appear in parentheses below observed frequencies

chological adjusting information provided. Results indicate that a majority (64 %) of this content was provided to the general audience during the first week of the amplification period, when the GRDA advisory went out. Audiences encountering this information after the Fourth of July would have less of a chance to receive these instructional messages which are meant to facilitate self-efficacy toward protecting their personal health and safety, including identifying possible signs and symptoms. Instructional actions audiences could employ to protect animal populations occurred in only 10 % of the analyzed reports.

As the second week of the crisis unfolded over the Fourth of July holiday period, over 60 % of the statements discounted or diminished the threat. It was not until the third week following the GRDA warning that instructional information regarding the possible symptoms of BGA reemerged in the media reports; this was followed shortly by the official announcement lifting the GRDA's BGA warning.

12.5 Discussion, Implications, and Remedies

A key limitation of this study is the subjectivity in pinpointing the precrisis, crisis, and postcrisis stages of the event, but this is expected in cases such as this. In order to frame the pragmatic value of the findings, this discussion section will focus on applied crisis communication strategies that can be utilized to achieve strategic message control and produce well-balanced information outcomes in the future. Five strategies and techniques for bolstering crisis communication effectiveness are detailed below.

12.5.1 Understanding Controlled Versus Uncontrolled Media

Crises increase the need for organizations to control their messages. Traditional media tactics such as media advisories and press releases are often utilized to convey crisis-related messages to media for the distribution to audiences so they may become knowledgeable and proactive in limiting and mitigating their personal susceptibility to identified threats. These communication tactics are considered to be *uncontrolled* to organizations because media gatekeepers (e.g., editors, producers) control the timing, distribution, and placement of the content. In contrast, websites, internal memorandums, white papers/position statements, and organizational websites are considered to be *controlled* because the organization retains authority in the presentation, content, and the timed release of the information.

Based on the media reports analyzed in this case, the GRDA (2011) spent more time clarifying their initial, abstract warning "discouraging bodily contact with the water" (para. 2) during the amplification period of the crisis rather than explicitly and concisely providing instructional and adjusting information as the crisis unfolded. The GRDA's reliance on uncontrolled media as the primary mechanism for the release of risk information during the precrisis phase resulted in increased uncertainty.

Some audiences received information that specified known HAB locations, while others received information on how to avoid contact with BGA, and others received information on BGA's threat to animal populations. Reporters and editors controlled presentation of BGA-related information in traditional media, where the GRDA's warning was diluted. The media pattern that emerged from this BGA outbreak is consistent with An and Gower's (2009) findings that the most common crisis frames media employ to convey crisis narratives are *responsibility*, *economic*, *conflict*, *human interest*, and *moral*. Policymakers at the local level (e.g., state representatives) and state level (e.g., Oklahoma Governor), along with local business owners repeatedly endorsed and framed this crisis as one of economic impact, opposed to an actual public health risk. In doing so they undermined the scientifically based warning information the GRDA was attempting to provide.

Millner et al. (2011) label third-party sources of critical information during crises as *proxy communicators*. "Although proxy communicators can fill the void created when an organization in crisis chooses to remain mute, this substitution is not without problems" (Ulmer et al. 2014, p. 157). Proxy communicators in this case undermined the communication efforts meant to limit or mitigate public harm. For example, Grove's state representative was quoted in the *Tulsa World* on July 2, 2011, stating, "I'm more concerned about the urine level than I am the blue-green algae level" (Morgan 2011, para. 12).

The issuance of the GRDA public health warning was characterized by the publisher of The Chronicle of Grand Lake as "sensationalized journalism… an irresponsible act of passive terrorism," who further alleged that "not one single case at any medical facility in the Grand Lake area related to BGA" (Ruth 2011, p. 1). These diminishing and deceptive statements occurred during the crisis, and despite initial media reports, which featured the first-known victim of BGA in the Grand Lake area on June 24, 2011 (Andes 2011).

This attitude toward threat diminishment by arguably highly credible sources was also shared among business owners within the Grand Lake community. The president of Arrowhead Investment and Development Co. stated "This has been a major overreaction, in my opinion,…It's very damaging economically," (Morgan 2011, para. 7). The aversion to accepting the risk BGA posed by a variety of proxy communicators, including policymakers and business owners, was problematic. Researchers found that not only were these voices integrated into the framing of the threat, but also the severity of the risk.

It is notable that within the 62 articles included in this sample, at no time did any business entity or policymaker associated with Grand Lake mentions or references the health and welfare of the current residents or holiday visitors prior to their own economic interests. Instead, the media frames that incorporated these voices focused almost solely on the economic impact opposed to the public health threat or environmental risk.

Controlling the statements of proxy communicators in crisis events is a difficult if not impossible task; however, one method for controlling crisis-related information is through the development of ghost sites. Ghost sites are a particularly effective strategy that can be strategically designed in precrisis stages with collaborative efforts from management and computer technology personnel. Ghost sites are webpages that can mask the front page of an organization's website but remain in the dark, or in a ghost state, unseen by public eyes until the time a crisis occurs. Once activated, the strategy allows organizations to convey reliable and timely information to mass audiences throughout the crisis life cycle. This strategy is a formal, controlled means of communication and can often serve as a valuable information resource center for media.

12.5.2 Controlling the "Image" of a Crisis

Attempting to control environmental crisis messages in uncontrolled media environments is a daunting task. Just as the media frame information for audiences, scientists and organizations have a role in creating an interpretative framework of environmental crises for the media. Online media environments contain both text-laden information (e.g., blogosphere) and image-laden information (e.g., imagesphere), which create the need for crisis communicators to control more than words. Current social media users are posting more than 300 million photos per day to Facebook, and 70 % of all actions on social media involve images (Lisbonne 2012).

The rise of the social media site Pinterest© is just one example of the growing emphasis on images in our contemporary visual culture. Pinterest, a site completely composed of images, hit ten million monthly unique visitors faster than any independent website in history, yet most crisis management plans do not include a process of gathering, clearing, and disseminating *official* images during crisis events. In the Grand Lake case the media incorporated a variety of outdated stock imagery and environmental pictures that were not native to the Grand Lake region (Scripps Media 2011). One publication staged a photo op featuring the Oklahoma Governor standing alongside the president of Arrowhead Investment and Development Co., a local business, overlooking the calm, non-green waters of Grand Lake and juxtaposed with a story encouraging people not to cancel their travel plans to the Grand Lake area (Canfield 2011). Strategic communication efforts to provide and disseminate relevant and appropriate crisis-related imagery can be achieved by developing a portfolio of imagery as well as policies and processes for collecting, selecting, and distributing that imagery. Digital media kits are often created to store and package the imagery in preparation for future distribution.

12.5.3 Engaging New Media

It is important to note that no formal responsive engagement from the GRDA or DEQ was found within the data set representing the blogosphere. This silence allowed streams of patently deceptive information from unreliable sources to flow freely, thus creating a reputational threat to the very subject matter experts (e.g., scientists, public health officials, GRDA) who sought to serve public interest. As mentioned earlier, the provision for clear, concrete instructional information and psychological adjusting information for general audiences was virtually absent in the GRDA's warning strategy. Audiences were left questioning the need for precautionary measures as to the protection of their personal safety and animal welfare, and this opened the door of scrutiny from unregulated new-media bloggers who emerged as dominant voices related to this issue. The lack of a coordinated online public notification strategy addressing the threat's severity and the public's susceptibility was problematic. Unfortunately, this is not all that uncommon and frequently results from a lack of professional training as to the value of strategic online crisis communication and the need for monitoring and managing of new media communication systems for both process and evaluative feedback.

As in the Grand Lake case, informal online communication can undermine formal attempts to deliver proactive public health and environmental information. Environmental organizations cannot rely on the media to handle their reputation management in crisis situations. Software technology systems such as Social Mention®, Radian6®, and Sysomos® can assist with media analysis, issues management, reputation monitoring, and social networking engagement during environmental crisis situations. These platforms are valuable in selecting which nontraditional media channels are best to reach stakeholders with timely, reliable, and instant information or refute false, unverified, and unreliable information.

12.5.4 Postcrisis Communication: Emphasis on Learning, Recovery, and Renewal

One of the cardinal tenets in postcrisis management is the need for organizations to deliver postcrisis messages of renewal and recovery that inform and educate members on the steps being taken to ensure the crisis does not reemerge, or, if it does, the steps that are being taken to reduce or mitigate the damage. In this case, no follow-up postcrisis communication related to the event was noted. Only four of the media reports included in this analysis provided any postcrisis response. All were general announcements concerning the removal of threat warnings, and none specifically reported measures meant to prevent or curb the development of future HABs. This finding may have occurred because the data set only extended three weeks past the warning removal date, and a longer time period may have yielded different results. Still, to this date, we have not seen proactive steps, similar to what was deployed in the Marion Reservoir case, to prevent reoccurrence by addressing nutrient loadings or additional state-level grant funding to support the preservation and protection of water quality.

More information has been learned since the spotlight of the media dissipated. In August 2011 sales tax numbers indicated the BGA outbreak had minimal economic impact on the Grand Lake business community (Young 2011). In the postcrisis stage we have also learned the severity of the public health impact (PHI) resulting from the Grand Lake BGA outbreak. According to the Acute Disease Service (ADS) August 22, 2011 report, the PHI was much larger than anticipated. The ADS conducted phone interviews with 49 out of 54 possible BGA exposures occurring between June 5 and August 20, 2011. Of the 37 cases reported in Oklahoma, 20 were directly and negatively impacted by harmful algal blooms (HABs) on Grand Lake. The most common documented symptoms by the ADS were upper respiratory (coughing, shortness of breath, wheezing), gastrointestinal (nausea, vomiting, diarrhea), dermatological (itchiness, rash, blisters), and neurological (numbness, vertigo, vision disturbance). No fatalities were reported.

In May 2012 policymakers enacted Oklahoma Senate Bill 259, which stripped the GRDA's authority to notify the general public about BGA risks. The legislation transferred the authority instead to Oklahoma's Department of Tourism and Recreation, the state agency that promotes the use of state lakes and rivers. SB 259 prohibits *any* Oklahoma state agency or *any* county in Oklahoma from posting BGA advisories or warnings.

In addition to the change in agency duties, the law established a new public disclosure threshold for blue-green algae blooms, eliminating the previous threshold of 20,000 cells per milliliter of water to a new, higher standard of 100,000 cells per milliliter. World Health Organization (WHO) guidelines indicate a moderate probability of adverse health effects from exposure to water containing between 20,000 and 100,000 cyanobacteria cells. That level increases to a high probability of adverse health effects when the cells per milliliter go beyond the 100,000 mark (Carter 2012). Beyond just increasing the amount of BGA that has to be present in the water, the legislation also requires the microcystin, or toxicity, of that BGA to be equal to or greater than 20 parts per billion (ppb) prior to any public warning. Because of this inclusion, Oklahoma's SB 259 is the most stringent piece of environmental legislation restricting public notification of BGA outbreaks in the United States today. In this case there were no stakeholder mobilization, no advocacy groups, no investigative media coverage, and no public outcry prior to the implementation of the law.

12.5.5 Crisis Planning and Preparation

Most generally, stakeholders who collectively face environmental risks share an interest in environmental quality and public health; therefore, these forms of environmental crisis events provide opportunities to motivate community involvement from stakeholders with a wide variety of backgrounds and interests. Bandura (1997) defined this concept *collective efficacy* as a "group's shared belief in its combined ability to undertake courses of action to achieve a goal" (p. 291). Sampson et al. (1997) alternatively defined collective efficacy as the "level of social cohesion among neighbors, and their willingness to take collective action for common good" (p. 292). Depending on the issue, forms of collective efficacy and mobilization may range from letter-writing campaigns, educational outreach, public meetings, protests, and demands on public officials to formulate grassroots efforts.

Through strategic collaboration in postcrisis states, community leaders possess the capacity to foster community involvement from key stakeholders who are likely to be affected in the future. These stakeholders should be involved in the prevention thinking processes. A proactive approach can help foment community-based goals and objectives to aid in bolstering the quality of future crisis communication efforts. This process of collaborative engagement is vital to facilitating the community-building processes that result in collective efficacy.

12.6 Conclusion

The Grand Lake Watershed Alliance Foundation (GLWAF), a nonprofit organization dedicated to preserving, protecting, and improving water quality within the four-state watershed, contends the threat BGA poses remains, as it began raising a red flag in 2004, publishing tests showing elevated nutrient levels polluting the Grand Lake Watershed, but it wasn't until July 1, 2011, that "it made headline news" (Turner 2011, para. 6). Unlike the Marion Reservoir case that limited public access to contaminated water and the mum effect demonstrated by the Lake of the Ozarks DNR, the GRDA issued a notice that allowed residents and holiday visitors to make a personal choice based on known information at the time. Researchers hold that the Oklahoma GRDA used a conservative interpretation of the precautionary principle, which was in accord with public interest; however, beyond just relying on uncontrolled media channels, there was a subsequent failure to provide ongoing instructional and psychological adjusting information as to how key stakeholder groups (e.g., residents, holiday visitors, businesses) could still have fun *at* Grand Lake opposed to *in* Grand Lake, and this in turn increased public outrage.

This analysis found there was an utter failure for government agencies and political officials to speak in one voice in the interest of the environment and public health. Long-term accountability for the policymakers, business owners, and media outlets which placed a priority on economic impact rather than the public health of their consumers, audiences, and voting base is beyond the scope of this case analysis. It is clear that SB 259, the resulting legislation, was designed to substantially reduce the ability of those agencies responsible for health and safety to protect the public in order to manage future potential economic impacts.

Environmental crises are opportunities—opportunities to engage, recover, learn, grow, and renew. Naturally occurring environmental threats will not change unless the environmental conditions that contribute to their emergence change; however, the manner in which key stakeholders respond to crisis events can be augmented. This is where we can use crises, such as this, as case-based exemplars to inform community-based crisis management planning and preparation efforts, as well as improve future crisis communications.

References

An SK, Gower KK (2009) How do the news media frame crises? A content analysis of crisis news coverage. Public Relat Rev 35:107–112

Andes N (2011) Bacteria warning for lake swimmers. KTUL.com. http://www.fox23.com/news/local/story/Blue-Green-Algae-Scare Impacts-Grand-Lake-Business/PRXKhpmJvE6sqeMKUdizzg.cspx. Accessed 4 Oct 2011

Bandura A (1997) Self-efficacy: the exercise of control. Freeman, New York

Blackman P (2009) The Marion Reservoir story. Grand Lake Watershed Alliance Foundation. http://glwaf.org/?p=223. Accessed 3 Nov 2011

Canfield K (2011) Despite advisory, Fallin spending holiday weekend at Grand Lake. Tulsa World. http://www.tulsaworld.com/news/article.aspx?subjecid=12&articledid=20110702_12_0_KE. Accessed 4 Oct 2011

Carter S (2012) Expert: new law for monitoring blue-green algae inadequate for public protection in Okla. The J record. http://www.questia.com/newspaper/1P2-33093760/expert-new-law-for-monitoring-blue- Accessed 12 Mar 2014

Casaletto D (2011). We strongly discourage any body contact with the water at this point. Ozarks water watch. http://www.archive.constantcontact.com/fs075/1102224436468/archive/1106483306152.html. Accessed 4 Oct 2011

Coombs WT (2008) Conceptualizing crisis communication. In Heath R, O'Hair D (eds) Handbook of risk and crisis communication, pp 99–118

Cox R (2007) Nature's "crisis disciplines": does environmental communication have an ethical duty. Environ Comm J Nat Cult 1:5–20

Entman RM (1993) Framing: toward clarification of a fractured paradigm. J Commun 43(4):51–58

EPA (2014) Climate change and harmful algal blooms. http://www2.epa.gov/nutrientpollution/climate-change-and-harmful-algal-blooms. Accessed 14 May 2014

Freestone D, Hey E (1996) Origins and development of the precautionary principle. In: Freestone D, Hey E (eds) The precautionary principle and international law. Kluwer Law International, The Hague, pp 3–15

Goering L (1997) Death by pond scum: blue-green neurotoxin -- algae killed 69 people in Brazil, and the problem could spread. The Chicago tribune. http://community.seattletimes.nwsource.com/archive/?date=19971207&slug=2576812. Accessed 1 Jun 2014

Goffman E (1974) Frame analysis: an essay on the organization of experience. Harper & Row, New York

GRDA (2011) GRDA discouraging swimming in Grand Lake. http://www.grda.com/grda-discouraging-swimming-in grand-lake/. Accessed 4 Oct 2011

Gregory R (1989) Improving risk communications: questions of content and intent. In: Leiss W (ed) Prospects and problems in risk communication. University of Waterloo Press, Waterloo, pp 98–132

Heath RL, Palenchar MJ, Proutheau S, Hocke T (2007) Nature, crisis, risk, science, and society: what is our ethical responsibility. Environ Comm 1:34–38

Kasperson RE (1992) The social amplification of risk: progress in developing an integrative framework. In: Golding D, Krimsky S (eds) Social theories of risk. Praeger, Westport, pp 153–178

Levin IP, Gaeth GJ, Schreiber J, Lauriola M (2002) A new look at framing effects: distribution of effect sizes, individual differences, and independence of types of effects. Organ Behav Hum Decis Process 88(1):411–429

Lisbonne S (2012) Just picture it: the imagesphere. http://techcrunch.com/2012/09/08/just-picture-it-the-imagesphere/. Accessed 2 May 2013

Maguire S, Ellis J (2009) The precautionary principle and risk communication. In: Heath RL, O'Hair D (eds) Handbook of risk and crisis communication. Erlbaum, Mahwah, pp 119–137

Mason A, Wright KB (2011) Framing medical tourism: an examination of appeal, risk, convalescence, accreditation, and interactivity in medical tourism web sites. J Health Commun 16:163–177

McComas K, Shanahan J (1999) Telling stories about global climate change: measuring the impact of narratives on issue cycles. Commun Res 26(1):30–57

Miller A (2010) E. coli at the Lake of the Ozarks: one year later. Vox Magazine. http://www.vox-magazine.com/stories/2010/05/13/e-coli-. Accessed 9 Nov 2011

Millner AG, Veil SR, Sellnow TL (2011) Proxy communication in crisis response. Public Relat Rev 37(1):74–76

Morgan R (2011) Lakegoers urged to stay out of Grand. The Tulsa World. http://www.tulsaworld.com/news/state/lakegoers-urged-to-stay-out-of-grand/article_d6f06b31-6d9b-55b9-9393-ffe2cf967dcc.html. Accessed 1 Nov 2011

Nelkin D (1995) Selling science: how the press covers science and technology. WH Freeman, New York

O'Riordan T, Jordan A (1995) The precautionary principle in contemporary environmental politics. Environ Values 4(3):191–312

Peter ML (2011) BGA poses threat to animals and humans. http://www.petandkennelsupply.com/blue-green-algae. Accessed 10 Mar 2012

Potter W, Levine-Donnerstein D (1999) Rethinking validity and reliability in content analysis. J Appl Commun Res 27:258–284

Richardson B (1994) Socio-technical disasters: profile and prevalence. Disaster Prevent Manage 3(4):41–69

Ruth B (2011) The chronicle of Grand Lake. http://www.grand-chronicle.com. Accessed 10 Aug 2011

Sampson RJ, Raudenbush SW, Earls F (1997) Neighborhoods and violent crime: a multilevel study of collective efficacy. Science 277:918–924

Schwarze S (2007) Environmental communication as a discipline of crisis. Environ Comm J Nat Cult 1:87–98

Scott C (2009) Grounded theory. In: Littlejohn S, Foss K (eds) Encyclopedia of communication theory. Sage, Thousand Oaks, pp 448–451. doi:10.4135/9781412959384.n166

Scripps Media (2011) Blue green algae found on Grand Lake in northeast Oklahoma. Tulsa, Oklahoma. KJRH-TV. http://www.kjrh.com/dpp/news/local_news/blue-green-algae-found-on-grand-lake-in. Accessed 4 Oct 2011

Tesser A, Rosen S (1972) Similarity of objective fate as a determinant of the reluctance to transmit unpleasant information: the MUM effect. J Pers Society Psychol 23(1):46–53. doi:10.1037/h0032881

Turner B (2011) BGA outbreak brings awareness of lake population. GTR newspaper. http://gtrnews.com/greater-tulsa-reporter. Accessed 4 Oct 2011

Ulmer RR, Sellnow TL, Seeger MW (2014) Effective crisis communication: moving from crisis to opportunity. Sage, Thousand Oaks

Ungar S (1992) The rise and (relative) decline of global warming as a social problem. Sociol Q 33(4):483–501

Wilson KM (1995) Mass media as sources of global warming knowledge. Mass Comm Review 22:75–89

Young C (2011) Sales tax numbers show algae had little effect on Grand Lake revenue. Tulsa World. http://www.tulsaworld.com/news/state/sales-tax. Accessed 2 Jan 2012

Zehr SC (2000) Public representations of scientific uncertainty about global climate change. Public Underst Sci 9:85–103

Chapter 13
Characteristics of Extreme Monsoon Floods and Local Land Use in the Lower Mekong Basin, Cambodia

Naoko Nagumo, Sumiko Kubo, and Toshihiko Sugai

Abstract Rivers in the Lower Mekong Basin, Cambodia, receive additional water and waterborne sediments annually during the monsoon. In the 2011 monsoon season, the region was inundated by extreme flood events. This chapter identifies inundation characteristics and explains fluvial landforms developed over a long-term flood history near Phnom Penh and Kampong Thom in Cambodia, based on land classification and field surveys. Understanding the development of fluvial plains and promoting land use in accordance with river behavior can minimize flood risk and allow local populations to live sustainably in flood-prone areas.

Keywords Land classification • Fluvial plain • Lower Mekong Basin • 2011 flood event • Cambodia

13.1 Introduction

Fluvial plains are relatively flat and large areas of land with plenty of secure water resources, meaning they have been utilized for the construction of dwellings and cities over the course of human history. However, fluvial plains are very vulnerable to floods because they are formed by repeated sediment transportation and accumulation during floods. Populations of fluvial plains have used trial and error to balance efficient water resource use and flood risk mitigation. Since many major cities

N. Nagumo, Ph.D. (✉)
International Centre for Water Hazard and Risk Management,
Public Works Research Institute, Tsukuba, Ibaraki, Japan
e-mail: n-nagumo55@pwri.go.jp

S. Kubo, D.Sc.
School of Education, Waseda University, Bunkyo, Tokyo, Japan

T. Sugai, D.Sc.
Graduate School of Frontier Sciences, The University of Tokyo, Kashiwa, Chiba, Japan

J.L. Drake et al. (eds.), *Communicating Climate-Change and Natural Hazard Risk and Cultivating Resilience*, Advances in Natural and Technological Hazards Research 45, DOI 10.1007/978-3-319-20161-0_13

worldwide are currently located on fluvial plains, understanding the fluvial plain formation and the associated risks and benefits is important for future safe, secure, and sustainable living.

The fluvial plains in the Indochina region are more vulnerable to floods than other climatic regions, as rainfall is concentrated during the monsoon season and the topography has a relatively low gradient. Temperature variations are small over the course of a year, but seasonal precipitation levels vary greatly. The monsoon season lasts for several months and river water levels stay high, whereas very little precipitation falls in the dry season.

In the lower large river basins of Indochina, such as the Chao Phraya River in Thailand and the Mekong River in Cambodia and Vietnam (see Fig. 13.1), construction of continuous levees and drainage systems to mitigate flood damage has been insufficient. Therefore, these rivers approximate natural behavior, and the local population occasionally experience severe floods as well as annual seasonal floods. Nevertheless, floodwater is an important resource for local agriculture, fisheries, and everyday life: The population receives not only the risks but also the benefits of frequent floods. Fluvial plains in these areas also store floodwater and retard the flow of excess water, mitigating the flood risk to downstream areas.

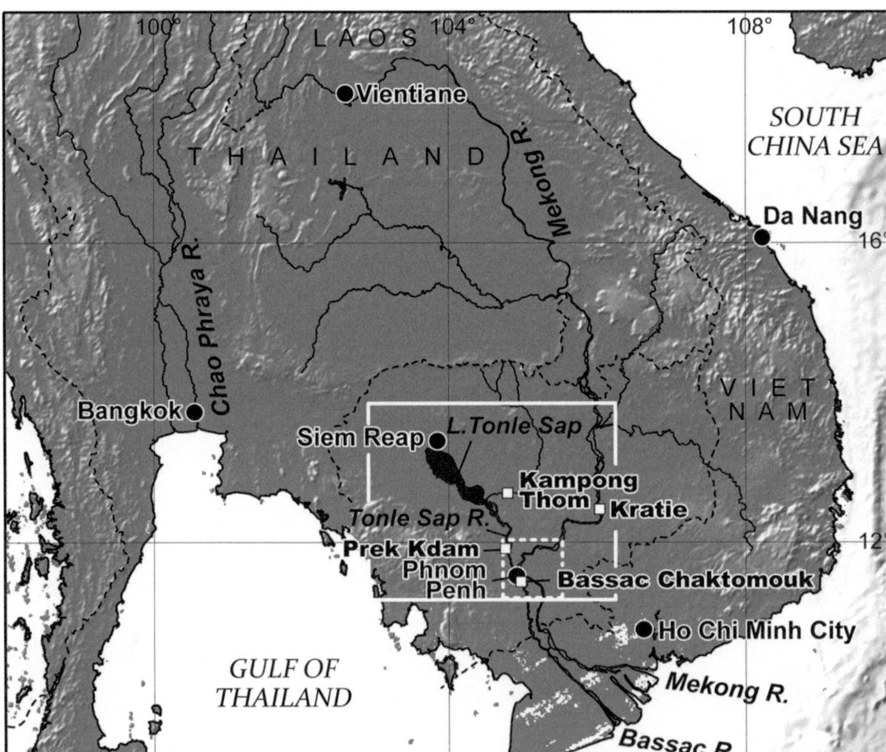

Fig. 13.1 Location map of the study area. *Solid box* exhibits the extent of Fig. 13.2, and *dashed box* means the extent of Figs. 13.4 and 13.5. *Black circles* indicate major cities, and *white squares* mean gauge stations explained in Fig. 13.3

This chapter focuses on the monsoon-affected fluvial Lower Mekong Basin in Cambodia and the 2011 flood event. Discussions are conducted from geographic and geomorphic standpoints to illuminate the relationship between geomorphic characteristics and inundation depth and illustrate how local people have adapted to a monsoon-affected fluvial environment with the associated flood risks and benefits. These discussions could be useful in improving quality of life in flood-prone fluvial areas, and in considering how we should prepare for and respond to hazardous flood events, particularly given the expected future climate changes.

Based on previous geomorphic research by the authors (e.g., Kubo 2006, 2008; Nagumo et al. 2011, 2013), flood characteristics are examined by a field survey of the Phnom Penh and Kampong Thom areas carried out in March 2012. Related data, such as hydrological data, satellite images, and official reports, were also collected. During the field survey near Phnom Penh, floodmarks found on electricity poles and building walls were measured to identify maximum inundation depths during the 2011 flood. These marks are due to floodwater containing suspended sediments that can leave floodmarks for several months, indicating the maximum inundation depths of the previous monsoon season.

13.2 The Lower Mekong Basin, Cambodia

The Lower Mekong Basin includes the area downstream of Yunnan, China, to the South China Sea. Downstream of Yunnan, the Mekong flows through Myanmar, Lao, Cambodia, and Vietnam. The majority of Cambodia is located in the basin. The Mekong River floodplain starts to develop at Kampong Cham, some 70 km upstream from Phnom Penh, and the delta system begins at Phnom Penh flowing toward the South China Sea. At Phnom Penh, the Mekong meets the Tonle Sap River, which flows from a large freshwater lake (Lake Tonle Sap) in Southeast Asia. Soon after the confluence, the distributary Bassac River begins to branch toward the South China Sea (see Fig. 13.1). This X-shaped river confluence and branch system has been termed *Quatre Bras*, which translates as four arms (Kubo 2006).

During the monsoon season (May to October), the region is affected by a southwesterly monsoon, and annual precipitation is concentrated in this season. In response to the marked seasonal precipitation change, the Tonle Sap River has unique flood properties involving the Mekong and Lake Tonle Sap. Although the river flows into the Mekong in the dry season, its flow direction reverses in the monsoon season and transports a huge volume of water into the lake. As a result, the water level and surface area of the lake change dramatically in each season. Masumoto et al. (2007b) found that between 2003 and 2005 the lake surface area increased from c. 1600 to 12,000 km^2 in the monsoon season and that the lake level at Chong Khneas, near the northern coast of the lake, increased to from a minimum of c. 1 m to c. 9 m above sea level. This system ensures reliable rice production and fishery yields for the surrounding areas, and the Mekong–Tonle Sap region is a vital resource for the Cambodian people (Hortle et al. 2004; Mekong River Commission/WUP-FIN 2003, 2007, 2011).

13.3 General Conditions in the 2011 Flood Event

13.3.1 Weather Conditions and Water Level Changes

Between late August and early October in 2011, exceptionally heavy rainfall promoted increased flooding of the fluvial plains in the Indochina region. The monsoon season in Cambodia starts when the Intertropical Convergence Zone (ITCZ) moves northward. In 2011, convective activity of the ITCZ became active between early June and September, and observed precipitation was 1.2–1.8 times greater than average for the monsoon season (Japan Meteorological Agency 2011). Additionally, tropical storms and typhoons frequently visited the South China Sea; in particular, the tropical storms Haima in late June and Nock-ten in late July provided heavy rain to the Lower Mekong Basin (Mekong River Commission 2011). As a result, the area along the Mekong and Lake Tonle Sap were inundated with above-normal monsoon season levels (see Fig. 13.2), and a flood event of a similar scale to the extreme flood in 2000 (e.g., Kubo 2006: Umitsu et al. 2004) resulted.

Water levels for the Mekong in April (dry season) change by c. 15 m at Kratie by October (peak monsoon season) (see Fig. 13.3). However, in 2011, water levels became higher than average with intense precipitation at the gauge stations of the main rivers around Phnom Penh and reached alarm levels. In particular, the water level of the Tonle Sap River at the Preak Dam station reached flood level in October, which is close to the maximum reached in the 2000 event in the latter half of September. At Kampong Thom, along the Stung Sen River, water levels were also higher than normal for the season and reached c. 14 m.

13.3.2 Social and Economic Damage

In response to the heavy rain and increasing water level, local newspapers, such as Phnom Penh Post and Cambodian Daily, posted many articles about the abnormal flood situation and associated damage since the end of August. Articles on these

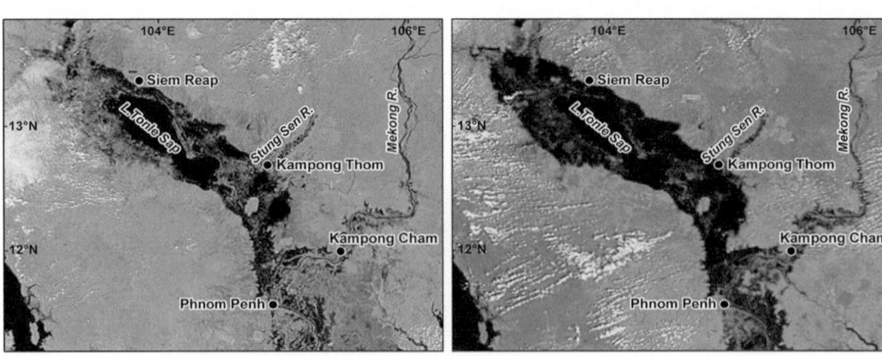

Fig. 13.2 Inundation areas of the 2011 flood event (Original Terra images are from NASA/GSFC, Rapid Response (*left*: 30 Oct 2009, *right*: 30 Oct 2011))

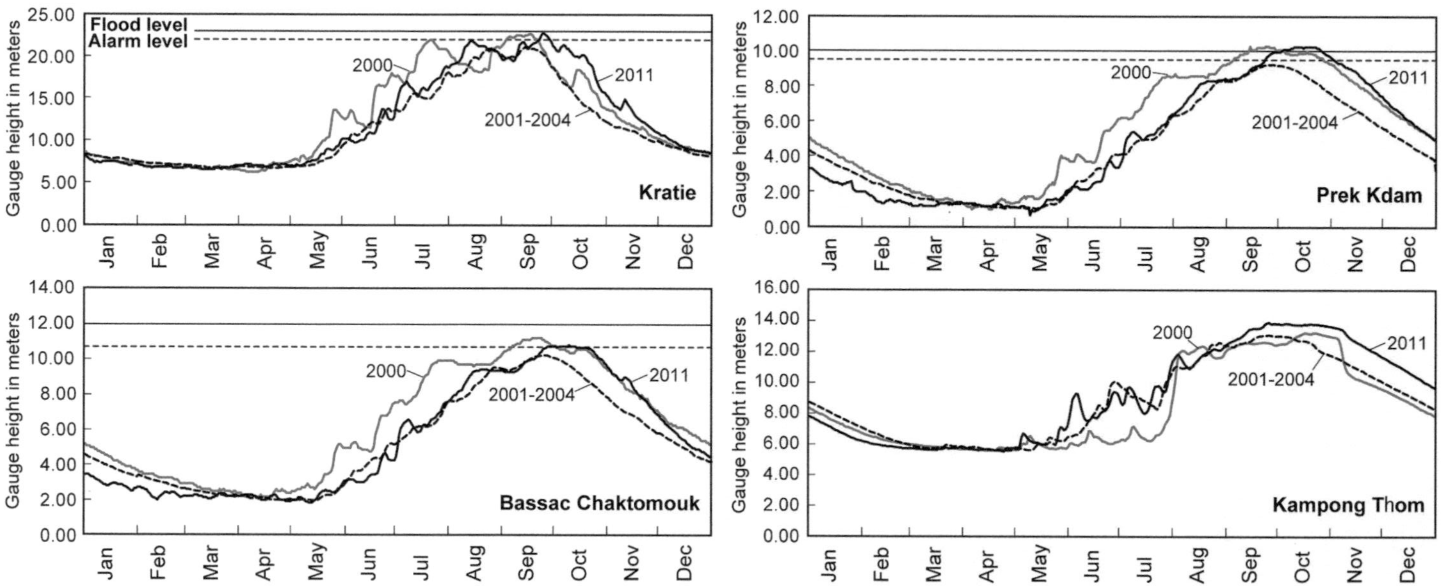

Fig. 13.3 Daily water levels of year 2000, 2011 and average levels between 2001 and 2004 at gauge stations of the Mekong (Kratie), Tonle Sap (Prek Dam), Bassac (Bassac Chaktomuk), and Stung Sen (Kampong Thom) Rivers. The location of those stations is shown in Fig. 13.1. Flood and alarm levels at Kampong Thom are unknown (Original data is provided by Mekong River Commission)

topics were posted until mid-October, when the focus of the articles shifted to the recovery situation. This suggests that flood damage gradually settled down after mid-October.

The National Committee for Disaster Management of Cambodia declared a state of emergency for all provinces and municipalities on 22 September, especially those along the Mekong, Tonle Sap, and Bassac rivers. These areas were put on high alert and requested to conduct flood emergency responses (Asian Development Bank 2012). An announcement on 28 October indicated that 18 out of 24 provinces were affected, 247 fatalities were recorded, and 350,000 households were evacuated.

Additionally, more than 400,000 ha of paddy fields were affected, c. 10.7 % of total crops were destroyed, and various roads and bridges were damaged by the flood (see Table 13.1 for information from Kawamura 2002; United Nations in Cambodia 2011). Foreign tourists visiting the Banteay Srei archaeological site in a rural area near Siem Reap could not return to the city because of the flood and had to be rescued by Cambodian army helicopters.

The total economic damage to Cambodia caused by the 2011 extreme flood was calculated at $620 million (Asian Development Bank 2012). However, this amount

Table 13.1 Major social and economic damage by the 2011 flood event in Cambodia compared to the damage in 2000

Items	Year 2011	Year 2000
Provinces (out of 24)	18[a]	18[b]
Fatalities	247[a]	347[b]
Damaged households	350,274[a] (over 1.64 million people)	750,618[b] (over 3.44 million people)
Evacuated households	51,594[a]	85,000[b] (387,000 people)
Rice fields	423,449 ha[a] (10.7 % of total crops were destroyed)	616,750 ha[b] (including 347,107 ha crop damages)
Schools	1396[c]	Primary schools: 860[b] Secondary schools: 128[b]
Health centers	115[c]	158[b]
Rural water supply	Wells: 77,544[c] Community ponds: 579[c]	Bridges and pipelines: 19[b] Toilets: 5,404[b] Wells, pipes, and pumps: 11,967[b]
Irrigation	Schemes: 329 (partly)[c] Canals: 54 km[c] Reservoir embankment: 122 km[c]	Hydrological stations: 123[b]
Roads	National and provincial roads: 363 km[c] Bridges and culverts: 177[c] Rural roads: 4,470 km[c]	National, provincial, and other roads: 1,500 km[b] Bridges: 115[b] Railways: 1,521 km[b]

Note
[a]United Nations in Cambodia 2011
[b]Kawamura 2002
[c]Asian Development Bank 2012

is considerably smaller than the economic damage to Thailand (World Bank 2012), since infrastructure and industrial sectors in Cambodia are not as developed. Thailand was heavily affected by the 2011 flood in the Chao Phraya River Basin. The flood inundated Bangkok and paralyzed the capital city.

During the flood, people in the Tonle Sap riverside area of Phnom Penh stockpiled sandbags because water levels nearly came to reach the top of the riverbank. Sewage overflowed because of poor rainwater drainage in many places. Also, relatively low areas, such as those around the former Boeung Kak Lake (reclaimed and extinct), were inundated. As a result of the heavy damage and critical situation, Cambodian schools postponed the start of the new semester on 1 October, and the annual boat race of the water festival in the Tonle Sap River in early November was cancelled.

13.4 Geomorphic Characteristics and the 2011 Flood Event in the Phnom Penh Area

13.4.1 Land Classification

Geomorphological features around Phnom Penh are characterized by the three major rivers: the Mekong, Tonle Sap, and Bassac (see Fig. 13.4). These rivers construct a large fluvial plain, with thinner sediment accumulation in the western portion and thicker sediment accumulation in the eastern portion, above the basement rock. The area is roughly divided into five geomorphic groups based on the aerial photograph interpretations and satellite image observations combined with field surveys. Each group has distinct flood characteristics (Kubo 2006, 2008).

Uplands and Alluvial Fans The western section of the fluvial plain is higher than the Mekong River floodplain. A gentle alluvial fan is found along the Preak Tnaot River, and Phnom Penh city is located on the eastern margin of this fan. Terrace-like uplands border the floodplain to the north and south of this fan. The Preak Tnaot River occasionally floods, but only locally, and floodwater from the Tonle Sap and Bassac rivers never affects the alluvial fan. Residual hills (monadnocks) and surrounding pediments are also recognized in the area.

The Mekong River Floodplain Upstream of Kampong Cham, higher uplands and basalt plateaus narrow the floodplain. Moving downstream from this narrowing, the Mekong River channel changes from a braided to a meandering pattern. Channel shifts and the consequent bank erosion are remarkable between Kampong Cham and Phnom Penh, as confirmed by comparing the present channel locations with those shown on old topographic maps. Many abandoned meandering channels are indicated by natural levee distributions along the Preak Kang Chak and Preak Mukh Kampul rivers, which show traces of the former Mekong channel. Around Phnom Penh, combinations of natural levees and back marshes are identified along the

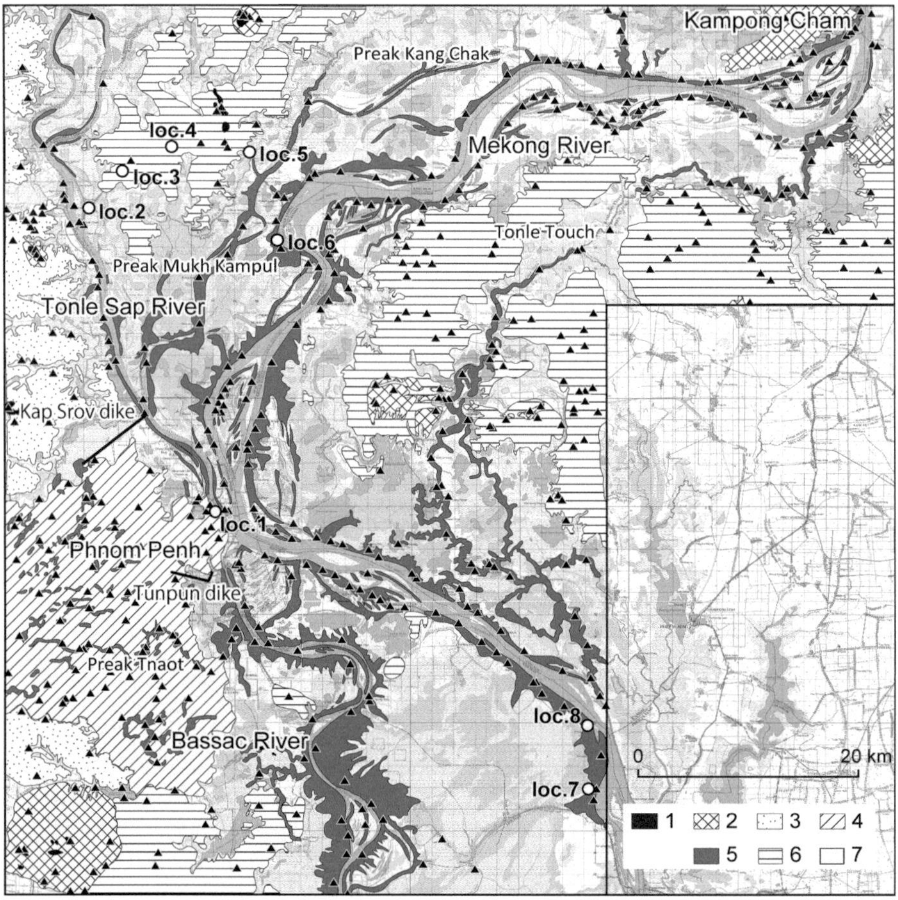

Fig. 13.4 Land classification map and the location of Buddhist temples around Phnom Penh. *White circles* indicate the location of investigated sites of Fig. 13.6, and *triangles* mean the location of Buddhist temples. *1* hill, *2* pediment, *3* upland, *4* alluvial fan, *5* natural levee, *6* higher alluvial surface, *7* back marsh (Modified from Nagumo and Kubo 2013. Base map is black-and-white 1:100,000 topographic maps, and land classification is by Kubo (2006, 2008))

Mekong, but the natural levees are comparatively small (less than 1000 m wide). Major roads and villages are located on these narrow natural levees, and available land that is free from flood risk is limited.

Tonle Sap River Floodplain The Tonle Sap River flows along the western margin of the fluvial plain, and the flow direction reverses between the dry and monsoon seasons. Although the channel is stable, natural levees are poorly developed, and back marshes found between the Tonle Sap and the Mekong are lower and swampier than those of the Mekong River floodplain. In the Phnom Penh city area, the river channel bank is artificially covered by concrete, and the Kap Srov dike is located in the northeast of Phnom Penh to prevent flooding of the city area.

Bassac River Floodplain The Bassac River diverges from the Mekong at Phnom Penh. It has extensive natural levee-like features that radiate from the main channel. These slightly higher features are developed along a number of irrigation canals, called *colmatage*, connecting the Bassac to the back marsh behind it. Monsoon floods transport deposits into the back marsh from the river and construct slightly higher geomorphic surfaces along the canal. Floodwater from the Bassac River rarely overflows because the water is led into the back marsh through the canal. The artificial Tunpun dike is located on the right bank of the river, and it has an important role in protecting Phnom Penh city from floods when combined with the Kap Srov dike in the Tonle Sap floodplain.

Higher Alluvial Surfaces Alluvial surfaces that are slightly higher than the surrounding floodplain are found on the left bank of the Mekong between the Mekong and Tonle Sap rivers. The border between these surfaces and the floodplain is indistinct and is generally defined by the land use transition from paddy fields to swampy areas. Major roads and villages are located on this surface. Along the Mekong and Tonle Sap, characteristic small dikes, called *tomnup*, are constructed around the edge of this geomorphic surface to store water for rice cropping during the early stage of the dry season.

13.4.2 Inundated Areas

The Tonle Sap Lake located in northwest Cambodia has the function of a natural retarding basin. When the lake receives floodwater in a normal monsoon season, it expands several times larger than its area in dry season, but the inundation area rarely reaches to Siem Reap and Kampong Thom. However, in the 2011 flood event, larger areas were inundated when compared to images from 2009 (see Fig. 13.2), and floodwater almost reached both Siem Reap and Kampong Thom.

The majority of the Tonle Sap floodplain that is narrowed by uplands was inundated, because floodwater flowed into higher alluvial surfaces that are rarely inundated during a normal monsoon season (see Fig. 13.5). Along the Mekong River, floodwater overflowed and covered the floodplain downstream of Kampong Cham. Natural levees remained above the water surface, but the majority of back marshes and higher alluvial surfaces on the left bank were inundated. Natural levees along the Preak Kang Chak and Preak Mukh Kampul rivers escaped the flood.

Downstream from Phnom Penh, floodwater inundated back marshes on the left bank, but continuous natural levees prevented the inflow of floodwater on the right bank. Additionally, floodwater did not reach natural levees along the Bassac River or the right bank of the Tonle Sap River. Uplands and gentle fans, combined with the Kap Slov and Tunpun dikes, were effective in preventing flooding of the city area.

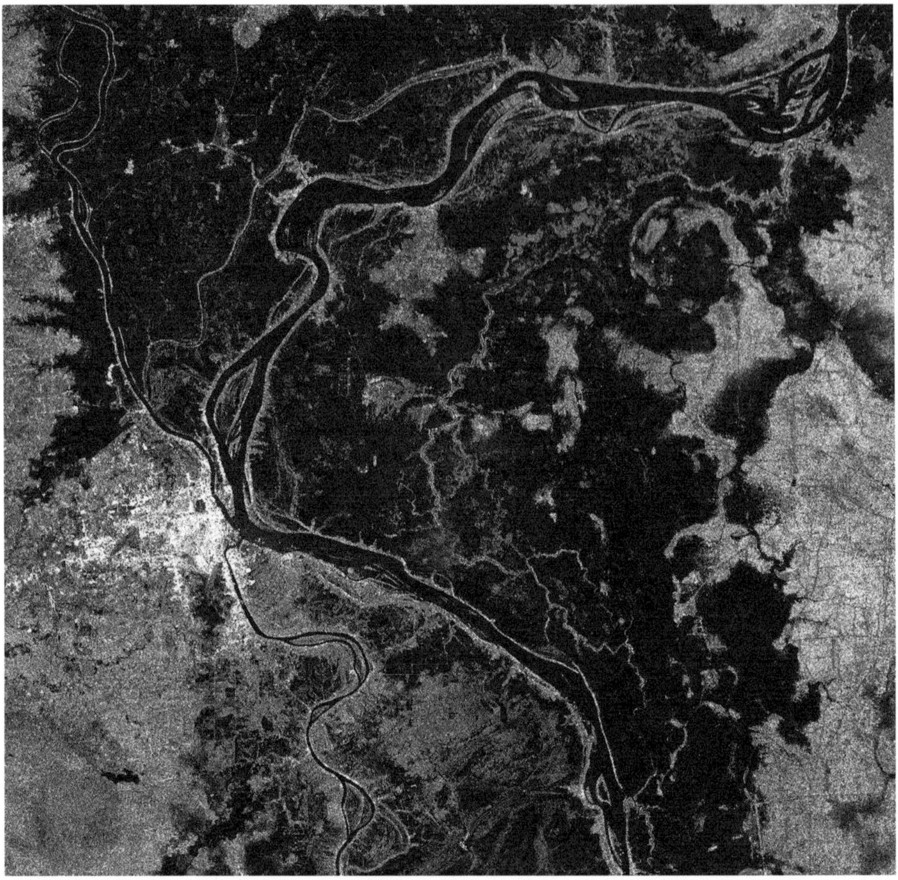

Fig. 13.5 Radarsat-2 imagery of 8 October 2011. The dark areas are all inundated (Original data provided by the Mekong River Commission)

13.4.3 Geomorphic Characteristics and Inundation Depths

The majority of the Cambodian plain was formed by erosional processes, and fluvial sediments are relatively thin. Monsoon floodwaters and sediments flow into the basin are stored temporarily in Tonle Sap Lake and are gradually drained downstream from the end of the monsoon season through the dry season. Although the Mekong River floodplain becomes wider around Kampong Cham to the downstream, the upland borders fluvial plain in the western side and the Preak Tnaot River approaches the Mekong, constructing an alluvial fan. Therefore, the eastern area of fluvial plain plays a major role in the storage of floodwaters. Quaternary deposits are considerably thicker in the eastern side of the fluvial plain and have reached >130 m (Kubo 2008).

Inundation depths, observed by field surveys, depend principally on the geomorphic surface (see Fig. 13.6). For example, inundation depths of natural levees (loc. 6, loc. 8) are very small (c. 1 m), implying that natural levees are seldom affected by normal monsoon floods. Higher alluvial surfaces (loc. 3, loc. 4), which are rarely inundated, showed inundation depths of 2–3 m. This is in agreement with the water level recorded at Preak Dam, which unusually reached flood level.

The back marsh of the Tonle Sap River (loc. 2) showed the largest inundation depths of more than 4 m, whereas the Mekong back marsh (loc. 7) was around 2 m. This difference is due to uplands on the right bank of the Tonle Sap River approaching the river channel and flared alluvial surfaces and hills on the left bank narrowing the floodwater path (see Fig. 13.2). Although much of the floodwater from the Mekong and Tonle Sap rivers flowed into the lake, the narrowed area blocked some water, increasing the flood plain inundation depth.

In addition, natural levees on the concave bank along the Mekong (loc. 6) were heavily affected by bank erosion. The maximum horizontal erosion rate of this bank was indicated at more than c. 20 m/year (Oketani 2008). On concave banks of natural levee areas along the Mekong, bank erosion also should be considered as the primary risk.

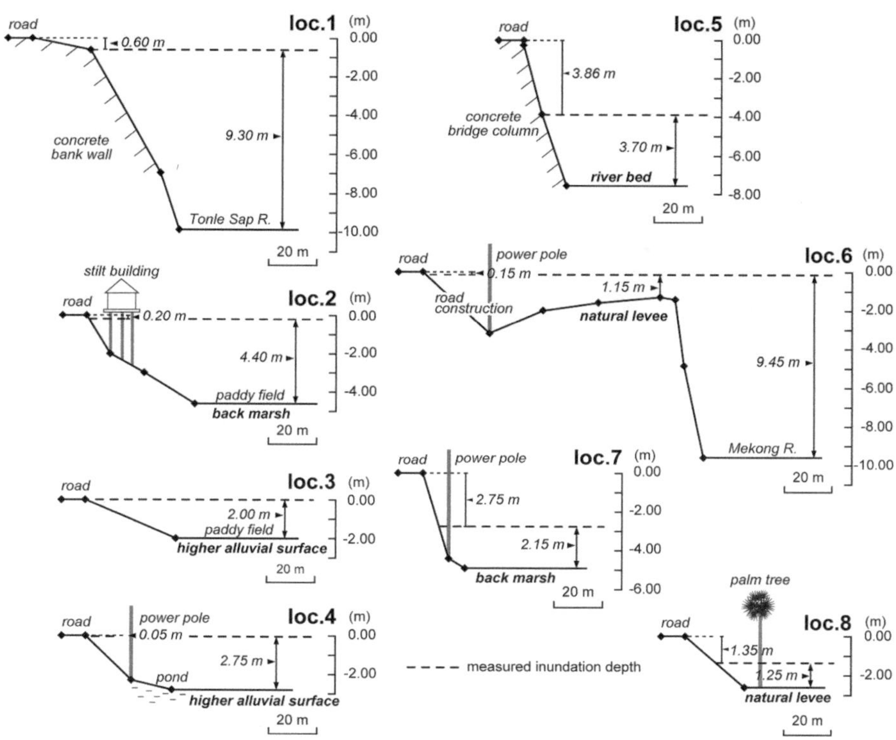

Fig. 13.6 Geomorphic cross sections and inundation depths (Modified from Nagumo and Kubo 2013). Laser range finder is used for field measurements

Although much of Phnom Penh coincides with the confluence point of the Mekong and Tonle Sap rivers (see Figs. 13.4 and 13.5), flooding did not occur much. Phnom Penh is located at the edge of a gentle fan (Preak Tnaot River), which means flooding was relatively limited and caused by rain-fed water and drainage failures. The Kap Srov and Tunpun dikes also played an important role in preventing floodwater flow into the city.

As described above, correspondence between geomorphic characteristics and inundation depth was observed. When a large flood event occurs, such as the 2000 or 2011 floods, floodwater flows along the Mekong River floodplain and inundation depths in back marshes increases along the Tonle Sap River. Furthermore, higher alluvial surfaces that are not affected by normal floods are at risk of inundation. Populations along the Tonle Sap River should be aware of the high inundation risk, and along the Mekong River, the erosion risk at concave banks should be considered. Phnom Penh was relatively safe because it is located at the edge of an alluvial fan. However, if the failure of the drainage system continues and the sprawl of the city into the surrounding back marsh is accelerated, the city may face an unprecedented flood hazard.

13.5 Geomorphic Characteristics and the 2011 Flood Event in the Kampong Thom Area

Kampong Thom is one of the major regional centers in central Cambodia and is located on the lower Stung Sen River basin. This river is the largest tributary of the Tonle Sap system, which drains an area of c. 16,000 m^2 and flows into Lake Tonle Sap (see Figs. 13.1 and 13.2). In the upper sections, tributary streams combine to increase the flow volume in a mountainous area; in the middle sections, the river meanders through an incised valley with decreased gradient; in its lower sections, it has a 7-km-wide floodplain, flows into the lake plain, and has a meandering channel, 7 m deep, with a rectangular cross section and a very flat (0.06/1000) longitudinal profile. Water levels at Kampong Thom typically reach c. 13 m, approximately twice the dry season level of c. 6 m (Nagumo et al. 2013).

The floodplain around Kampong Thom is roughly divided into back marsh, meander scrolls, and abandoned channels (see Fig. 13.7). Back marshes are further divided into back marsh I, II, and III. Back marsh I is the highest surface and is rarely inundated even in the monsoon season; back marsh II is partly inundated by monsoon floodwaters; and back marsh III is fully submerged in the monsoon season and remains sporadically swampy in the dry season. Meander scrolls and abandoned channels are well developed, which implies that strong traction promotes the replacement of deposits and channel migration, whereas natural levees are rare. Accumulation of back marsh is relatively slow, with increments of 0.1 mm/year during the Holocene (Nagumo et al. 2011, 2012, 2013).

In the 2011 event, floodwater overflowed from the river and suspended materials spread from the northeast to the southwest over the floodplain, inundating back marsh II as well as back marsh III (see Fig. 13.7). This was exceptional as floodwater overflowed at the lower portion of the floodplain and rarely reached back marsh II during usual events.

The furthest downstream area is affected by both lake expansion and upstream floodwater from the Stung Sen River. In 2011, as a result of the unusual volume of floodwater from both upstream and downstream, efficient drainage was prevented and large volumes of excess water were stored near Kampong Thom.

During large flood events, floodwater spreads over the floodplain causing significant inundation over a long period. Additionally, floodwater allows suspended sediments to accumulate, which is consistent with the observed constant accumulation rate regardless of the distance from the river channel (Nagumo et al. 2011). However, overflow rarely occurs and floodwater seldom spreads over the floodplain, so back marsh accumulation rate is very low. Nevertheless, the back marsh accumulation probably increases during exceptional flood events such as that seen in 2011, and the high concentration of sediments associated with this event has raised the surfaces of back marshes II and III.

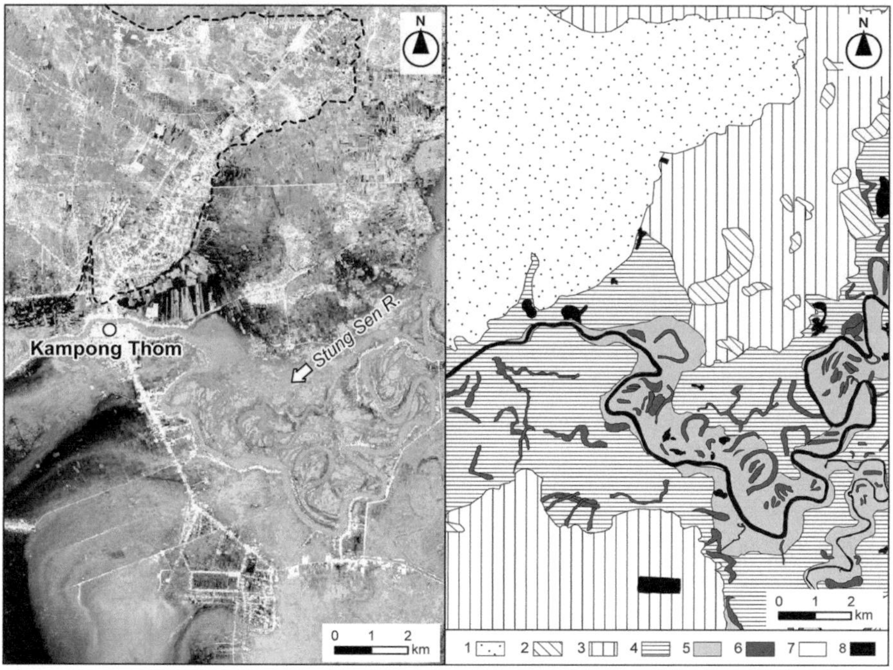

Fig. 13.7 Inundation areas and land classification map around Kampong Thom. *1* upland, *2* back marsh I, *3* back marsh II, *4* back marsh III, *5* meander scroll, *6* abandoned channel, *7* sandbar, *8* water surface (Modified from Nagumo et al. (2012). *Left*: black-and-white GeoEye-1 imagery of 30 October 2011, *right*: land classification map)

13.6 Local Lifestyles with Seasonal and Extreme Flood

As we have seen so far, monsoon floods occur annually, whereas extreme flood events often happen owing to climatic and geomorphic conditions in the Lower Mekong Basin. Therefore, effective land use to minimize the flood risk and maximize benefits is essential for this area.

In Fig. 13.4, the distribution of Buddhist temples is shown overlaying a land classification map of the area around Phnom Penh. The distribution of temples is regarded as the distribution of traditional settlements, because the majority of Cambodians are Buddhists, and therefore most traditional settlements have a Buddhist temple.

Temple distribution is concentrated on natural levees along the Mekong and Bassac rivers, higher alluvial surfaces on the left bank of the Mekong, a few natural levees developed along the Tonle Sap River, and the gentle alluvial fan of the Preak Tnaot River. These locations escaped severe inundation during the 2011 event, but very few temples are located in the Mekong and Bassac back marshes, Tonle Sap back marshes, and higher alluvial surfaces, which were heavily inundated during the event. This distribution suggests that local people know experimentally which areas are safe from not only seasonal floods, but also exceptional flood events, and have consciously selected settlements with relatively high elevations.

Deposits in Cambodian Lower Mekong Basin transported by monsoon floods are mainly composed of sand and silt, and the erosion risk due to strong traction is relatively low except for some concave banks. Floodwater drains slowly over several months, providing water and fresh soil for paddy fields and rich fishery yields. Since the normal maximum water level is almost constant, stilt houses of sufficient height to clear the maximum water level (combined with the use of small boats) have popularly been constructed in the floodplain. Even if surrounding areas are inundated, the floodwater moves relatively slowly in this area, and stilt houses are rarely washed away. Therefore, the population lives comfortably during the monsoon season, with an abundance of water resources, showing that the stilt house may be the most sustainable house type in areas where seasonal long-term floods and occasional exceptional flood events occur.

Land uses in back marshes inundated for long periods along the Mekong and Tonle Sap rivers are not seen and retain natural conditions. In contrast, in other back marshes that surround higher alluvial surfaces and natural levees, floodwater is used for flood-recession rice planting at the end of the monsoon season. Fringing the edge of higher alluvial surfaces, small *tomnup* embankments of square or arc shape store water in the higher alluvial surface during the monsoon season. When the monsoon season finishes and floodwater retreats, part of the embankment is artificially cut and stored floodwater is used for rice production in surrounding strip-shaped paddy fields in the back marsh.

In contrast, rice production in higher alluvial surfaces and alluvial fans is rain-fed and conducted predominantly in the monsoon season. In some of these areas, grid *Pol Pot canals*, constructed during the Cambodian civil war, are easily found.

Along the Bassac River, characteristic *colmatage* canals have been constructed perpendicular to the river to bring floodwater from the river into the back marsh. These canals are said to have been developed during the French colonial period for irrigation and land reclamation and contribute to the transportation of deposits and the formation of levee-like surfaces. As a result, floodwater goes through the canal, and the land that is less susceptible to inundation is maintained along the canal. This land use is typical for this region (Kubo 2006; Masumoto et al. 2007a).

Although agricultural land use predominates and flood control measures are almost entirely absent in Cambodia, traditional land use, dwelling locations, and house structures in floodplain areas are well devised and adapted to cyclic environmental changes. Large areas of the floodplain become the path for floodwater, and some of this water is stored as a valuable resource for dry season food production and livelihood. Controlling floods in these areas by artificial embankments is not realistic. Suitable lifestyle selection for populations in flood-prone areas, as demonstrated in Cambodia, is important, particularly for those populations facing the effects of future environmental changes.

13.7 Conclusion: How Can We Coexist with Hazardous Floods?

This research shows geomorphic and flood characteristics of the Lower Mekong Basin in Cambodia, focusing on the 2011 flood event. From the results of field surveys and a review of previous research, clear relationships between inundation depth and geomorphic surface type were recognized. Additionally, land classification was shown to be effective in predicting flood behavior. Characteristic local land use and lifestyles were found to reduce flood risks and even benefit from cyclic flood events.

The possibility of extreme flood events is higher for fluvial plains than other geomorphic settings. Predictions of near-future flood threats are necessary to minimize risk through understanding past flood events and factors that affect floods. However, in highly developed fluvial plains, populations are excessively concentrated in vulnerable areas, and ongoing rapid developments are ignoring natural geomorphic conditions. Land conditions for inhabited sites have become more vulnerable and are often linked to increased flood potential.

For instance, Tokyo, the capital of Japan, extends on a fluvial plain facing Tokyo Bay. Annual precipitation for Tokyo is around 1600 mm, and the rainfall is concentrated from early summer to early autumn owing to frontal passage and typhoons. The length of rivers from mountain headwaters to the river mouths is comparatively short, and the longitudinal profiles are steep compared to continental rivers. Therefore, water and sediment levels increase rapidly when it rains, and mitigation of flood risk is challenging.

Since World War II, the population density in Tokyo has accelerated, and large-scale developments and the use of underground spaces have increased. At the same time, residential areas expanded in floodplains and even in former river channels and marshy areas. Artificial levees and drainage facilities are necessary to prevent flooding of these areas, and the function of urban rivers is specialized for the drainage of excess water. As severe floods with dike breaches have not happened over the past decades for the major rivers around Tokyo, people tend to be unconcerned about serious flood events.

Just as urbanized Thailand suffered more damage than Cambodia in 2011, densely populated and highly urbanized cities on fluvial plains, such as Bangkok and Tokyo, also have the potential to suffer significant flood damage. If people live expecting annual floods, as in Cambodia, a flood would not to be a disaster. Understanding the characteristics and development of fluvial plains, and promoting land use in accordance with river behavior, allows for sustainable living in flood-prone areas. In densely populated and highly urbanized cities on fluvial lowlands, is this way of life possible?

References

Asian Development Bank (2012) Flood damage emergency reconstruction project, preliminary damage and loss assessment. http://www.adb.org/projects/documents/flood-damage-emergency-reconstruction-project. Accessed 13 Nov 2012

Hortle KG, Lieng S, Valbo-Jorgensen J (2004) An introduction to Cambodia's inland fisheries, Mekong development series, no. 4. Mekong River Commission, Phnom Penh

Japan Meteorological Agency (2011) Heisei 23 nen 10 gatsu 12 nichi houdou happyou siryou. [News release on 12 Oct 2011] http://www.jma.go.jp/jma/press/1110/12a/world20111012.html (in Japanese). Accessed 10 Apr 2012

Kawamura M (2002) 2000 nen Mekong gawa ryuuiki kouzui saigai no fukkyuu jyoukyou to kongono kouzui taisaku [Restoration situation of 2000 flood event and future flood control in the Mekong River Basin]. Kokudo Kensetsu Bousai 12:92–109 (in Japanese)

Kubo S (2006) Mekong gawa karyuu heiya (Cambodia) ni okeru bichikei to kouzui tokusei, tochi riyou, misu riyou no tokushoku [Microtopography and flood characteristics, land use, and water use in the Lower Mekong Basin (Cambodia)]. Gakujutsu Kenkyu, Geogr Hist Soc Sci 54:1–9 (in Japanese)

Kubo S (2008) Geomorphological features and subsurface geology of the Lower Mekong Plain around Phnom Penh city, Cambodia (Southeast Asia). Revista Geográfica Acadêmica 2:20–32

Masumoto T, Ogawa S, Horikawa N, Rikimaru A, Kubo S, Somura H, Pham TH, Kamal R (2007a) Development of prediction models of the change of agricultural water use in the Asia monsoon region. Bull Natl Ins Rural Eng 46:67–90 (in Japanese with English Abstract)

Masumoto T, Tsujimoto K, Somura H (2007b) Hydro-meteorological observation and analysis of observed data at Tonle Sap Lake and its environs, urban and paddy areas. Tech Rep Natl Ins Rural Eng 206:219–236 (in Japanese with English Abstract)

Mekong River Commission (2011) Flood situation report 2011. Mekong River Commission technical paper 36. Phnom Penh

Mekong River Commission/WUP-FIN (2003) Draft final report—modelling Tonle Sap for environmental impact assessment and management support. WUP-FIN Phase 1—Modelling of the Flow Regime and Water Quality of the Tonle Sap. Finnish Environment Institute, Helsinki

Mekong River Commission/WUP-FIN (2007) Final report—Part 2: Research findings and recommendations. WUP-FIN Phase 2—hydrological, environmental and socio-economic modelling tools for the lower Mekong Basin impact assessment. Mekong River Commission and Finnish Environment Institute Consultancy Consortium, Vientiane, Laos

Nagumo N, Kubo S (2013) 2011 flooding and fluvial micro landforms in the Cambodian lower Mekong plain. E-Journal GEO 8:141–152 (in Japanese with English Abstract)

Nagumo N, Sugai T, Kubo S (2011) Characteristics of fluvial lowland in lower reach of the Stung Sen River, central Cambodia. Trans Japan Geomorph Union 32:142–151

Nagumo N, Sugai T, Kubo S (2012) 2011 nen Cambodia kouzui to Sen gawa karyuu heiya no bichikei [Year 2011 Cambodian flood and microtopography in the lower Stung Sen River]. Chiri 57:18–24 (in Japanese)

Nagumo N, Sugai S, Kubo S (2013) Late quaternary floodplain development along the Stun Sen River in the lower Mekong Basin, Cambodia. Geomorphology 198:84–95

Oketani S (2008) Cambodia, Mekong karyuuiki no dosha no shinshoku, taiseki to hitobito no sei-katsu [Sedimentary erosion and accumulation, and people livelihood in the Lower Mekong River basin, Cambodia]. Chiri 53:110–115 (in Japanese)

Umitsu M, Nguyen VL, Ta TKO (2004) Geo-environment and flood 2000 in the Mekong River delta. J Facult Lit, Nagoya University History 50:57–69 (in Japanese with English Abstract)

United Nations in Cambodia (2011) Cambodia-flood season situation report #8, Monday 5 Dec 2011. http://reliefweb.int/report/cambodia/flood-season-situation-report-8. Accessed 24 Jan 2012

World Bank (2012) Thai flood 2011, rapid assessment for resilient recovery and reconstruction planning. https://www.gfdrr.org/sites/gfdrr.org/files/publication/Thai_Flood_2011_2.pdf. Accessed 19 Apr 2013

Chapter 14
The Value of Earth Observations: Methods and Findings on the Value of Landsat Imagery

Holly M. Miller, Larisa O. Serbina, Leslie A. Richardson, Sarah J. Ryker, and Timothy R. Newman

Abstract Data from Earth observation systems are used extensively in managing and monitoring natural resources, natural hazards, and the impacts of climate change, but the value of such data can be difficult to estimate, particularly when it is available at no cost. Assessing the socioeconomic and scientific value of these data provides a better understanding of the existing and emerging research, science, and applications related to this information and contributes to the decision-making process regarding current and future Earth observation systems. Recent USGS research on Landsat data has advanced the literature in this area by using a variety of methods to estimate value. The results of a 2012 survey of Landsat users, a 2013 requirements assessment, and a 2013 case studies of applications of Landsat imagery are discussed.

Keywords Landsat imagery • Value of information • Socioeconomic benefits • Remote sensing • Earth observations

14.1 Introduction

The US Geological Survey (USGS) provides a variety of Earth observation information to the public at no cost, such as data from the streamgage and Landsat programs. These data are used extensively in managing and monitoring natural resources, natural hazards, and the impacts of climate change. Given the budget difficulties and potential data gaps faced by Earth observing systems around the

H.M. Miller (✉) L.O. Serbina • L.A. Richardson
US Geological Survey, Fort Collins, CO, USA
e-mail: millerh@usgs.gov

S.J. Ryker • T.R. Newman
US Geological Survey, Reston, VA, USA

J.L. Drake et al. (eds.), *Communicating Climate-Change and Natural Hazard Risk and Cultivating Resilience*, Advances in Natural and Technological Hazards Research 45, DOI 10.1007/978-3-319-20161-0_14

globe, understanding the socioeconomic and scientific value of these data will be useful to decision-makers for the next decade and beyond.

The process of valuing these data provides a better understanding of the existing and emerging research, science, and applications related to the information provided by these programs. In order to value information, the first step is to identify how the information is used and by whom. This information can be used to target user needs, communicate with users, and, ultimately, improve the provision of the information and identify which applications should be supported. Recent USGS research on Landsat data has advanced the literature in this area by using a variety of methods to estimate value.

Landsat satellite imagery has long been recognized as unique among remotely sensed data due to the combination of its extensive archive, global coverage, and relatively high spatial and temporal resolution. Since the imagery became downloadable at no cost in 2008, the number of users registered with USGS has increased tenfold to more than 50,000 registered users, while the number of scenes downloaded annually has increased more than a hundredfold to over 5 million scenes. It is clear that the imagery is being used extensively, and understanding the benefits provided by open access to the imagery can help inform decisions involving its provision.

While there have been some studies on the value of Landsat, it is not fully understood. To further this understanding, we have used a variety of approaches, both qualitative and quantitative, to evaluate the benefits obtained from the use of Landsat imagery. Three projects will be discussed in this chapter:

1. a 2012 online survey of Landsat users registered with USGS
2. a 2013 requirements assessment to identify operational or decision-supportive uses of Landsat
3. a 2013 case studies of the application of Landsat in water resources

14.2 Challenges in Valuing Landsat Imagery

There are a variety of challenges in assessing the value of Landsat imagery. Landsat is used in a wide variety of applications by hundreds of thousands of people, implying substantial value. For example, Landsat imagery is used extensively in Google Earth,[1] a program that has been downloaded over a billion times. However, a full valuation of the imagery would require a comprehensive knowledge of all the users and uses of the imagery. This knowledge is difficult to attain, given the public-good nature of the imagery.

Landsat imagery has characteristics of a public good because multiple users can use the same image at the same time, and, once an image is made publically available, there is little or no cost associated with providing that same image to additional users. Once the imagery is downloaded, it can be distributed to other users without

[1] Any use of trade, firm, or product names is for descriptive purposes only and does not imply endorsement by the US Government.

attribution or incorporated into derived products then distributed to secondary users. The free and open data policy enacted in 2008 resulted in rapidly increasing use accompanied by new types of analysis and use of the imagery in new applications. New users and applications of the imagery are constantly identified (e.g., Miller et al. 2011, 2013), but it is impossible to identify all the downstream users.

In addition, USGS and other private and public entities are developing value-added products such as the LandsatLook Viewer, a full-resolution browsing tool, that make the data accessible to an even wider variety of users. These trends of increasing users, uses, and products indicate the value of Landsat is also increasing. Monitoring and assessing the constantly changing use and value of Landsat imagery represent a unique challenge. Another challenge stems from assigning dollar figures to the value of a public good such as Landsat imagery.

Frequently, monetary value is the most desired and requested type of value by policymakers, but estimating such a value can then be used to argue that the data should no longer be available at no cost. However, the enormous increase in use and applications (and subsequently, in the value generated by the use of the imagery) could not have occurred without the free and open data policy. Despite the barriers to assessing the value of Landsat imagery, the imagery is characterized by a number of positive attributes, which can be used to estimate or demonstrate the value. The more-than-40-year global archive of imagery is unique. When compared to other remotely sensed data, Landsat stands out as the only data available to perform time series analysis over decades almost anywhere in the world. Landsat is used by many people in many different applications (Miller et al. 2013).

For some people, this widespread use clearly demonstrates value; others need more evidence. Landsat users are often willing to demonstrate the value of the imagery. Pursuing case studies and other qualitative work can be successful since the community already recognizes the importance of communicating the value of Landsat. While estimating the value of information from sources such as Landsat imagery continues to be a challenge, there is a range of existing qualitative and quantitative methods that can be used to communicate the value of data and information sources that do not have a market price. There is a growing interest among researchers, practitioners, and policymakers regarding the valuation of data and information sources such as Landsat. For example, the Socioeconomic Benefits Community is a recently formed community of practice focused on assessing the benefits of geospatial and environmental information (http://www.socioeconomicbenefits.org) and the theme of the 2013 Geospatial World Forum was '*Monetizing geospatial value and practices*.' The Socioeconomics Benefits Community and the Geospatial World Forum are not connected in any way

14.3 Valuations of Landsat in the Literature

There is a small but growing literature on the value of Landsat that utilizes a variety of approaches. Broadly, value can be assessed quantitatively or qualitatively. Quantitative methods use numerical units, which are measurable, comparable, and well defined, such as dollars, hours, or lives saved. Monetary values are easy to compare to the costs of the Landsat program, making a cost-benefit analysis potentially possible. Numerical values are also easy to understand, and the value is often immediately apparent. However, it is not possible to quantitatively estimate the total value of the Landsat program. The free and open data policy makes it difficult to track the users and uses of the imagery, which in turn makes it almost impossible to account for all of the benefits generated by Landsat. Additionally, quantitative values are not always the most effective way to engage an audience, particularly if the value was obtained using complicated analyses.

Qualitative value is typically demonstrated through narratives or stories that focus on difficult to quantify benefits, such as the provision of baseline information, the facilitation of decision-making processes, or the ability to ask and answer previously unsolvable research questions. Some may consider qualitative data more readily obtainable than quantitative data. For instance, a case study of a single organization or specific application can be achieved via a few in-depth interviews. This type of research can provide a good story and context for quantitative data. Numbers often mask a richer narrative that can be compelling on its own. However, qualitative data are harder to compare and often stand alone, though they can be presented in one package. It can also leave people asking for a more quantitative measurement of value. A good story can be useful for communication purposes, but for policy and funding purposes, a quantitative value is frequently requested as well.

There have been few quantitative valuations of Landsat imagery, but those that have been conducted have shown significant monetary benefits from the use of the imagery. The most common approach to valuing Landsat has been to demonstrate the cost savings that occur from using the imagery. The Mapping EvapoTranspiration at high Resolution with Internalized Calibration (METRIC) model was developed by the Idaho Department of Water Resources and uses Landsat thermal data, along with data from other sources, to measure agricultural water use as an alternative to traditional methods, such as power consumption coefficients. METRIC is used extensively by western states (Serbina and Miller 2014), either in lieu of or along with traditional water monitoring practices (e.g., measuring headgate flows, traveling to a location to see if it has been irrigated or not). METRIC was originally estimated to cost 80 % less than traditional methods (Morse et al. 2008), but the actual savings have been closer to 90 % (Landsat Advisory Group 2012).

The Landsat Advisory Group (LAG 2012) found substantial potential cost savings from the use of METRIC across the western United States, estimating that \$20–\$73 million could be saved annually. LAG found similarly large cost savings across a variety of Federal and State governmental applications of Landsat imagery. They estimated a savings of \$178–\$235 million annually over 10 applications, ranging from consumptive water use and forestry to agriculture and flood mitigation. Using this approach, costs are relatively easy to calculate and to compare, resulting in defensible numbers. This approach does not take into account any benefits created by the use of the imagery above and beyond that of the cost savings. For instance, an improvement in data quality could lead to a more accurate analysis and thus myriad resultant benefits.

Another approach to valuing Landsat is a cost-benefit analysis. Booz Allen Hamilton (BAH 2012) conducted a cost-benefit analysis using a variety of case studies from many applications of Landsat imagery. Each case study addressed a major use of the imagery and used existing data to calculate the benefits provided by Landsat in that use. A cost-benefit analysis using a Bayesian network approach was used to estimate benefits. BAH determined the return on investment for Landsat to be more than $119 billion over a 15-year period (2003–2018). Cost-benefit analysis is a common approach that tends to be well understood. It covers a lot of different application areas but still results in an aggregate value and can be an efficient approach if existing data are used to determine the benefits of Landsat. There is no available estimate of the contribution of the value of information provided by Landsat to existing economic activity. This analysis used the 1 % rule based on an estimate by Nordhaus (1986), but there has been no research to determine if this is appropriate to apply to the use of Landsat imagery in multiple applications.

The potential revenue from the use of Landsat can also be valued. Bernknopf and his colleagues (2012) used a variety of data, including Landsat thermal data, to determine where the greatest risk of nitrate contamination of drinking water wells was in northeast Iowa. They then used this information to determine the optimum pattern of crop distribution (soybeans and corn) to avoid well water contamination and maximize yield. The integrated assessment approach incorporated models for agricultural production, nitrate leaching, groundwater nitrate dynamics, groundwater protection, and economic optimization. They found that if the optimum crop distribution was applied, the maximized yield would result in an increased profit of $858 (±197) million annually, with a net present value of $38.1 (±8.8) billion into the indefinite future. This approach demonstrates the relative value of Landsat imagery when it is combined with other data, which is most often how Landsat is used. It provides multiple levels where economic benefit could be accrued. For instance, the value of a map that assesses the probability of well contamination could also be quantified based on its contribution to the prevention of health problems. In order to realize the economic benefits of the crop distribution application, new policy would have to be enacted to encourage farmers to plant certain crops (e.g., incentives for planting corn instead of soybeans or leaving land fallow). As the authors state, it is difficult to know the extent to which crop optimization could be obtained, but it will most likely not be perfect, thus reducing the estimated benefits.

In addition to estimating the benefits created by the use of Landsat, the impacts of the loss of Landsat can also be used to assess value. In 2006, the American Society for Photogrammetry and Remote Sensing (ASPRS) conducted a survey of remote sensing professionals (Green 2008). ASPRS estimated an annual loss to respondents of $936 million if Landsat imagery were not available. It is often easier for people to estimate losses rather than calculate benefits, though the value questions asked may have been difficult for some users to answer due to their broad nature.

In addition to the quantitative valuations of Landsat imagery, there are myriad examples of qualitative approaches. These typically are case studies focusing on a specific use or user of Landsat. AmericaView, an organization that supports the use of Landsat through research and education at the state level, produces fact sheets describing the use of and benefits from Landsat (find examples at http://www.americaview.org/downloadable-fact-sheets). A recent NASA (2013) publication describes the benefits of using Landsat in fire, land use and land cover change, water, agriculture, ecosystems, and forestry applications. Qualitative benefits include providing baseline data, increasing accuracy, improving long-term mitigation and planning, and detecting change over time. These qualitative case studies are useful in communicating the applications and direct benefits of Landsat relevant to individual decision and policymakers, as well as providing context to quantitative data.

14.4 Recent USGS Studies of the Value of Landsat

In a continuing effort to assess the value of Landsat imagery, the USGS has undertaken several different projects. The three projects described here each take a different approach to estimating the value of Landsat and provide a mix of qualitative and quantitative data. They each have their advantages and disadvantages and provide different types of complementary valuations, which can be used in diverse policy situations.

14.4.1 Survey of Landsat Users

In April 2012, a survey was sent to 44,731 Landsat users registered with the USGS Earth Resources Observation and Science (EROS) Center (Miller et al. 2013). The response rate to the survey was 30 % with 11,275 current (having used Landsat in their work the year prior to the survey) and 2,198 past Landsat users responding. The results reported here apply to current users registered with USGS. The value of Landsat imagery was measured in a variety of ways. The importance of, dependence on, and impacts of the hypothetical and actual loss of the imagery were explored. The monetary value of Landsat was estimated using the contingent valuation method, a nonmarket approach to valuing public goods.

On average, respondents used Landsat in 46 % of their work. Three-quarters of the users stated that Landsat imagery was somewhat or very important in their current work, and three-quarters also indicated that they were somewhat or very dependent on Landsat to complete their work (Miller et al. 2013). Importance and dependence were found to be related to the level of use of Landsat imagery in work. More than 60 % of heavy users of the imagery (using Landsat in 71 % or more of their work) were very dependent on the imagery, compared to a quarter of light users (using Landsat in 30 % or less of their work; see Fig. 14.1). More than 65 % of heavy users considered Landsat to be very important to their work, compared to a quarter of light users. Dependence also correlated to the percentage of work using Landsat that was operational (continuous or ongoing work that either relies on the consistent availability of Landsat imagery or is mandated). Very dependent users

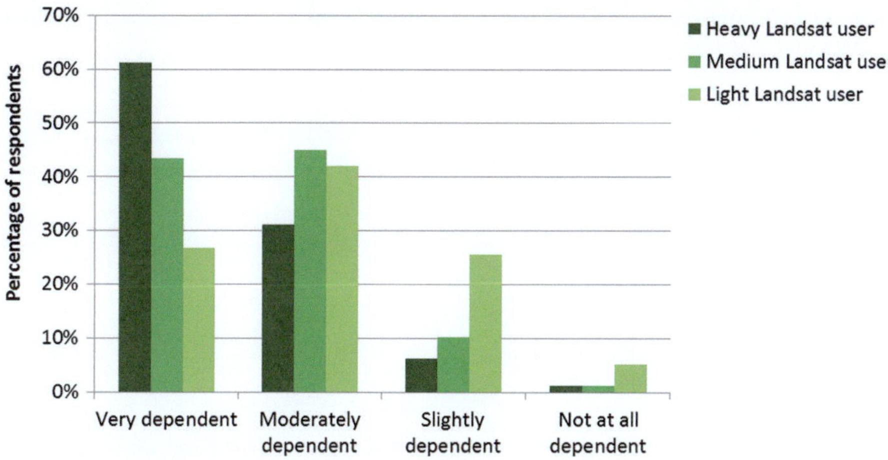

Fig. 14.1 Dependence on Landsat imagery among users with varying levels of use of Landsat imagery (Reproduced from Miller et al. 2013, p. 22, Fig. 14)

estimated a higher percentage of their work to be operational (40 %) compared to users who were not at all dependent (24 %).

Users were also asked to respond to a hypothetical situation in which all Landsat imagery (new and archived) was no longer available. More than 65 % of the users reported that they would have to discontinue at least some of their work; on average, these users would discontinue half of their work, indicating a strong dependence on the imagery (Miller et al. 2013; see Table 14.1). The majority (86 %) would use substitute information for some of their remaining work. Satellite imagery was the most common substitution, but other data sets and fieldwork would also be used by more than half of those who would use substitute information. The use of often expensive and time-consuming fieldwork as a substitute data source seems to indicate that fieldwork might be the only viable alternative for certain types of data.

During the time the survey was being administered, one of the two Landsat satellites, Landsat 5, was not acquiring new imagery. Landsat 7 was acquiring new imagery at the time but delivering images with missing data due to a technical issue. This unique situation provided an opportunity to determine the actual responses of users to the loss of one segment of Landsat imagery. The majority of users (79 %) had used Landsat 5 imagery in the year prior to the survey, and more than two-thirds of Landsat 5 users (69 %) believed their work had been impacted negatively by the loss of Landsat 5 imagery, whether due to reduced quality or scope of work or more time spent on work (Miller et al. 2013). A quantitative measure of the loss was the amount spent on imagery or other data to replace Landsat 5 data. Though only 8 % of the Landsat 5 users spent money on substitute information in the 30 days prior to the survey, they spent an average of $3,765. The population of users registered with EROS spent an estimated $9.4 million in the 30 days prior to the survey to obtain imagery or other data to replace Landsat 5 data.

The contingent valuation method (CVM) was used to estimate the economic value derived from the use of Landsat imagery, known as consumer surplus.

Table 14.1 Percentages of current Landsat users who would take certain actions if Landsat imagery were no longer available

Action taken if Landsat was no longer available	Current Landsat users	US users	International users
For work that uses Landsat, percentage of users who would...	*n≥5,903*	*n≥1,703*	*n≥4,200*
...discontinue some of work	66 %	66 %	66 %
...use substitute information in some of work	86 %	83 %	87 %
...continue some of work without substitute information	57 %	46 %	61 %
For those who would use substitute information, percentage of users who would use...	*n≥4,779*	*n≥1,301*	*n≥3,478*
...different imagery	98 %	97 %	98 %
...other data sets	59 %	49 %	63 %
...on-the-ground fieldwork	57 %	46 %	61 %

Reproduced from Miller et al. 2013, p. 24, Table 8

Consumer surplus provides the best measure of societal benefits resulting from government programs (Office of Management and Budget 1992). CVM is a survey-based stated preference or intended behavior technique recommended for use by Federal agencies (US Environmental Protection Agency 2000; US Water Resources Council 1983). Users were presented with a hypothetical, though probable, scenario in which both Landsat 5 and 7 became inoperable before Landsat 8 was launched, creating a data gap. As mentioned previously, Landsat 5 was, at the time, not acquiring new imagery, so this situation had already partly come to pass. If a data gap were to occur, users may have had to obtain imagery elsewhere. Users were asked if they would be willing to pay a certain dollar amount (each respondent received a predetermined amount ranging from $10 to $10,000) for a scene equivalent to a Landsat scene in the event of a data gap. They were instructed to assume they were restricted to their current budget, and the money to pay for this imagery would have to come out of that budget.

Differences between certain user groups in the benefits they received from the imagery were hypothesized. A similar user survey conducted in 2009 (Miller et al. 2011) revealed significant differences in benefits among sectors, and similar differences were expected for this survey. Many users were new users (43 %), or those who had never used Landsat before it became available at no cost in 2008, and these users were expected to receive fewer benefits than established users (41 %), or those who used Landsat consistently both before and after the no-cost data policy was enacted (Miller et al. 2013). There was also a group of returning users (16 %) who were similar to the new users; the two groups were combined for the CVM analysis. Lastly, the users were divided into US and international users to better understand the benefits accrued by each group.

The analysis looked at four groups (US and international established and new/returning) to determine the economic benefits for each one. The differences in

Table 14.2 Annual aggregate economic benefits to Landsat users registered with the US Geological Survey from Landsat imagery distributed by the Earth Resources Observation and Science (EROS) Center in 2011

Landsat user group	Number of scenes obtained in 2011 from EROS	Average economic benefit per scene	Annual economic benefit (millions)	Lower bound (millions)
US users				
Established	1,687,600	$912	$1,539	$1,399
New/returning	692,508	$367	$254	$236
US total	*2,380,108*		*$1,793*	*$1,635*
International users				
Established	320,522	$930	$298	$270
New/returning	218,196	$463	$101	$93
International total	*538,718*		*$399*	*$363*
Total	**2,918,826**		**$2,192**	**$1,998**

Reproduced from Miller et al. 2013, p. 36, Table 13

benefits among sectors were accounted for in each group. The annual economic benefit from Landsat imagery obtained from EROS in 2011 was just over $1.79 billion for US users and almost $400 million for international users, resulting in a total economic benefit of $2.19 billion (see Table 14.2; Miller et al. 2013). This estimate does not represent the entire societal benefit from Landsat imagery because it accounts only for the benefits received by direct users (i.e., those that download scenes directly from USGS). Any benefits that users receive from derived or value-added products that include Landsat imagery were not estimated. This estimate is also a reflection of economic value received by users under the free and open data policy. If a price is charged, it is expected that the number of users will decrease, as well as the number of scenes each remaining user obtains. This decrease in the number of users and the amount of imagery used would reduce the societal benefits of the imagery while not creating any additional benefits, resulting in a net loss of benefits.

14.4.2 Requirements Assessment

In 2012, under the direction of the White House Office of Science and Technology Policy (OSTP), a National Earth Observations Task Force conducted an assessment of 362 Earth observation systems (space, air, land, and sea platforms) with regard to their contributions within a framework of 13 societal-benefit areas (National Science and Technology Council 2014). Among 132 satellite systems considered, Landsat ranked second highest in impact, surpassed only by the Global Positioning System. Findings from this assessment informed the National Plan for Civil Earth

Observations (National Science and Technology Council 2014), which was released from the White House on July 18, 2014.

In a related 2013 effort, a USGS National Land Imaging Requirements Pilot Project elicited 151 distinct, representative Federal-agency applications where Landsat data are used routinely to produce consistent services or information products. The study identified operational or decision-supportive uses of Landsat in a broader variety of fields than was previously recognized. The associated USGS report (Vadnais and Stensaas 2014) highlights key requirements such as the need for an eight-day (or fewer) satellite site-revisit cycle, 30-m pixel resolution, and simultaneous Visible to Short Wavelength InfraRed (VSWIR) and thermal measurements. For example, the assessment found that 60 % of elicited threshold user requirements call for eight-day or more frequent revisit.

14.4.3 Case Studies

Serbina and Miller (2014) examined 25 applications of Landsat imagery in water resources to shed more light on the benefits accrued from the imagery and to gain a better understanding of the program's value in water resources applications. The users included Federal, State, and county governments; private companies; and non-governmental organizations. Users applied Landsat in consumptive use mapping, water budgeting and accounting, irrigation improvements, resolving water rights cases, humanitarian aid, flood mapping and monitoring, and land cover mapping. The case studies tell a more descriptive story of how Landsat imagery is used and what its value is to different private and public entities, but include quantitative valuations where possible. A few case study findings are outlined below.

In the private sector, E. & J. Gallo, the largest winery in the world, is using Landsat imagery in a modified METRIC model to improve yield and grape quality while decreasing the amount of water applied by 20–30 %, depending on the region (Serbina and Miller 2014). More precise water monitoring and application has enabled Gallo to increase the quality of the grapes and thus improve the quality of its wine. Gallo has also spent less money on irrigation and pruning (too much water causes excess foliage which must be removed). In Chile, the use of Landsat and METRIC (see Fig. 14.2) has led to cost savings of $80/acre in 3,700 acres of olive orchards, and a 30–60 % reduction in water applied to 6,000 acres of vineyards. Grape quality has also improved by 30–35 %.

The US Bureau of Reclamation uses Landsat extensively in estimating evapotranspiration and evaporation, as well as identifying types, locations, and acreages of crops, irrigated lands, and riparian vegetation (Serbina and Miller 2014). The agency saves approximately $40,000 each year by using Landsat instead of traditional methods for estimating evapotranspiration on agricultural and riparian lands. Unlike traditional on-the-ground methods, Landsat is able to provide complete coverage of the lands, which need to be monitored; if this was to be replicated through fieldwork, it would cost around $580,000 each year. Aside from the cost savings to

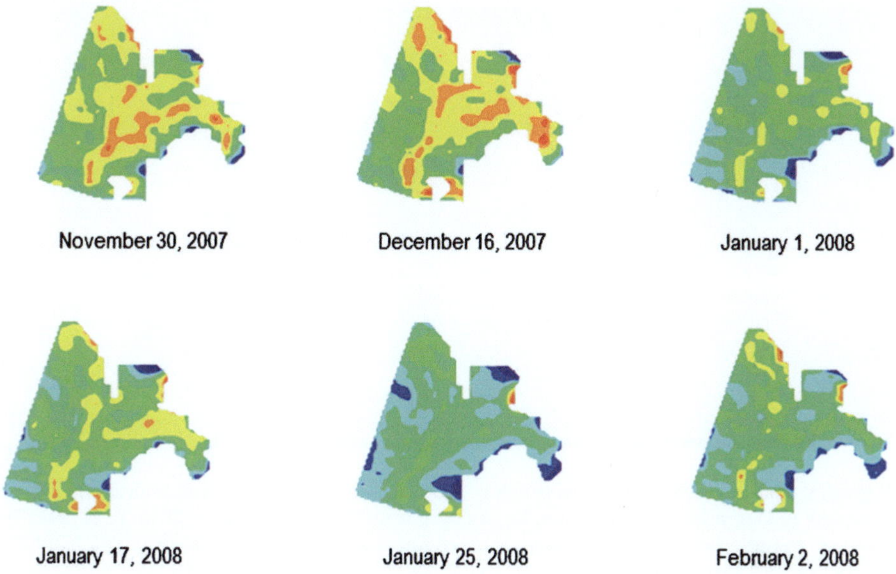

Fig. 14.2 Crop coefficient mapping of a drip-irrigated vineyard in Chile using images from Landsat 7 ETM+. *Blues* and *greens* indicate more evapotranspiration and *yellows*, *oranges*, and *reds* indicate less evapotranspiration (Courtesy of Universidad de Talca. Reproduced from Serbina and Miller 2014, p. 28, Fig. 20). From L-R, top to bottom: 30 Nov 2007, 16 Dec 2007, 1 Jan 2008, 17 Jan 2008, 25 Jan 2008, 2 Feb 2008

the agency, the reports and data sets they produce are used by myriad other entities, including Federal and State agencies, universities, and private businesses. Without this information derived from Landsat imagery, these organizations would have to spend money to produce or find alternative data.

Multiple western states use Landsat imagery to monitor consumptive water use (Serbina and Miller 2014). The State of Wyoming is using Landsat imagery in the METRIC model to assess evapotranspiration so they can more efficiently meet their annual consumptive use reporting requirements under the Upper Colorado River Basin Compact. The use of Landsat and METRIC has yielded consistent savings from a third to as much as one-half of the total costs compared to on-the-ground methods. Colorado, Idaho, and Nevada (see Fig. 14.3) are also using Landsat for water management.

Landsat imagery has proven critical in preserving minimum needed stream flows, as well as settling water rights disputes (Serbina and Miller 2014). In the Klamath Basin in Oregon, Landsat and METRIC were used to determine which water rights were being used to their fullest extent. The Klamath Basin Restoration Agreement Water Use Retirement Program could then make offers to purchase water rights which would result in the largest increase in in-stream flows. In Idaho, the A&B Irrigation District, a senior groundwater user, claimed that junior groundwater users were pumping too much water. Landsat and METRIC analysis of the

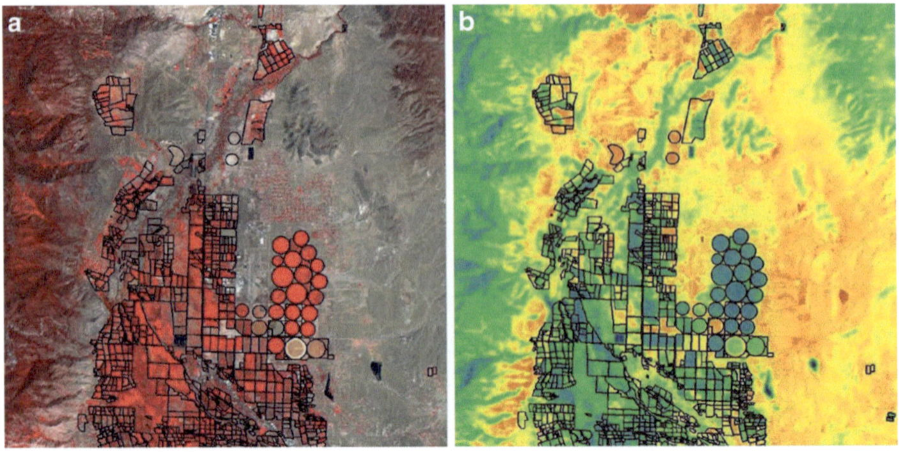

Fig. 14.3 Images of Northern Mason Valley, Nevada, showing irrigated and nonirrigated lands (Courtesy of Desert Research Institute. Reproduced from Serbina and Miller 2014, p. 12, Fig. 7). From left to right: (**a**) Landsat false-color infrared image showing irrigated lands in *red*, Northern Carson Valley, Nevada, August 3, 2009. (**b**) Landsat image showing evapotranspiration using energy balance (irrigated lands in *blue*), Northern Carson Valley, Nevada, August 3, 2009

current evapotranspiration in the district did not show a water shortage in the area (see Fig. 14.4), and A&B's claim was dismissed.

Landsat has also been used for water exploration in developing nations. Amid the Darfur Crisis in 2004, more than 250,000 Sudanese refugees were forced to relocate to camps in the desert landscape of eastern Chad (Serbina and Miller 2014). Many refugee camps did not have access to water on site. Every passing day without an adequate supply of water meant the loss of 200 children's lives in the camps, and water trucking cost millions of US dollars per day. Radar Technologies International used Landsat imagery in their WATEX System to help identify and drill 1,800 water wells, with a drilling success rate of 98 %, contributing to the survival of hundreds of thousands of people (see Fig. 14.5).

14.5 Conclusions

Landsat satellite imagery is highly valued by users in a wide range of application areas and sectors. Many users would be negatively impacted by the loss of Landsat imagery; some were already affected during the gap in Landsat 5 provision. The value of Landsat can be measured monetarily, but it can also be measured by lives saved, by improvements in accuracy or quality, and by advancements in scientific research. Though valuing a public good such as Landsat imagery can be difficult, using a variety of methods provides different types of valuations. The valuations can

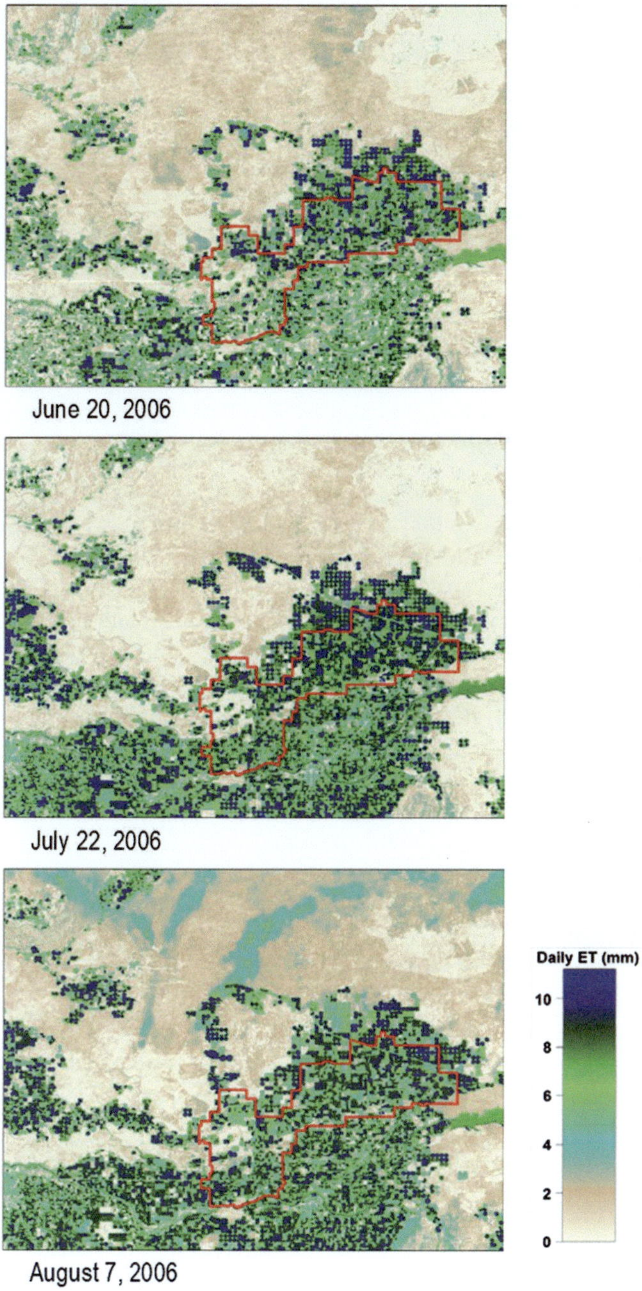

Fig. 14.4 METRIC evapotranspiration images of A&B Irrigation District (outlined in *red*), Idaho. Courtesy of Idaho Department of Water Resources. *ET* evapotranspiration, *mm* millimeters (Reproduced from Serbina and Miller 2014, p. 40, Fig. 29). From top to bottom: 20 June 2006, 22 July 2006, 7 August 2006

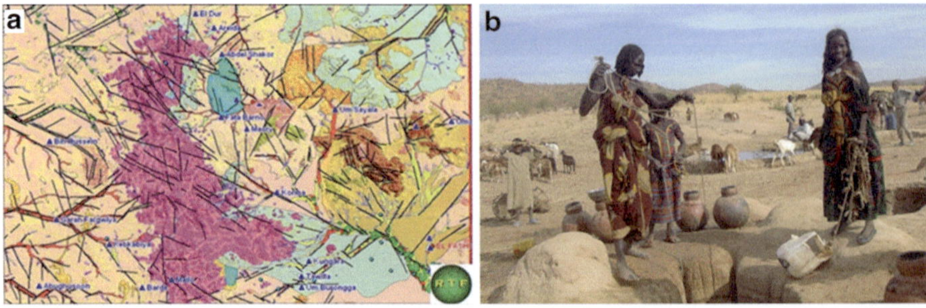

Fig. 14.5 Successful water well drilled in Eastern Ouaddai, Chad, located based on information provided by Landsat images processed in the WATEX model (Courtesy of Radar Technologies International. Reproduced from Serbina and Miller 2014, p. 42–43, Figs. 34 and 35). From left to right: (**a**) Potential water drilling sites map for West Darfur, Sudan, produced by WATEX. (**b**) Successful water well in Wadi Gaga Campsite, Eastern Ouaddai, Chad

then be tailored for use in varying communication situations with different audiences. The approaches used for valuing Landsat imagery can also be used to estimate the value of other Earth observations and information with public-good characteristics. While determining the full value of Landsat imagery and similar data sources may not be possible, the use of multiple valuation approaches can provide ample evidence of the benefits accrued from such information.

References

Bernknopf RL, Forney WM, Raunikar RP, Mishra SK (2012) Estimating the benefits of land imagery in environmental applications: a case study in nonpoint source pollution of groundwater. In: Laxminarayan R, Macauley MK (eds) The value of information. Amsterdam, Springer, pp 257–299

Booz Allen Hamilton (BAH) (2012) USGS land satellites data system (LSDS) cost benefit analysis. U.S. Geological Survey, Reston

Green K (2008) Results from the ASPRS moderate land imagery survey and next steps forward. Paper presented at the civil commercial imagery evaluation workshop, Fairfax

Landsat Advisory Group (2012) The value proposition for ten Landsat applications. Washington, DC: National Geospatial Advisory Committee. http://www.fgdc.gov/ngac/meetings/september-2012/ngac-landsat-economic-value-paper-FINAL.pdf. Accessed 6 Aug 2014

Miller HM, Sexton NR, Koontz L, Loomis J, Koontz SR, Hermans C (2011) The users, uses, and value of Landsat and other moderate-resolution satellite imagery in the United States: executive report (open-file report 2011–1031). US Geological Survey, Fort Collins

Miller HM, Richardson L, Koontz SR, Loomis J, Koontz L (2013) Users, uses, and value of Landsat satellite imagery: results from the 2012 survey of users (Open-File Report 2013–1269). US Geological Survey, Fort Collins, CO

Morse A, Kramber WJ, Allen RG (2008) Cost comparison for monitoring irrigation water use: Landsat thermal data versus power consumption data. Paper presented at Pecora 17 (17th William T. Pecora memorial remote sensing symposium), Denver

NASA (2013) Landsat: continuing to improving everyday life. NASA, Goddard Space Flight Center, Greenbelt. http://landsat.gsfc.nasa.gov/pdf_archive/LandsatImprovingLife.pdf. Accessed 6 Aug 2014

National Science and Technology Council (2014) National plan for civil earth observations. Executive Office of the President, Washington, DC. http://www.whitehouse.gov/sites/default/files/microsites/ostp/NSTC/2014_national_plan_for_civil_earth_observations.pdf. Accessed 25 Aug 2014

Nordhaus WD (1986) The value of information. In: Krasnow R (ed) Policy aspects of climate forecasting. Resources for the Future, Washington, DC, pp 129–134

Office of Management and Budget (1992) Guidelines and discount rates for benefit-cost analysis of federal programs (office of management and budget circular A-94, revised transmittal memo #64). Executive Office of the President, Washington, DC. http://www.whitehouse.gov/omb/circulars_a094. Accessed 29 Aug 2014

Serbina L, Miller HM (2014) Landsat and water: case studies of the uses and benefits of Landsat imagery in water resources (open-file report 2014–1108). US Geological Survey, Fort Collins

U.S. Environmental Protection Agency (2000) Guidelines for preparing economic analyses (EPA 240-R-00-003). U.S. Environmental Protection Agency, Washington, DC

U.S. Water Resources Council (1983) Economic and environmental principles and guidelines for water and related land resources implementation studies. US Government Printing Office, Washington, DC

Vadnais C, Stensaas G (2014) National Land Imaging Requirements (NLIR) pilot project summary report: summary of moderate resolution imaging user requirements (Open-file report 2014–1107). US Geological Survey, Reston. http://dx.doi.org/10.3133/ofr20141107. Accessed 25 Aug 2014

Part IV
Learning Forward: Communicating Climate Change Among Diverse Audiences

Chapter 15
Carbon Offsets in California: Science in the Policy Development Process

Barbara Haya, Aaron Strong, Emily Grubert, and Danny Cullenward

Abstract Natural and social scientists are increasingly stepping out of purely academic roles to actively inform science-based climate change policies. This chapter examines a practical example of science and policy interaction. We focus on the implementation of California's global warming law, based on our participation in the public process surrounding the development of two new carbon offset protocols. Most of our work on the protocols focused on strategies for ensuring that the environmental quality of the program remains robust in the face of significant scientific and behavioral uncertainty about protocol outcomes. In addition to responding to technical issues raised by government staff, our contributions—along with those from other outside scientists—helped expand the protocol development discussion to include important scientific issues that would not have otherwise been part of the process. We close by highlighting the need for more scientists to proactively engage the climate policy development process.

Keywords Carbon offsets • Climate change policy • Carbon markets • Science and policy

15.1 Introduction and Background

Natural and social scientists in the field of global climate change are increasingly stepping out of purely academic roles to inform and support policy that is science-based. This chapter explores the roles that science and scientists play in climate policy development using an example from the California climate policy process. Beginning in the spring of 2013, we participated in the public process for

B. Haya, Ph.D. (✉)
Stanford Law School, Stanford, CA, USA
e-mail: bhaya@berkeley.edu

A. Strong, M.A. • E. Grubert, M.S.
Stanford University, Stanford, CA, USA

D. Cullenward, Ph.D.
University of California, Berkeley, Berkeley, CA, USA

J.L. Drake et al. (eds.), *Communicating Climate-Change and Natural Hazard Risk and Cultivating Resilience*, Advances in Natural and Technological Hazards Research 45, DOI 10.1007/978-3-319-20161-0_15

developing two new carbon offset protocols in California. We relay our experiences as scientists in these processes with two main goals. First, we describe the types of input we and other natural and social scientists provided to regulators, in order to shed light on how scientific issues emerge in policy development and the associated role scientists play in practice. Second, we hope this example will encourage interested scientists to engage the climate policy process more directly. Fundamentally, we believe that scientists' active participation in climate policy development can improve policy outcomes and generate useful research agendas.

The primary theme of our work is supporting the robustness of California's offsets policies, a topic on which most of our efforts focused. As used in discussions of global climate change, another term—*resilience*—most commonly refers to the ability of communities or nature to adapt to the uncertain impacts of climate change. In the context of climate change policy, *robustness* offers a similar framing. It refers to the ability of a policy to reliably meet its goals despite substantial uncertainty in predicting or measuring its outcomes (Lempert and Schlesinger 2000).

The concept of policy robustness is particularly relevant in the context of policies concerning carbon offsets because of the deep scientific and behavioral uncertainties involved in calculating accurate emission reductions from offset projects. Because greenhouse gas emitters in a climate policy system that recognizes offsets—such as California's carbon market—use offset credits to justify increased emissions within the policy system's boundaries, it is critical that offsets accurately represent true emission reductions. Meeting this standard is no simple matter, however, as it requires scientifically complex and inherently uncertain methodologies.

The uncertainty stems from the need to calculate emission reductions by comparing an offset project's emissions against an inherently unknowable counterfactual scenario: the emissions that would have occurred without the offset project. Both estimates are subject to uncertain physical, social, and economic drivers. In light of this uncertainty, ensuring that offset credits represent true emission reductions requires conservative decisions about project and baseline emissions to ensure that protocols actually reduce the credited emissions reductions. Accordingly, our participation in California's public policy development processes focused on ways to preserve the robustness of the two offset protocols on which we worked.

The chapter is organized as follows. We begin with an overview of California's climate mitigation policies, describing how offsets fit into the state policy system, as well as the key challenges offsets pose for policy-makers. Next, we describe our activities as stakeholders in the public process for developing new offset protocols. We illustrate our work with a handful of examples that highlight scientific issues that emerged in the policy process, including issues that the regulatory agency identified for public input, as well as those issues we raised in our independent capacity. In the final section, we offer some concluding thoughts about our experience and the various roles we and other scientists played in these policy processes. Finally, we encourage other environmental scientists to explore proactive models of policy engagement.

15.1.1 California's Climate Policy

In 2006, California passed the Global Warming Solutions Act (AB 32), launching the state's comprehensive approach to climate mitigation policy. Its key feature is a legally binding requirement to reduce statewide greenhouse gas (GHG) emissions back to 1990 levels by the year 2020. To accomplish this goal, state law delegated broad authority to the California Air Resources Board (CARB), which developed a suite of climate policy instruments over the last several years (CARB 2008, 2014a). The most prominent is California's cap-and-trade program. This program applies to California's electricity, industrial, and fuels sectors, covering about 85 % of statewide emissions.

Briefly, cap-and-trade carbon markets set an overall limit (or *cap*) on anthropogenic greenhouse gas emissions within the covered sectors. The regulator then issues tradable emissions allowances, with the total number equal to the cap. Each emissions allowance credit confers the right to emit one tonne of GHG pollution (measured in tonnes of CO_2 equivalent, tCO_2e). Covered entities must submit one allowance per tCO_2e of pollution they emit. Since allowances are tradable, if a regulated emitter can reduce emissions more cheaply than the price of a permit, it can do so, freeing up permits to sell to others who face costlier mitigation opportunities. This lowers compliance costs compared to a system in which each emitter must meet an established standard without trading.

Carbon offsets extend the flexibility of this approach by allowing covered entities to seek lower-cost emission reduction opportunities outside of the carbon market—for example, in another state or in an economic sector not covered by the cap—instead of reducing emissions within the capped sectors. The financial benefits to regulated emitters are straightforward: expanding the range of mitigation opportunities outside the capped system through offsets reduces compliance costs. Since climate change is driven by the global stock of GHGs in the atmosphere, reducing one tonne of emissions has the same effect regardless of location.[1] As we discuss below, however, accurately calculating the net emissions reductions raises new challenges.

15.1.2 Offsets in California

Companies subject to the cap-and-trade market can use offset credits to cover up to 8 % of their total emissions. This limit on the use of offsets appears significantly more generous when expressed as a percentage of the total mitigation required in the carbon market: if all regulated parties use the maximum amount allowed, offsets

[1] Though other pollution impacts that are coincident with the greenhouse gas emissions may have important local and regional effects, including on public health

would contribute about half of the total emission reductions expected under California's climate policy through 2020 (Haya 2013).

Carbon offsets in California work as follows. CARB issues offset credits for projects that follow approved protocols. The protocols themselves determine what project activities are eligible and define the methodologies by which projects estimate their emission reductions. Thus, offset protocols must be designed to anticipate all of the emissions-related drivers that apply in a given sector—a task that typically involves complex issues of environmental and social science.

Although the decision to develop a new protocol lies entirely at CARB's discretion, offset protocol methodologies must meet certain standards. State law and market regulations both require that emission reductions from offsets be "real, additional, quantifiable, permanent, verifiable, and enforceable."[2] Each of these terms has a formal legal definition. The most challenging requirement has been *additionality*, defined in AB 32 as crediting only those emission reductions that are made "in addition to any greenhouse gas emission reduction otherwise required by law or regulation, and any other greenhouse gas emission reduction that otherwise would occur."[3] CARB's climate regulations provide more context on how additionality is to be tested, requiring the use of a "conservative, business-as-usual scenario."[4]

The regulations also directly address uncertainty and risk management, defining conservative scenarios as those whose "project baseline assumptions, emission factors, and methodologies that are more likely than not to understate net GHG emission reductions or GHG removal enhancements for an offset project to address uncertainties affecting the calculation or measurement of [net GHG reductions]."[5]

Finally, it is important to recognize that political perspectives on offsets vary widely. Many stakeholders, including most major emitters in the market, are strongly supportive of offsets as a mechanism to keep compliance costs low. After all, the supply of offset credits is widely expected to meaningfully reduce carbon market prices relative to a market without offsets (Borenstein et al. 2014; EPRI 2013). In contrast, several nonprofit stakeholders have expressed concerns about whether California's offsets truly represent reductions in GHG emissions. For example, two environmental groups sued CARB, claiming that the agency's decision to evaluate additionality using a performance standard at the protocol level does not satisfy the requirements of AB 32. The trial court rejected the plaintiffs' claims, finding that CARB had the necessary legal authority to adopt its performance standard approach. The court then applied a highly deferential standard to review CARB's treatment of additionality in each of its existing protocols (*Our Children's Earth Foundation v. CARB* 2015). Beyond highlighting the political opposition to offsets, this decision suggests that future legal challenges to CARB's protocol methodologies would face a difficult legal test under which the regulator is likely to prevail.

[2] Cal. Code Regs. tit. 17, § 95802(a)(14); see also Cal. Health & Safety Code § 38562(d)(1)-(2).

[3] Cal. Health & Safety Code § 38562(d)(2).

[4] Cal. Code Regs. tit. 17, § 95802(a)(4).

[5] Cal. Code Regs. tit. 17, § 95802(a)(76).

15.1.3 Critical Issues for Carbon Offsets

Offsets raise a number of technical challenges, and CARB's two new protocols are no exception. A carbon market maintains its environmental integrity only if the offset credits it recognizes represent actual net reductions in greenhouse gas emissions. In practice, however, uncertainty about those reductions requires detailed scientific input and is often the subject of significant controversy.

A critical task for policy-makers is establishing a robust standard for offset additionality. An offset project is considered additional only if it occurred because of the financial investment made in return for offset credits. In other words, an offset program should only credit those emission reductions it causes and should not credit reductions that would otherwise have occurred. This standard is necessary to ensure that any climate policy system that accepts offsets achieves its intended emission reductions. But additionality is difficult to achieve in practice. Several studies have shown that a large portion of credits generated by the Clean Development Mechanism (CDM, the Kyoto Protocol's offsets program) were non-additional projects that would have occurred without the financial incentive of offset credits and thus do not represent net emission reductions (Cullenward and Wara 2014; Haya 2009; Haya and Parekh 2011; Wara 2008). As a result, their use by countries to meet Kyoto Protocol targets came at the expense of real reductions in greenhouse gas emissions.

Two issues further complicate the basic question of establishing whether offset credits represent real additional emission reductions. First, uncertainty analysis is particularly important for offset projects in the land-use and agricultural sectors, where emissions vary widely across location, crop, and ecosystem types. Second, there is the risk that offset program incentives cause emissions to increase outside of offset project boundaries. The most egregious example involves offset credits in the CDM awarded for the destruction of hydrofluorocarbons (HFCs), a potent family of greenhouse gases emitted as byproducts in the production of certain refrigerants. Manufacturers realized they could earn greater profits from destroying HFCs than from the sale of the refrigerant itself. There is strong evidence that they increased their production as a result of this incentive, creating surplus HFC byproducts that they subsequently destroyed to earn offset income (Wara 2008). Beyond enticing non-additional credits, the income from HFC-related offsets might have discouraged national governments from directly regulating HFC emissions, in order to maintain offset project eligibility—an effect that has been documented for a range of other project types (Figueres 2006).

Although the problems observed in past offset systems remain relevant, it is important to recognize that CARB's approach to additionality is different than that of its predecessor, the Kyoto Protocol's CDM. The CDM requires individual offset project applicants to evaluate their counterfactual emissions scenarios and demonstrate additionality for each individual project. In contrast, the California system makes these determinations at the protocol level by defining project eligibility criteria. Once CARB has approved a protocol, a project applicant needs only to demonstrate compliance with the protocol's eligibility criteria in order to earn credit. Given the use of up-front project eligibility criteria, robust protocol design is particularly critical to ensuring that California's offset credits represent real emission reductions.

Finally, we note the importance of CARB's early offset protocols as institutional precedents in American climate policy. As one of the first legally binding climate policies in the United States, California's cap-and-trade system has already become a standard point of reference for climate policy design. In turn, CARB's treatment of complex and uncertain scientific issues in its offset protocol development process will surely set an important example for others.

15.1.4 Proposed Mine Methane Capture and Rice Cultivation Protocols

By the beginning of 2013, CARB had approved four offset protocols covering projects in the following areas: (1) forestry, (2) urban forestry, (3) livestock waste management, and (4) destruction of ozone-depleting substances. We participated in the policy development process for two new protocols: (1) mine methane capture and (2) rice cultivation, which we describe briefly here for background.

CARB approved the Mine Methane Capture (MMC) protocol in April 2014 (CARB 2014b), following a year of development and stakeholder engagement. The protocol awards credits to projects that capture methane that otherwise would have been released into the atmosphere from coal and trona[6] mining activities. CARB's MMC protocol recognizes two types of projects. Methane can be captured for use as a fuel, such as by injecting captured gas into natural gas pipelines or using it to fire an on-site power plant. Alternatively, MMC projects can destroy methane without putting it to productive use through flaring or oxidation. In any of these cases, methane (CH_4) is converted to carbon dioxide (CO_2), a much less potent greenhouse gas.

At the time that this chapter was written, CARB was in the process of developing a rice cultivation protocol and responding to comments submitted on a discussion draft of the protocol released in March 2014. This protocol would credit reductions in methane emissions from changes in rice cultivation practice in California and the South Central United States. Rice cultivation produces methane emissions because production fields are submerged under water for a large portion of the year. This causes biomass to decompose without oxygen, producing CH_4 rather than CO_2. Methane emissions can be reduced if the fields are submerged for less time or if less biomass is left on the field to decompose anaerobically.

[6]Trona is a mineral mined as the primary source of sodium carbonate in the United States.

15.2 Science in the Policy Development Process

In April 2013, CARB established technical working groups to bring together stakeholders to inform the development of two new offset protocols. The working groups included offset project developers, project verifiers (who verify that project developers have met the protocol's requirements), representatives from industries facing compliance obligations in the carbon market (i.e., offset buyers), environmental nonprofit staff, academic research scientists, representatives from organizations that develop offsets standards for voluntary carbon markets, and state and federal officials from outside agencies. Each working group convened approximately once every three months, though additional discussion continued between meetings.

15.2.1 The Interdisciplinary Nature of Climate Change Policy Development

As a preliminary matter, we note that the scientific and technical expertise needed to ensure the environmental integrity of carbon offset protocols spans a wide range of disciplines. For example, the MMC and rice cultivation protocols drew on experts—including a number of outside scientists, in addition to our group—who provided advice on statistical uncertainty assessment, biogeochemical and ecological modeling, field measurements of gas fluxes, economic analysis, life-cycle analysis, basic mineralogy, engineering of mine construction, wildlife ecology, insect population dynamics, the sociology of agricultural crop production practices, modeling hydrological connectivity above- and belowground, state and federal water law, land-use law, environmental law, and organizational theory. As this list indicates, there are many opportunities for a variety of scientific experts to proactively engage the climate policy process—no agency has all of the necessary experts on staff.

15.2.2 What Did We Do?

Our participation in the offset protocol development process included a wide range of activities. We interfaced with a variety of stakeholders, including CARB staff, CARB board members, offset project developers, and nonprofit groups. Similarly, our communications ranged from informal conversations in person to formal written comment letters. As members of the technical working groups for each protocol, we attended meetings at the agency's headquarters in Sacramento and brought attention to issues we viewed as critical to the environmental integrity of the draft protocols as they developed, based on detailed independent analysis.

We provided our assessments to CARB staff as informal communications and later submitted formal comment letters during public comment periods in the administrative process. At times when we believed that CARB was not adequately addressing critical concerns, we spoke with individual CARB staff and board members outside of the formal working group process, occasionally with the participation of other stakeholders; we also raised our concerns through public testimony at formal board meetings.

The overarching goal of our involvement was to apply our research team's interdisciplinary expertise to helping ensure the environmental quality of the protocols. We did not use a single set of methods in our contributions, but rather, each of us brought methods from our respective disciplines to our shared goal. Below, we offer examples of scientific issues that highlight the kinds of input we offered in an effort to ensure that California's offset protocols reflect the best available science and are robust in the face of significant uncertainty.

Our examples are organized according to different ways that scientific issues arose in the policy development process—at the agency's request or according to our independent review of the protocols—rather than by protocol or chronology. In this way, we hope to illustrate both how science was used in developing the protocols and what roles scientists can expect (or be expected) to play in such processes.

15.2.3 Scientific Issues Raised by the Agency

Our first category of scientific engagement in the policy development process focuses on those issues that CARB proactively identified, either via agency staff asking stakeholders directly for input or by inclusion on agency-drafted meeting agendas. We review one such example in this section.

15.2.3.1 Scale of Uncertainty Assessment in Model-Estimated Emissions from Rice Cultivation

If the proposed rice cultivation protocol is adopted, it will become the first California protocol to use a computer-based model to estimate emission reductions. Using a model is necessary in this case because direct field measurements of emissions are technically challenging, costly, and time-consuming. The proposed protocol relies on a mechanistic biogeochemical model, the DeNitrification-DeComposition (DNDC) model, originally developed at the University of New Hampshire (2012).

The DNDC model is used to estimate offset project emissions and emission reductions. Through the technical working group, we—along with other scientists, including DNDC model developers, biogeochemists, and agricultural experts—addressed questions about model uncertainty and validation, the model's ability to estimate emissions of the potent GHG nitrous oxide (N_2O), and specific biogeochemical parameters used in the model.

Models are by definition simplifications of complex processes and are not perfectly accurate. Accordingly, the draft protocol applies a *deduction* that reduces the model-estimated emission reductions to conservatively account for any model error. Early drafts of the protocol included this deduction, but applied only one value for all eligible projects. Since DNDC must be field-calibrated to particular crop types, however, we were concerned that a blanket assessment of an uncertainty deduction for model error was too general and would not reflect the uncertainty of the model as it would be applied in the rice cultivation protocol—specifically, to fields in different ecosystems, with different cultivars, and in different regions around the country.

We focused our attention on how finely to parse assessments of model uncertainty, raising this issue in both formal and informal comments. Ultimately, the draft protocol included separate uncertainty deduction calculations for each of the rice-growing regions, rather than a single uncertainty deduction for all applications of the model. Furthermore, CARB decided to update the uncertainty deduction coefficients on an annual basis, a feature that will make the protocol more robust in light of new information. On the other hand, there is no formal mechanism for updating the model itself in response to newly published scientific information that directly affects relevant calculations. In the end, the potential for model structures and inputs to change highlights the profound challenge of integrating active scientific research into a fixed policy structure. Inevitably, there will be trade-offs between the adaptability of the protocol to new information and the stability of compliance rules that offset project developers desire.

15.2.4 Scientific Issues We Raised

A second category of scientific engagement describes our independent evaluation of issues that emerged during the protocol development process, as opposed to the assessment of issues on which CARB specifically requested input. In this section, we discuss examples of issues we raised about the conservative estimation of emission reductions from individual projects, additionality assessment, and the risk of unintended consequences caused by interactions between offset protocols and other policies. In some cases, we raised questions that were not being addressed at the time, and in others, we advanced new perspectives on issues that were already under agency consideration.

15.2.4.1 Statistical Bias in the Rice Cultivation Emissions Model

Statistical bias occurs when a prediction repeatedly over- or underestimates real-world outcomes. A model is unbiased if its outcomes are equally likely to over- and underpredict actual emissions as determined by direct field measurements. An unbiased model may still over- or underestimate the reductions achieved by an individual offset project, but the uncertainty deduction factor (discussed above) ensures that over-crediting is still avoided with a high degree of certainty. However, a model that has not been validated as statistically unbiased for the project types credited under the protocol may result in an overestimation of the emissions reduced by those project types, even after the uncertainty deduction factor is applied.

During the rice protocol development process, CARB staff referred to hundreds of field measurements that had validated the DNDC model, finding no trend in the estimates. Thus, they concluded that the model was not biased. We were concerned, however, that some of the project types eligible under the protocol were not included in the data used to validate the model. Noting this gap, we argued that an assessment of bias at the level of the entire DNDC model was insufficient, and that project-type specific assessment of model bias was warranted. To avoid over-crediting, we suggested that CARB approve the eligibility of a project type under the protocol only if the DNDC model has been validated to have no statistical bias for the type of activities credited by that project type. As of this writing and to the best of our knowledge, CARB staff provided the technical working group with only a list of published references, not the actual data from the model runs used in the bias assessment.

As CARB continues to collect field data to validate the model, we hope to view the complete dataset on which CARB validates the DNDC model. This example illustrates the important role scientists play in reviewing the technical basis of policy—in this case, the methods used to assess statistical bias in an emissions model, in order to avoid over-crediting. It also illustrates the importance of transparency and access to data, both of which are necessary to enable scientific review.

15.2.4.2 Additionality of Methane Capture at Abandoned Mines

Our second example in this category concerns the treatment of additionality in the MMC protocol. CARB determines the additionality of different project types by assessing whether the project activity is *common practice* among a relevant population; a project type is considered additional if it is not common practice. Applying this approach to methane capture at abandoned mines under the MMC protocol, CARB staff studied abandoned underground mines in the United States, finding that "few currently capture and destroy mine methane. Methane capture and destruction is therefore deemed not to be business-as-usual at these mines" (CARB 2013, p. 7). This language suggests that CARB was prepared to deem all abandoned mine methane control projects additional under the MMC protocol.

The case of methane capture at abandoned mines demonstrates the importance of assessing additionality for subcategories of project types and not just for the entire population of possible projects as a whole. It also highlights the value of performing a conservative quantitative assessment to examine compliance with the protocol level additionality standard. While only 38 of the more than 10,000 abandoned mines in the United States have implemented methane capture projects, these 38 mines emit one third of all methane released from abandoned mines in the country (Ruby Canyon Engineering 2013a). Thus, existing methane capture projects at abandoned mines are correlated with high rates of methane emissions—exactly as one would expect, given that the costs of capturing methane decrease as the rate and concentration of methane emissions at mines increase.

If all abandoned mines were eligible for MMC offset credits, the protocol could generate non-additional credits from projects that would have proceeded regardless of the financial incentives offsets provide. Indeed, if methane capture project development trends at abandoned mines from the last two decades were to continue, the volume of non-additional credits enabled by CARB's initial common practice assessment would likely far exceed methane capture from truly additional projects enabled by the financial incentive created by the offsets program as assessed by Ruby Canyon Engineering (2013b).

A more detailed analysis of abandoned mines suggested a path forward. Currently, most methane capture at abandoned mines occurs at mines that captured methane for pipeline injection when they were active. In fact, all mines that captured methane and were closed within the last ten years continued to capture methane after being abandoned. Methane capture at this subcategory of mines is undoubtedly common practice. Accordingly, CARB narrowed its eligibility criteria in the final protocol it adopted in April 2014, excluding those abandoned mines where methane had been captured and injected into pipelines when the mine was active (CARB 2014b, p. 14).

Our calculations showed that this approach excludes most, but not all, of the non-additional crediting that would conceivably be generated under CARB's initial definition of common practice at abandoned mines. While most non-additional methane capture is excluded from crediting by the narrowing of CARB's eligibility criteria for abandoned mines, past trends suggest that a smaller amount of methane capture may still be cost-effective on its own. We performed a quantitative analysis on the narrowed pool of eligible projects.

We found that if past trends in the development of new methane capture projects at abandoned mines that never previously captured methane were to continue, the expected generation of credits from non-additional projects is likely to be small compared to the expected effect of the protocol on new project development. Our analysis further indicated that under-crediting from conservative methodologies used to estimate emission reductions from abandoned mines under the protocol can reasonably be expected to counterbalance this non-additional crediting.[7] In other words, even though it is likely that some abandoned mines that would have chosen to implement methane capture technology regardless of the offset credit could generate credits under the protocol, the total quantity of offset credits generated by the protocol is unlikely to exceed the net emission reductions enabled by the protocol.

[7] For a more detailed description of this assessment, please see comments submitted by Barbara Haya on behalf of our research team dated February 14, 2014, "RE: Comments on the informal draft of the Mine Methane Capture (MMC) Projects Compliance Offset Protocol released 31 January 2014" available on California Air Resources Board's Workshop Comments Log: http://www.arb.ca.gov/lispub/comm2/bccommlog.php?listname=discussion-draft-ws.

As a result, we concluded that the protocol is expected to meet the additionality requirement defined under AB 32.

In addition to describing how the regulator's approach to a particular technical issue evolved during the MMC protocol development process, this example illustrates a methodological issue that speaks to the broader architecture of California's offsets policy. CARB's common practice approach appears to be designed to avoid the subjectivity of other eligibility metrics by referring to objective measurements of the frequency of emission-reducing activities. Nevertheless, we believe that this approach belies a persistent analytical subjectivity. As the abandoned mine issue shows, how CARB defines the population of project types against which it makes its common practice determination has important implications for the additionality of the offset protocol as whole. This example illustrates the importance of performing additionality assessments on subcategories of projects and conservatively excluding subcategories that could be considered common practice. More broadly, it also shows that the decision to use a common practice standard does not avoid the need for careful risk assessments of possible outcomes; these assessments remain necessary to identify appropriate project eligibility criteria that contain the risk of over-crediting.

15.2.4.3 Potential Conflicts with Clean Air Act Implementation

Our final example concerns a prospective impact that could occur beyond offset project boundaries. Here, our analysis focused on the potential for California's MMC protocol to interfere with other states' implementation of regulations under the federal Clean Air Act. The problem is this: although California's offset regulations exclude as ineligible those offset projects whose emission-reducing activities are separately required by law, they do not consider the incentive California's offset protocols create to keep legal standards in other jurisdictions low.

Under the Clean Air Act, any major new source of greenhouse gases is required to apply for a Prevention of Significant Deterioration (PSD) permit from its state environmental agency. In turn, the state agency is required to determine the best available control technology (BACT) for that particular project. State agencies have broad discretion in setting each project's BACT, with limited room for the federal Environmental Protection Agency (EPA) to review their findings. We expressed concern that California's MMC protocol would create incentives for out-of-state agencies to keep GHG BACT standards for mines artificially low. After all, were an out-of-state regulator to require methane destruction under the BACT determination for a PSD permit that methane destruction project would become ineligible for offset credits (and revenues).

In order to mitigate this risk, we recommended a do-no-harm precaution, temporarily excluding from the MMC protocol those mines that would require a PSD permit under the Clean Air Act. Once a specified number of PSD permits were issued to comparable mines, however, we suggested the MMC protocol could then expand its eligibility to mines that required PSD permits—so long as the early BACT determinations indicate that this course would be appropriate. Ultimately, these issues were not addressed in the adopted protocol and will be monitored informally.

15.3 Conclusions

The development of two new carbon offset protocols in California provides a rich case study in science-based policy-making. As public members of the technical working groups established by the California Air Resources Board, we both observed and contributed to the scientific discussions that arose during the course of protocol development. In addition to responding to the issues and questions raised by CARB directly, we—along with other outside scientists—played an essential role in expanding the protocol development discussion.

Most importantly, our engagement focused extra attention on the robustness of the protocols, providing strategies to avoid over-crediting despite substantial uncertainty in predicting protocol outcomes. Robustness is critical in the development of carbon offset protocols because of the significant scientific and behavioral uncertainty involved in accurately calculating emission reductions from individual projects. Fundamentally, this uncertainty stems from the challenge of estimating emission reductions (and the number of offset credits awarded) against an inherently unknowable counterfactual scenario of what would have happened without the offset program. Because offset credits are used in place of emission reductions within existing climate policy systems, methodological decisions must be made conservatively and guided by scientific risk assessments in order to avoid weakening these systems. Protocols should also be responsive to new scientific information and changes in the socioeconomic drivers of emissions. By conducting independent analyses of these kinds of issues, we aimed to increase the agency's capacity to evaluate key risks and improve the robustness of the offset protocols.

Finally, we hope the examples in this chapter encourage more members of the scientific community to seek ways to actively engage the development of climate policies. Although the offset protocols on which we worked were certainly informed by traditional scientific publications, our experience shows how the full treatment of scientific issues in the policy process occurs more through direct participation than literature reviews. Many of the critical policy questions involving science and uncertainty analysis would be difficult, if not impossible, to anticipate from a detached distance. In addition, their successful resolution depends on professional relationships built through iterative interactions in the policy process. Collectively, these factors suggest the need for more academics to explore ways to actively engage the climate policy process in the future.

References

Borenstein S, Bushnell J, Wolak FA, Zaragoza-Watkins M (2014) Report of the market simulation group on competitive supply/demand balance in the California allowance market and the potential for market manipulation. Energy Institute @ Haas working paper #251. https://ei.haas.berkeley.edu/papers.html. Accessed 2 Mar 2015

California Air Resources Board (2008) Climate change scoping plan: a framework for change. Sacramento. http://www.arb.ca.gov/cc/scopingplan/document/adopted_scoping_plan.pdf. Accessed 2 Mar 2015

California Air Resources Board (2013) Staff report: initial statement of reasons, proposed regulation to implement the California cap-and-trade program, appendix A, staff report and compliance offset protocol, mine methane capture projects (4 Sept 2013), Sacramento. http://www.arb.ca.gov/regact/2013/capandtrade13/capandtrade13isorappa.pdf. Accessed 2 Mar 2015

California Air Resources Board (2014a) First update to the climate change scoping plan: building on the framework. Sacramento. http://www.arb.ca.gov/cc/scopingplan/2013_update/first_update_climate_change_scoping_plan.pdf. Accessed 2 Mar 2015

California Air Resources Board (2014b) Compliance offset protocol mine methane capture projects: capturing and destroying methane from US coal and trona mines. Sacramento (25 Apr 2014). http://www.arb.ca.gov/cc/capandtrade/mmc_oal_april2014.pdf. Accessed 2 Mar 2015

Our Children's Earth Foundation v. CARB, 234 Cal. App. 4th 870 (Cal. Ct. App. 2015)

Cullenward D, Wara M (2014) Carbon markets: effective climate policy? Science 344:1460–1461

Electric Power Research Institute (2013) Exploring the interaction between California's greenhouse gas emissions cap-and-trade program and complementary emissions reduction policies. EPRI report #3002000298. http://www.epri.com/abstracts/Pages/ProductAbstract.aspx?ProductId=000000003002000298. Accessed 2 Mar 2015

Figueres C (2006) Sectoral CDM: opening the CDM to the yet unrealized goal of sustainable development. Int J Sustain Dev Law Pol 2(1):5–26

Haya B (2009) Measuring emissions against an alternative future: fundamental flaws in the structure of the Kyoto protocol's clean development mechanism. Energy and Resources Group working paper ER09-001, University of California, Berkeley. http://papers.ssrn.com/sol3/papers.cfm?abstract_id=1562065. Accessed 2 Mar 2015

Haya B (2013) California's carbon offsets program: the offsets limit explained. http://bhaya.berkeley.edu/docs/QuantityofAB32offsetscredits.xlsx. Accessed 2 Mar 2015

Haya B, Parekh P (2011) Hydropower in the CDM: examining additionality and criteria for sustainability. Energy and Resources Group working paper ER11-001, University of California, Berkeley. http://papers.ssrn.com/sol3/papers.cfm?abstract_id=2120862. Accessed 2 Mar 2015

Lempert R, Schlesinger P (2000) Robust strategies for abating climate change. Clim Change 45:387–401

Ruby Canyon Engineering (2013a) Abandoned coal mine methane offsets protocol: background information on performance standard and additionality. http://www.arb.ca.gov/cc/capandtrade/protocols/mmc/rce_amm_background.pdf. Accessed 2 Mar 2015

Ruby Canyon Engineering (2013b) Letter from Ronald C. Collings to Jessica Bede. Subject: California Air Resources Board: proposed compliance offset protocol mine methane capture projects, dated August 19, 2013. http://www.arb.ca.gov/regact/2013/capandtrade13/capandtrade13.htm. Accessed 22 Oct 2013

University of New Hampshire (2012) DNDC user's guide version 9.5. Institute for the Study of Earth, Oceans, and Space. University of New Hampshire. http://www.dndc.sr.unh.edu/model/GuideDNDC95.pdf. Accessed 2 Mar 2015

Wara M (2008) Measuring the clean development mechanism's performance and potential. UCLA Law Rev 55:1759–1803

Chapter 16
Fostering Educator Resilience: Engaging the Educational Community to Address the Natural Hazards of Climate Change

Minda R. Berbeco and Mark McCaffrey

Abstract Climate change and the natural hazards associated with it are some of the greatest environmental, economic, and political challenges of our time. Given how important students are to the future of addressing these issues, it is surprising that those who are tasked with educating them, teachers, have been so greatly overlooked. This chapter will address the challenges and opportunities for creating a resilient, supported, and informed educational community that can work with students to plan for the future.

Keywords Natural hazards • Climate change • Climate education • Climate literacy • Next Generation Science Standards • Common Core State Standards • National Climate Assessment • Science denial

16.1 Introduction

Climate change and the natural hazards associated with it are some of the greatest environmental, economic, and political challenges of our time. Complex, often nonlinear, in their causes and effects, both are often overlooked and at times deliberately avoided because of their overwhelming implications. Both have nonintuitive aspects that may baffle those unfamiliar with the science and mathematics behind them. A 500-year flood could in theory occur 2 years in a row, and even as global average temperatures continue to rise, record-breaking cold temperatures still are being set. When students are not afforded the opportunity to learn about these vital topics in their formal education, their alternative or naïve concepts may well continue into adulthood.

M.R. Berbeco, Ph.D. (✉) • M. McCaffrey, M.A.
National Center for Science Education, Oakland, CA, USA
e-mail: berbeco@ncse.com

J.L. Drake et al. (eds.), *Communicating Climate-Change and Natural Hazard Risk and Cultivating Resilience*, Advances in Natural and Technological Hazards Research 45, DOI 10.1007/978-3-319-20161-0_16

As the world prepares for future disruptions associated with a changing climate, many political, scientific, and social groups are working to understand the potential impacts and associated costs. There is still confusion, however, within the public sphere about the legitimacy of the science that indicates human activity is responsible for climate change as well as the economic and political implications. Though in the recent past less than half of Americans believed climate change was human caused, and a third believed there was a lot of disagreement within the scientific community as to whether or not it was happening (Leiserowitz et al. 2010), recent research suggests more general acceptance of human causation but a shallow understanding (Leiserowitz et al. 2013).

In an attempt to enlist the public in a more comprehensive discussion of the natural hazards associated with climate change, many researchers have been working on the best way to engage with the public: how to discuss the topic and how to get the public interested (Hamilton et al. 2012; Hassol 2008; Leiserowitz et al. 2010). Attempts have also been made to track, analyze, and improve media coverage of this topic (MT Boykoff and Boykoff 2007; Huertas and Kriegsman 2014), including a recent focus on utilizing weathercasters in particular as spokespeople for discussing these issues. When the third National Climate Assessment was released in the spring of 2014, President Obama invited weather forecasters to the White House to discuss the implications and encourage them to address these issues with their viewers (Davenport 2014).

As future citizens and policy makers themselves, students are another group that has been heavily researched (Hamilton College 2007; Leiserowitz et al. 2011). Attempts have been made to discover how to best reach this audience without unduly alarming them about the consequences of natural hazards or discouraging them from thinking about mitigation and adaptation (Hicks and Bord 2001; Ojala 2012). Indeed, students are in some respects a perfect group with which to work. They are already immersed in an environment conducive both to learning and to action, and though many admit to know little about climate change, they report they are interested in learning more (Leiserowitz et al. 2011).

In light of how important students are to the future of addressing the challenges of climate change and its associated natural hazards, it is perplexing that teachers, who are tasked with educating them, have been largely overlooked. Research has been limited on understanding how educators address these topics with students, the best practices for climate change education, or even the opinions and concerns of the educators themselves. Moreover, though science educators are typically trained in the sciences and pedagogical practices, there is a risk, with a topic such as climate change that is politically contentious, that educators will be too concerned about political implications to address the topic forthrightly.

As a reliable and trusted source for accurate scientific information, science educators are depended on as the first and often the last time many students engage with scientific concepts. Ensuring that science teachers have a clear and accurate understanding of the science is central to a scientifically literate citizenry. In this article, the value of educators in addressing both climate change and its implications related to natural hazards will be addressed. Through a thorough discussion of the challenges and opportunities for engagement, including an emphasis on informed responses, we can create a more resilient educational community ready to confront the challenges ahead.

16.2 Unique Challenges for Educators

There are many challenges to addressing climate change and its associated natural hazards. These include science denial and politicization related directly to climate change (Rosenau 2012), student misconceptions of both the causes and impacts of climate change, and the associated natural hazards (Boyes et al. 1993; Harrington 2008; McCaffrey and Buhr 2008), as well as psychosocial issues attendant on bringing potentially upsetting topics into the classroom (Hicks and Bord 2001; Ojala 2012). These challenges have been expounded upon in the past, particularly as they relate to student understanding of these concepts (Berbeco and McCaffrey 2014). Here, though, the particular challenges that educators face will be addressed. Also discussed here will be how educators' misconceptions and personal beliefs trickle down into substandard teaching practices and how the scientific community can better support educators in teaching the core science in an accurate and confident manner.

There is much that can be garnered from the social science research that attempts to understand the challenges educators face when teaching a socially contentious topic. Prior to climate change being integrated into the classroom, evolution was perhaps the most socially controversial topic to arise in science courses in the United States. A 2007 National Survey of Public High School Biology Teachers (Berkman et al. 2008) found that few educators steered away altogether from teaching general evolutionary processes and less than 20 % avoided the discussion of human evolution. However, a quarter of the educators reported that they devoted at least an hour to creationism or intelligent design in addition to evolution and, of those, nearly half felt that creationism was "a valid scientific alternative to Darwinian explanations for the origin of species" (p. 922).

Of at least equal concern, however, was what the researchers subsequently dubbed "the cautious 60 %" of science teachers who, while not preaching creationism, nevertheless fail to be "strong advocates for evolutionary biology" (Berkman and Plutzer 2011, p. 404). It would be easy to blame these educators for shirking academic and professional responsibilities as their creationist colleagues do. This, however, would be a simplification of the challenges such teachers face and dismissive of the realities of teaching topics that can be emotionally challenging both to the students and to the teachers themselves.

In a separate study by Griffith and Brem (2004), researchers surveyed biology teachers about the sorts of pressures they face when addressing evolution. They found three categories of teachers: *scientist*, *selective*, and *conflicted* (p. 791). The *scientist* teachers felt there was no place for controversial social issues in their classroom. This is not to say they avoided teaching evolution or did not acknowledge that evolution was perceived by many in the public sphere to be controversial, but rather that they felt no internal conflict about teaching the topic in a straightforward manner and insisting on its scientific value. They felt a deep love of the science and a desire to share that with students, reporting few classroom stresses and not worrying about incidents or conflicts arising in the classroom.

The *selective* teachers were concerned with harmony in the classroom, specifically selecting topics that would be less controversial to avoid conflict. Moreover, they altered their teaching structure from more open to more closed when they addressed evolution (e.g., fewer opportunities for questions and more of a lecture style). The last group, *conflicted* teachers, experienced stressors internally as well as through interactions with their students. They had grave concerns regarding the consequences for teaching evolution, and their concerns were mostly religious in nature, causing them much personal stress. This group felt the least comfortable with the content as well and was more likely to present the content as controversial. Griffith and Brem's (2004) categories correlate, roughly, with the three groups identified by Berkman et al. (2008): the 28 % who present evolution as a unifying theme of biology, the cautious 60 %, and the 13 % who present creationism as scientifically credible.

These findings are not unique to educators who teach evolution but can be extrapolated to include those who address any scientific topic that has a social or political controversy associated with it, such as climate change. Rather than religious in nature as with evolution, the challenges to climate change are of a more political nature (Rosenau 2012). Still, with climate change, one would expect a similar breakdown of the educational community, with some teachers (such as those in the *scientist* group) choosing to present the scientific consensus forthrightly, others (the *selective* group) focusing on those areas with the least bit of controversy, and a third group choosing to misrepresent the science as controversial rather than to present the scientific consensus.

Regarding climate change, far more educators may fall into this last group than with evolution. Rather than reflecting their ideological preferences, though, it may be a misguided pedagogical choice on their part. In a study of science teachers in Colorado, Wise (2010) found that a large percentage (85 %) of earth science teachers who taught about climate change felt that they should be teaching *both sides* of the public controversy (p. 297). Their reasoning varied, however; 25 % reported that they thought their *both sides* perspective actually represented the science well, 50 % reported they used it as a critical thinking activity, and 25 % reported they taught *both sides* but emphasized the scientific consensus. Few of the teachers reported having encountered pressure not to teach about climate change, so their concerns about conflict presumably arose internally or through anticipation of possible controversy.

Though pedagogical approaches focusing on scientific controversy have been touted over the years (Metz 2013; Osborne 2010), there are many pitfalls that educators have not been trained to avoid (Berbeco et al. 2014). First and foremost, there is the question of teacher content preparation. In the aforementioned Colorado study, few teachers reported learning about climate change from college- or graduate-level courses, professional development, or school in-service opportunities. Rather, the majority reported that they learned about the issue from climate-change-specific websites (68 %) or magazines (58 %). This is not unique, as others have found both students and educators utilize the Internet in preference to many other conventional sources such as textbooks to learn about climate change (Berbeco et al. 2013; Leiserowitz et al. 2011). This is also troubling due to the variety of Internet sources with many different agendas—a website maintained by a federal scientific agency such as NASA or NOAA, for example, may give a different account of the science than a source with a more political agenda.

The consequence of the lack of teacher content preparation is clear. It has been noted in several texts that without proper training or sources of information, many current and preservice educators suffer from the same misconceptions as their students. These include holding misconceptions regarding the causes of climate change, often confusing other environmental hazards such as acid rain and ozone depletion with climate change (Groves and Pugh 1999; Papadimitriou 2014; Ratinen et al. 2012; Summers et al. 2001). Interestingly, in some cases preservice educators are more knowledgeable about these topics than educators who have been teaching longer either because it is fresher in their minds or because these topics are more frequently covered in science courses now than in the past (Summers et al. 2001). Whatever the causes of these misconceptions, the reality is that educators cannot possibly address these issues accurately without a better understanding themselves (Groves and Pugh 1999).

Though researchers have looked into the psychosocial impacts of addressing with students climate change and natural hazards and many have focused specifically on the potential to depress or discourage students (Ojala 2012), few researchers have looked into the effect on educators. If there is a class of educators who are concerned about the impact such discussions will have on their students (i.e., in the terms of the Griffith and Brem 2004 evolution study, the conflicted group), then perhaps a way to support them would be to help eliminate some of their internal conflict and associated concern. In a study of both pre- and in-service teachers, Lombardi and Sinatra (2013) looked at how teachers' emotions affected the perceived truthfulness of new information related to climate change. They found that the teachers' emotions were a significant predictor of their perceptions of the information. Moreover, the more eager educators were to reach a conclusion, the less likely they were to perceive climate change as plausible. This indicated that the individuals, in their desperation to come to a quick resolution, ended up relying too much on their own experiences or background knowledge than on new sources of information. Interestingly, educators who did not teach about climate change demonstrated greater anger and urgency to come to a decision than preservice teachers or in-service teachers who do teach about climate change—suggesting that the former group had a greater amount of anxiety related to the issue, which perhaps related to their willingness to address climate change in their classroom.

Based on this research, it would be easy to argue that it is simply the emotional position of educators that limits their ability to teach about climate change and natural hazards. But this argument would be far too simplistic, neglecting the evidence for a number of further sources of limitation. These findings are important to recognize when trying to engage educators. As with students, so with educators: there is not just a single problem such as lack of knowledge. Not only the lack of core knowledge, but also the lack of educational opportunities and resources and their own cognitive and emotional states coming to the topic, all tie together to make climate change and the related natural hazards a challenging topic for educators to address. This suggests that when educators use *teaching both sides* as a pedagogical approach to address climate change, it may be a way to circumvent some of these challenges of unfamiliarity with the science and discomfort with its implications. Teachers, being human, are not immune to their own feelings about and concerns with these difficult topics, and it may be easier from their perspective to credit the possibility that perhaps the science is still uncertain rather than face the harsh possibilities of an uncertain future.

16.3 New Leverage Points for Engagement with Educators

Rather than castigate educators for actions, behaviors, misunderstandings, and positions that are shared by the larger population, it is important for the larger educational and scientific communities to consider how best to support them in their activities. Though educators may be alone in the classroom, a stronger support network for all of these issues (from the scientific to the political to the emotional) would help to strengthen educators' confidence to present the scientific consensus, rather than throwing up their arms in a questionable pedagogical approach. There are many opportunities and leveraging points for both the professional development and scientific community to work with these educators in the coming years, a few of which will be discussed below.

National Climate Assessment The National Climate Assessment (NCA) is an interagency report representing contemporary research on the current and predicted impacts of climate change on the United States. First published in 2000, the third report came out in 2014, was written by over 300 authors, and went through a rigorous public and expert review process. The report is split into different sector and regional sections, including topics such as energy and agriculture (US Global Change Research Program 2014).

As a comprehensive review of the scientific literature, the NCA has the unique benefit of being a scientific document written for the lay audience, making it available to science teachers of all levels. It also addresses climate change on a regional, rather than global, scale, highlighting the natural hazards associated with each part of the country. This allows educators to use local information to discuss the associated impacts with students. The NCA, though, is the beginning of the conversation, not the end, for teaching students only about the impacts of climate change, specifically natural hazards, could be considered cruel, leaving them dismayed and disturbed by the prospect of such a challenging future ahead. The purpose of using such a document in the science classroom is not to terrify, but to engage—and engaging students in general with a serious dilemma and asking them then to devise solutions are an ideal learning opportunity. The third NCA is available online at http://nca2014.globalchange.gov/, and Learning Pathways developed to help unpack the regional chapters are available at http://climate.gov/teaching/2014-national-climate-assessment-resources-educators, NOAA's Climate.gov website.

State Science and Language Arts Standards There are many opportunities within the curriculum for educators to engage students in discussions both of climate change and natural hazards, and, with the adoption of new science, language arts, and mathematics standards across the country, the alignment with classroom activities will become even easier. Both the Next Generation Science Standards (NGSS 2013, available at http://www.nextgenscience.org/) and the Common Core Language Arts and Math Standards (http://www.corestandards.org/) provide opportunities for student inquiry of climate change and related topics.

While Common Core standards have been adopted by most states, the newer Next Generation Science Standards, released in 2013, are in the process of being adopted. Though this process may take several years, this set of K–12 science standards covers a wide range of topics relating to natural hazards, climate, energy, and risk assessment and reduction. The standards were developed through a state-led collaboration of 26 states and sponsored by the National Research Council, the National Science Teachers Association, the American Association for the Advancement of Science, and Achieve. Natural hazards and climate change fit directly or indirectly in multiple places within the standards (NGSS Lead States 2013). These standards are clear in their expectations of the inclusion of climate change and specifically how human activities have led to recent climate conditions.

For example, NGSS requires students to look at the evidence of warming, utilize models to simulate energy flows, and analyze authentic data related to global climate change. Furthermore, under the headings of *Human Impacts* and *Human Sustainability*, students are asked to analyze authentic data related to natural hazards to inform technology development for potential mitigation. Examples of natural hazards, of course, would include more obvious hazards such as volcanic eruptions and earthquakes, but severe weather events such as flooding and extreme precipitation events, which are expected to increase in frequency and severity due to climate change, would be covered, too. In high school, these themes are further developed to make a connection among natural hazards, climate change, and natural resources. Examples of these connections include how climate change will drive migration from changes in sea level rise, influence temperature and precipitation, and affect agriculture.

Importantly, rather than using this opportunity to alarm students about hazards and impacts, the standards ask for students to use scientific and engineering principles to monitor and minimize the problems, emphasizing responses and solutions rather than focusing only on the problem, which can be overwhelming. This allows educators to talk about management, technology, and policy options for addressing natural hazards related to climate change—not to encourage students to be blasé about the consequences, but encourage students to recognize the associated opportunities.

By engaging students in a discussion of the technical and/or policy opportunities and responses (without being dogmatic in approach), educators also have an opportunity to address the more thought-provoking question: What should we do? This allows educators to discuss a very challenging and potentially distressing topic with their students in an engaging manner—starting with the core science based on the scientific consensus and then enriching the experience by asking students to come up with potential mitigation or adaptation strategies for dealing with the consequences.

Perhaps surprisingly, the Common Core State Standards, K–12 English language arts and mathematics standards, also provide an opportunity to incorporate climate change content and problem-solving skills into the curriculum. As with the NGSS, the development of Common Core was initiated by education chiefs and governors and their representative organizations from 48 states, with teachers, administrators, and others providing input. Though their focus is math and English language arts, Common Core contains a set of standards devoted entirely to building science literacy, which can be leveraged to engage students in learning how to develop arguments that utilize the science to bring it out of the lab and into the public sphere.

In the early-year Common Core standards, students are asked to evaluate sources and distinguish among facts, judgments, and speculations. With a topic as political and rife with misinformation as climate change and its associated natural hazards, this is an ideal place for educators to help students wade through misinformation with a knowledgeable guide. In later years, students are asked to evaluate information from many sources to problem solve. This is a great opportunity to answer some of the more troubling issues related to natural hazards with a problem-solving rather than despairing focus (National Governors Association Center for Best Practices & Council of Chief State School Officers 2010).

Both NGSS and Common Core standards allow for a transition from a discussion of the scientific ideas to a more inclusive discussion of natural hazards and climate change, as students are asked to take the data and sources from their science classroom and integrate them into a cohesive discussion of potential resolutions. Through integrating subjects (like English language arts and the sciences), educators have a unique opportunity to deepen students' understanding of the science as well as to develop their scientific problem-solving skills to address larger public issues. Moreover, with new standards being implemented across the country, educators will be seeking out professional development opportunities to learn about these topics. Through leveraging these standards, scientists and other interested parties can encourage educators to use this opportunity to engage their students in a thoughtful and hopeful approach to the materials. Academically and emotionally, this would benefit not only the students but also the educators themselves.

Two important and authoritative resources for teachers that provide an overview of high-quality, vetted online resources relating to climate change and natural hazards are (1) CLEAN (http://cleanet.org), the Climate Literacy and Energy Awareness Network, funded through the National Science Foundation with support from NOAA, and (2) Climate.gov (http://climate.gov/teaching). Both include pedagogical guidance and annotations to hundreds of free, high-quality online resources, including videos, visualizations, and lesson plans, many of which are solution oriented.

16.4 Conclusions: Fostering a Resilient Educational Community

There are many challenges to creating a resilient and confident educational community capable of addressing natural hazards and climate change in a scientifically accurate and pedagogically appropriate manner. This community would benefit from greater support, both academic and emotional, on how best to address these issues through professional development and other training opportunities. In addition, since natural hazards and climate change are a part of the NGSS, educators will be seeking training on this topic, making it a natural place to give them more professional support. Similarly, with the release of the 2014 National Climate Assessment, educators have a unique opportunity to engage in a very user-friendly scientific document that is sector and region specific. This allows them to connect the concepts of climate change and natural hazards more closely to their curriculum, as well as to bring regionally specific issues to their students, making the science and potential solutions more relevant.

It is clear, though, that professional development and additional resources will not be enough to assist educators in bringing these ideas into their classroom. In order to get teachers to engage with these topics in a thoughtful and effective manner, their own feelings and concerns cannot be overlooked. Both their own potential discouragement and classroom conflicts are real challenges that need to be addressed for a teacher to feel confident in teaching a depressing and politically challenging topic in what otherwise would be a fairly straightforward and apolitical science course. The challenges to a resilient and capable educational community are not limited to mere knowledge but require a thorough investigation of all of the conflicts, internal and external, that can arise by addressing a scientific issue with complex political and social implications. Appropriate and meaningful responses to natural hazard and climate change risks based on a firm understanding of the science are key to engaging learners and countering feelings of being overwhelmed. A resilient community not only knows the science and how to use it, but also understands how to manage the personal, political, and social challenges when the topic intersects a public debate.

References

Berbeco M, McCaffrey MS (2014) Infusing climate and energy literacy throughout the curriculum. In: Drake JL, Kontar YY, Rife GS (eds) New trends in earth-science outreach and engagement. Springer, Cham, pp 155–163

Berbeco MR, Stuhlsatz M, White L, McCaffrey M (2013) Understanding global change needs assessment. Unpublished data

Berbeco M, McCaffrey M, Meikle E, Branch G (2014) Choose controversies wisely. The Science Teacher Apr/May:8–9

Berkman MB, Plutzer E (2011) Defeating creationism in the courtroom, but not in the classroom. Science 331:404–405

Berkman MB, Pacheco JS, Plutzer E (2008) Evolution and creationism in America's classrooms: a national portrait. PLoS Biol 6(5), e124. doi:10.1371/journal.pbio.0060124

Boyes E, Chuckran D, Stanisstreet M (1993) How do high school students perceive global climatic change: what are its manifestations? What are its origins? What corrective action can be taken? J Sci Educ Technol 2(4):541–557. doi:10.1007/BF00695323

Boykoff MT, Boykoff JM (2007) Climate change and journalistic norms: a case-study of US mass-media coverage. Geoforum 38(6):1190–1204. doi:10.1016/j.geoforum.2007.01.008

Davenport C (2014) Using weathercasters to deliver a climate change message. New York Times, 6 May 2014. http://www.nytimes.com/2014/05/07/us/politics/using-weathercasters-to-deliver-a-climate-change-message.html?_r=0. Accessed 5 Aug 2015

Griffith JA, Brem SK (2004) Teaching evolutionary biology: pressures, stress, and coping. J Res Sci Teach 41(8):791–809. doi:10.1002/tea.20027

Groves FH, Pugh AF (1999) Elementary pre-service teacher perceptions of the greenhouse effect. J Sci Technol Educ 8(1):75–81

Hamilton College (2007) Climate change and environmental issues poll, p 25. http://www.hamilton.edu/documents//news-sports-events/HCClimateChangePoll.pdf. Accessed 5 Aug 2014

Hamilton LC, Cutler MJ, Schaefer A (2012) Public knowledge and concern about polar-region warming. Polar Geogr 35(2):155–168. doi:10.1080/1088937X.2012.684155

Harrington J (2008) Misconceptions: barriers to improved climate literacy. Phys Geogr 29(6):575–584. doi:10.2747/0272-3646.29.6.575

Hassol S (2008) Improving how scientists communicate about climate change. Eos 89(11):106–107

Hicks D, Bord A (2001) Learning about global issues: why most educators only make things worse. Environ Educ Res 7(4):413–425. doi:10.1080/13504620120081287

Huertas A, Kriegsman R (2014) Science or spin ?. A report by the Union of concerned scientists, Washington, DC, p 12

Leiserowitz A, Maibach E, Roser-Renouf C, Smith N (2010) Climate change in the American mind : Americans' global warming beliefs and attitudes in June 2010. A report by the Yale Project on Climate Change, New Haven, pp 1–9

Leiserowitz A, Smith N, Marlon J (2011) American teens' knowledge of climate change. A report by the Yale Project on Climate Change, New Haven, p 63

Leiserowitz A, Maibach E, Roser-Renouf C, Feinberg G, Rosenthal S, Marlon J (2013) Climate change in the American mind: Americans' global warming beliefs and attitudes in November 2013. A report by the Yale Project on Climate Change, New Haven, p 65

Lombardi D, Sinatra GM (2013) Emotions about teaching about human-induced climate change. Int J Sci Educ 35(1):167–191. doi:10.1080/09500693.2012.738372

McCaffrey MS, Buhr SM (2008) Clarifying climate confusion: addressing systemic holes, cognitive gaps, and misconceptions through climate literacy. Phys Geogr 29(6):512–528. doi:10.2747/0272-3646.29.6.512

Metz S (2013) Let's argue. Sci Teach 80(5):6

National Governors Association Center for Best Practices & Council of Chief State School Officers (2010) Common core state standards for English language arts and literacy in history/social studies, science, and technical subjects. Washington, DC. http://www.corestandards.org/ELA-Literacy/. Accessed 5 Aug 2014

NGSS Lead States (2013) Next generation science standards: for states, by states. Washington, DC, pp 1–103. http://www.nextgenscience.org/next-generation-science-standardsAccessed. Accessed 5 Aug 2014

Ojala M (2012) Hope and climate change: the importance of hope for environmental engagement among young people. Environ Educ Res 18(5):625–642. doi:10.1080/13504622.2011.637157

Osborne J (2010) Arguing to learn in science: the role of collaborative, critical discourse. Science 328(5977):463–466. doi:10.1126/science.1183944

Papadimitriou V (2014) Prospective primary teachers' understanding of climate change, greenhouse effect, and ozone layer depletion. J Sci Educ Technol 13(2):299–307

Ratinen I, Viiri J, Lehesvuori S (2012) Primary school student teachers' understanding of climate change: comparing the results given by concept maps and communication analysis. Res Sci Educ 43(5):1801–1823. doi:10.1007/s11165-012-9329-7

Rosenau J (2012) Science denial: a guide for scientists. Trends Microbiol 20(12):567–569. doi:10.1016/j.tim.2012.10.002

Summers M, Kruger C, Childs A (2001) Understanding the science of environmental issues: development of a subject knowledge guide for primary teacher education. Int J Sci Educ 23(1):33–53. doi:10.1080/09500690116990

US Global Change Research Program (2014) National Climate Assessment. Washington, DC, p 827. http://nca2014.globalchange.gov/. Accessed 5 Aug 2014

Wise SB (2010) Climate change in the classroom: patterns, motivations, and barriers to instruction among Colorado science teachers. J Geosci Educ 58(5):297–309. doi:10.5408/1.3559695

Chapter 17
Communicating Uncertainty: A Challenge for Science Communication

Simon Schneider

Abstract Uncertainty is supposed to be the most underrated element of science communication. While being a crucial part of scientific research, uncertainty is perceived by the recipients of science communication as unwanted and prone to create mistakes and malfunction in everyday life. Nevertheless, uncertainty can and should play a significant role in science communication. A growing number of studies deals with different aspects of uncertainty and gives valuable ideas on how to implement the concept of uncertainty into science interpretation and communication. This article will draw the attention to the multidimensional characteristics of uncertainty. Since decision-makers have to consider variability, uncertainty, and probability into feasibility analysis, we show that the communication of the scientific surplus value of these concepts must be centered within modern science communication efforts. A special focus will be on the use of the concepts of uncertainty and variability within interpretive and educational programs in museum environments.

Keywords Science communication • Uncertainty • Science in museum • Communication concepts • Perception of science

17.1 Introduction

Today, by raising the awareness of scientific uncertainty as a driving force for scientific development and evolution, the media perspective on uncertainty is changing. Science journalists more and more realized that communicating uncertainty has become a crucial part of talking about science. Journalists tend to highlight the differences of various climate models, the uncertainty in the distribution of mineral raw materials, or the effects and results of volcanic unrest observed. While this process now is running within the media system, the value of uncertainty still is

---

S. Schneider, Dipl. (✉)
Department of Political and Social Sciences, Institut für Publizistik und Kommunikationswissenschafte, [Institute for Media and Communication Studies], Freie Universität Berlin, Berlin, Germany
e-mail: simon.schneider@fu-berlin.de

J.L. Drake et al. (eds.), *Communicating Climate-Change and Natural Hazard Risk and Cultivating Resilience*, Advances in Natural and Technological Hazards Research 45, DOI 10.1007/978-3-319-20161-0_17

underestimated in the public and in politics. Therefore, science communicators are in need for new and innovative ways to talk about scientific uncertainty.

One of the fundamental driving forces of scientific and technological evolution is uncertainty. Nevertheless, uncertainty has "many different dimensions …, and there is a lack of understanding [by decision makers] about their different characteristics, relative magnitudes, and available means of dealing with them" (Walker et al. 2003, p. 5).[1] Uncertainty in science is crucial to develop new questions and hypothesis, to search for processes and procedures, and to gain more knowledge about how our environment works. Uncertainty is a good thing—as long as one accepts uncertainty as a valuable part of science and research. Without uncertainty, science legitimacy and reputation would be at stake. Nevertheless, like nothing else, uncertainty supports the saying that the pen is mightier than the sword. Communicating science without sharing uncertainty about an issue may create "negative values, if it induces unwarranted confidence or is so hesitant that other, overstated claims push it aside" (Fischhoff 2013, para. 1). If the knowledge about uncertainty is used in an intentional miss informational way or – even worth – not mentioned at all, science communication can create massive destruction and foster the loss of science's credibility. By using uncertainty as part of propaganda, it has evolved into a weapon of Astroturf organizations as well as of lobbyist groups. To increase the recognition of the value of uncertainty by decision-makers and the public, science communication has to create new and more effective ways of communicating uncertainty.[2]

The key issue about scientific uncertainty is not to avoid uncertainty. Scientists know that there is nothing like true certainty within research. The only scientific discipline, which might be close to the elimination of uncertainty, is math. Whether it is economics, natural science, or social science, even the arts and humanities deal with uncertainty in their daily routine. So it is not the fact that there is a whole lot of uncertainty around; it is the way scientists treat uncertainty that constitutes good science and good scientific behavior. As Webster declared: First, science has to quantify the amount of uncertainty within its research and the effects of uncertainty for future outcomes. Then, if the amount of uncertainty is known, science has to communicate the quantified uncertainty (Webster 2003).

But the first step alone bears numerous traps and a multitude of possibilities for misinterpretation and misuse. Webster sees "empirical, methodological, institutional, and philosophical challenges" to identify and quantify uncertainty within science (Webster 2003, p. 1). This is grounded in the fact that uncertainty has innumerable sources. Therefore, a multitude of strategies, mostly adopted from risk management procedures, have been developed to get an idea about the uncertainty level. These strategies reach from sensitivity analysis to Monte Carlo simulations.

[1] Walker et al. (2003, p. 5) provide a "general definition of uncertainty as being any deviation from the unachievable ideal of completely deterministic knowledge of the relevant system."

[2] May (2001, p. 891) comments the necessity of communicating uncertainty in his column for Nature, in which he states that "it helps to recognize, and explicitly acknowledge, these uncertainties."

Despite its major role in science, uncertainty only plays a minor role in museums and most other channels of science communication. Strategies to communicate uncertainties have not yet integrated the value of uncertainty in its presentation of science and research. Museum administration and curators shy from sublime associations of uncertainty such as untruthfulness, unpredictability, delusion, and misdirection. But they also struggle with communicating accompanying concepts such as likelihood, probability, or plausibility. Nevertheless, a broadly based knowledge and familiarization with these terms is crucial for communicating uncertainty. Therefore, neuroscience and the evolutionary social sciences are currently searching for a new theory for the communication of uncertainty. Their efforts are focused on the risk communication, which is closely related, if not synonymous to communication about uncertainties. As Thompson (2002, p. 648) illustrates, risk assessment, risk management, and risk communication are strongly based on the "consideration of variability and uncertainty".[3]

By analyzing two case studies, Thompson (2002, p. 652) draws the conclusion that "… at a basic level these cases would suggest that variability and uncertainty in risks should be explicitly considered and addressed to ensure that ignoring them does not mislead either the risk manager or the public." Hence, results from research on risk communication can be transferred to the communication of uncertainty. Promising achievements were recently made within neuroscience (see, e.g., a case study by Schmälzle et al. 2013 or an elaboration by Slovic 2011). Here, a deep understanding of the processes and principals of "the way in which the human brain is evolved and organized to process information relevant to the perception of risk" (Tucker and Ferson 2008, p. 10) is conceptualized and awaits further evaluation.[4]

Over the last decade, the need to communicate uncertainty increased. Climate sciences and environmental sciences have faced massive propaganda campaigns by global industry and Astroturf organizations[5] such as the Heartland Institute in the USA or the IZ Klima (Information Center Climate) in Germany. These organizations use the deep societal mistrust in uncertainty to point out alleged unethical and intentional delusion of decision-makers and the public by scientists and their consultatory function. Scientists, who openly communicate uncertainty of climate model calculations, earthquake occurrence frequencies, or possible side effects of genetic manipulated semen, have to face massive campaigns against their research—and sometimes against their person or even their lives. Hence, new strategies to

[3] Thompson distinguishes between variability and uncertainty. Nevertheless, he shows that both concepts have similar impact on risk communication – therefore, it is justifiable that for the purpose of this article, both concepts are combined into the term "uncertainty."

[4] By interpreting numerous studies by anthropologists, psychologists, economists, and neuroscientists, Tucker reasons, "…that recent research in neuroscience and in the evolutionary social sciences is developing just such an explanatory and predictive theory." He does not explicitly quote such studies, by mentions, that they are "… providing broad-based experimental and observational support for this new perspective" (Tucker and Ferson 2008, p. 10).

[5] An interesting study about the impact of Astroturf communication has been conducted by Cho et al. (2011). They also give a detailed summary of recent research on astroturfing.

communicate uncertainty have to face the societal roots of the misunderstanding of the concept of uncertainty itself.

Evolutionary biology has shown that humans are empowered with "sensory structures well suited for practical decision making in uncertain environments" (Tucker and Ferson 2008, p. 10). Tucker and Ferson draw the conclusion that by presenting data "in formats the brain is adapted to understand," many of the irrational concepts about uncertainty are mitigated (Tucker and Ferson 2008, p. 10). The popular definition of uncertainty[6] is closely related to the definition of *vagueness* given by Frisch and Baron. They define vagueness as a "subjective experience of a lack of information that is relevant to a prediction" (Frisch and Baron 1988, p. 152). Thus, sociological models and theories about vagueness (also synonymously called ambiguity) are transferable to the communication of uncertainty.

Research on how geopolitical issues such as climate change policies are presented by the media (Cox 2010; Nisbet 2009) has shown that these issues are often framed in terms of potential gains or losses (e.g., Gifford and Comeau 2011; Reber and Berger 2005). Therefore, it seems that the understanding of the cognitive psychology of communicating uncertainty is in first order based on an overarching gain/loss framing effect. Kuhn (1997) analyzed the impact of uncertainty to the process of decision-making, which finally is dominantly driven by preconceptions about terms such as *uncertainty*, *vagueness*, or *probabilities*.[7] This leads to the common notion that most people adopt a cautious attitude when confronted with uncertainty. According to Kuhn (1997), "Vague probabilities of gains are underweighted relative to their anchors, and vague probabilities of losses are overweighed" (p. 57).[8]

17.2 Interpreting Uncertainty in New Ways

Take fracking, mountaintop mining, shale gases, sea level rise, or even natural hazard early warning as examples for scientific communication based on outstanding research while loaded with scientific uncertainty. Parallel to the increasing role of scientific uncertainty in strategic communication, science communicators developed a number of techniques to master the challenge of putting uncertainty in the focus. This meets the demand that "information on variability and on low-probability high-consequence events allows decision makers to account for society's risk-aversion" (Webster 2003, p. 1) within the decision-making process. Communication

---

[6] Uncertainty: something that is doubtful or unknown, retrieved from Merriam-Webster, www.merriam-webster.com (Merriam-Webster. www.merriam-webster.com. Accessed 1 Jul 2014).

[7] Other models for overarching framing effects include, for example, the concept of post-normal science by Funtowicz and Ravetz (1991).

[8] Kuhn refers to the prospect theory, developed by Kahneman and Tversky (1979), as well as to the decision-making model by Einhorn and Hogarth (1985) who introduced the individual attitude towards vagueness and its implications on the decision-making process (described as an overarching gain/loss framing effect).

offices of universities and research facilities today only allow scientific data to be published if probabilities or variations in results and models are shown. Scientists are advised to highlight uncertainty in press conferences and media releases. Nevertheless, by dealing with the media, science has to follow the logic of mass media,[9] leading to the observable phenomena called mediatization of science.[10] Seen from a press officer perspective, science still has to be spectacular and epoch-making.

On the other side, "media coverage of climate science in the policy context has often depicted it as an area of vigorous controversy to a degree that active climate scientists find unrecognizable" (Manning 2003, p. 9). In this example, climate science stands exemplary for other sciences like genetic engineering, the epidemiology of new diseases, and many others. Public relations specialists from science organizations are used to this. Professional science communicators adapt the media logic to communicate topics, which are of high scientific relevance, but shaped by uncertainty. As a result, the audience will read about spectacular findings while being confronted with abstract concepts such as likelihood, probability, or standard variations. By being scientifically correct, the inerrant nimbus of science is damaged. And at the end, the human mind makes the connection between uncertainty and failure. To initiate a paradigm change in interpreting scientific uncertainty, science communication faces one of its biggest upcoming challenges.

17.3 Museums Are a Fundamental Part of Science Communication

Cameron (2012) observed, "Climate change is a vast, complex phenomena impinging on biological and social life, economics, politics, and culture" that stretches a multitude of disciplines beyond current limits (p. 317). Referring to Hodge (2011), she draws the conclusion that climate change and its accompanying sciences invite "a new, critical examination of the roles and capacities of museums and science centers in these complex ecologies" (Cameron 2012, p. 317). Museums, while currently in the stage of reshaping from education-centered institutions to dialogue-encouraging venues, seek new topics. For example, presenting geology by showing off rock samples will be substituted by a concept that allows visitors to experience Earth and Space sciences as a profession, a fundament for society, and an entertaining discipline.

More and more, it is obvious that museums have mostly neglected Earth and Space sciences as one of the key disciplines for a sustainable future. This is especially true if focused on climate change sciences. Until now, only a few museums were dedicated to Earth and Space sciences; most natural museums focused on

---

[9] See Weingart (2005) for a detailed analysis of the increased orientation of science towards the media.

[10] See Rödder and Schäfer (2010) for a brief introduction into ongoing debates about mediatization.

natural history and taxonomy. They present minerals, volcanoes, or earthquakes, but usually from the phenomenological perspective. The utilization, economic value, hazard potential, and socioeconomic impact of Earth and Space sciences have not been on the agenda of museum curators. With the transition of the image of science from an aloof profession into a much-needed sociological function, museum administrators are changing their attitudes towards science.

Today, overarching themes like raw materials or sustainable energy are leading topics in new museum story lines. Accompanied by the increasing awareness that museums do play a major role in informing policy makers, a fundamental self-assessment currently triggers the transition from asymmetric museum communication to a place for dialogue-driven debate. Museums no longer simply provide information. They turn into venues for conversation, education, and entertainment and they "activate and broker discussions and decisions" (Cameron 2012, p. 317) around critical scientific topics.

Museums also tend to focus on the scientists rather than exclusively on the factual science and its results. While a large number of museums currently undergo this transition, museum professionals have adapted Earth and Space sciences as a leitmotif for their new exhibition concepts. Raw materials are no longer limited to presenting hundreds of colorful minerals. Modern museum concepts deal with the sociological conflict potentials in mining areas, with environmental friendly processing technologies and with recycling concepts. Always present is uncertainty. In this example is the uncertainty of resource quantity, of accessibility, and of political stability. Nevertheless, uncertainty is not a bold keyword in museum concepts. Uncertainty still has the aura of failure, errors, and mistakes. The value of scientific uncertainty is not yet fully recognized by communicators or museum professionals.

17.4 The Wegener Example

Now imagine the possibilities of communicating uncertainty. Think about a story of trial and error, finally leading to a successful and widely accepted new theory. Think, for example, about Alfred Wegener. He earned laughter and harsh critiques when first publicly talking about moving continents. After his speech about the new theory of plate tectonics in January 1912 in Frankfurt, Germany, all he got were comments along the lines of "Nonsense! What evidence do you have? None! Who do you think you are to question a long tradition of geological research?"

Now think about the following decades in Earth and Space sciences. Research vessels discovered the magnetic pattern of the ocean floor sediments in the Atlantic. Similar fossils were found in western Africa as well as in eastern South America. Rock samples from Brazil and Namibia showed striking similarities, leading to the idea of a single source for both samples. Decade after decade, new evidence was found that Wegener's theory was right, that continental movement has formed the appearance of Earth as we know it today, that plate tectonics was, in its principal idea, right. But still, there was uncertainty within the theory. What were the driving forces? How do continents behave if they collide? And a lot of other questions aroused with every piece of information that fitted the theory. Today, we do have a very profound theory of plate tectonics. We have an idea about the forces and the collision processes. But still, there is uncertainty. What role does water play, trapped in rocks in the deep interior of Earth? What will our planet looks like in ten, hundreds, or thousands of millions of years?

Uncertainty drives geoscientists to think about plate tectonics, leading to new insights into earthquake dynamics, volcanological processes, and the building mechanisms for raw materials. And all these are based on uncertainty and a theory pregnant with more questions than answers. Think about this story, the emotions, the meaning for our lives today and for the lives of future generations. This is the material museum professionals are looking for while creating new exhibition concepts.

17.5 Analyzing Uncertainty

There is a procedure that helps science itself to deal with uncertainty. This procedure alone is a great story for museum communication in itself. Earlier, we mentioned the methods scientists use to determine the quantity of uncertainty within their research. Sensitivity studies, for example, are created to observe the impact of small changes of single parameters to the results of experiments or model calculations. By choosing extreme, but nevertheless realistic model values for each parameter, scientists can identify the range of uncertainty of their model. In a following step, the model values are tested in relation to the likelihood of their appearance. This will create a feeling for the probability of extreme results (on both sides, minimum as well as maximum). Finally, the so-called Monte Carlo simulation is used to identify the frequency of occurrence for extreme results, leading to a robust statement about the uncertainty for specific results of the analyzed process based on the input assumptions. This procedure, while at first intimidating, is a simple and successful tool to get an idea about the uncertainty within a scientific observation.

Now think about an exhibition storyline based on this process. Think about climate change and how to communicate the impact of glacial meltdown to sea level rise. You can start by talking about the ice masses that are currently present on Earth. Present new ideas like how to measure the total mass as well as the melting rate of inland ice by satellite remote sensing analysis of the variation of the gravitational forces. Now you can start to calculate different scenarios: low-melting rates compared to high-melting rates. This will result in two different model predictions for a future sea level rise. Now highlight the real conditions and problems when measuring the actual ice-melt rates in Greenland. In situ measurements, satellite data changes in precipitation and so on. All these processes influence the data, leading to different scenarios for sea level rise. A narrative can be accompanied by human interest stories like the first all-female arctic researchers crew or the journeys of arctic explorers. You can show satellite images and remote sensing technology, a couple of in situ apparatuses for measuring the duration and intensity of sunshine in the arctic regions, outdoor gear for arctic scientists, and many other artifacts as well. This is a great interpretive story about science, based on uncertainty.

17.6 Social Precipitation of Uncertainty

By treating uncertainty as system immanent, as something that can be taken into account, uncertainty will lose its image as a parameter for unprofessionalism. If uncertainty is communicated with as much transparency as possible, the transition into a new assessment of uncertainty by society will be initiated. The public as well as political decision-makers will start to appreciate uncertainty as a valuable part of science communication. Still, communicating uncertainty requires that science builds an internal process "that tracks our understanding as it evolves and a policy process that can adapt and respond to new information" (Webster 2003, p. 8). And there is a second dimension to this transition: The professional misconduct of uncertainty as a weapon of propaganda will lose its ground. Hence, Astroturf organizations and lobbyist groups would no longer be able to use uncertainty to discredit research, if uncertainty is seen as an integral concept within science. As Webster (2003) states:

> The possibility of being wrong (as some of our results will inevitably prove to be) is not an argument for not doing the best we can to be explicit about what we know, what we don't know, and where we disagree for the moment. (p. 8)

In the end, communicating uncertainty highly depends on communication skills. The two most powerful tools are wording and the use of graphics. Both are also most prone to misinterpretation. For example, the potential of misinterpretation by an imprudent use of subjective word-based classification of uncertainty is widely underestimated by scientists. Fischhoff points out, "we do not intend the term *unlikely* to imply an event will not happen. We use *probably* and *likely* to indicate there is a greater than even chance" (Fischhoff 2013, section "What we mean when we say: an explanation of estimative language," para. 3). The subjective relationship of terms used in science communication to provide information about uncertainty and probability reaches from *remote* to *almost certain*, meaning an objective probability from nearly zero (remote) to close to 100 (almost certain).

But there is a large difference in how people choose to operationalize these terms. Communication science highlights that the message is interpreted by the receiver. Terms used to subjectively characterize uncertainty will be perceived and interpreted differently by each individual depending on personal interests, knowledge, and many other factors. Furthermore, the use of words to describe *uncertainty* and *probability* by scientists might be different from the way lay readers typically do.[11] Thus, scientists and science communicators should find a consensus on how to

[11] Patt and Schrag showed this in an experiment, which used the Third Assessment Report by the Intergovernmental Panel on Climate Change (IPCC). They finally asked if the IPCC intentionally

use a standardized set of uncertainty-related words, especially if these words are used in communicating risk and uncertainty.

The other important technique to communicate uncertainty is using graphic visualizations. In a study about risk and uncertainty communication in respect to seismic hazards, Bostrom et al. (2008) mentioned that "studies of visual representation of risk to date have focused on statistical graphics and symbols" (p. 30). While at the same time the utilization of animations, graphics, and pictures to communicate uncertainty and risk are not analyzed in detail, social and communication scientists are convinced that graphic visualizations do have a higher impact as language on the perception of said issues.[12] There is robust knowledge about the effects and interpretation of color (e.g., that red is—at least in western societies—associated with danger[13]). However, the effects of charts, graphs, maps, interactive displays, or animations are still not fully understood.[14]

17.7 The Montauk Guidance

At the 2006 Montauk meeting, scientists from neuroscience, psychometrics, and evolutionary biology and risk scientists developed a portfolio of *quick response* tools for communicating risk. These guidelines are transferable to communicating uncertainty, since both issues are tightly connected to each other. Next to many other strategies for communicating uncertainty,[15] these guidelines can help to integrate scientific uncertainty into science communication.

The participants of the 2006 meeting claimed that the use of natural frequencies will help to communicate uncertainties to a lay public. They criticize that single-event probabilities, relative frequencies, and percent values are far too abstract concepts that will lead to misinterpretation and misunderstandings about uncertainty and risk (see Kurz-Milcke et al. 2008). Natural frequencies such as weekly, yearly, or others are strongly connected to the daily experience of humans. Therefore, these values do have meaning to the audience and should therefore be preferred in communication with nonexperts.

Another recommendation is to raise the communicator's awareness of differences in numeracy and mental models of risk and uncertainty. By dealing with numbers and terms like *probability*, *likelihood*, or *distribution*, the expert learns that

---

misused wording to lead readers to underestimate the probability of high-magnitude possible outcomes (Patt and Schrag 2003).

[12] A critical review on the visual representation of information can be found in Bresciani and Eppler (2008). A case study with an elaborated review of extant literature about the use and perception of text versus graphics is presented by Hochpöchler et al. (2013).

[13] A comprehensive review of extant literature, for example, on the relation between color and psychological functioning can be found in Elliot et al. (2007).

[14] For a brief summary of various researches on the effects of graphic visualization for risk and uncertainty communication, see Bostrom et al. (2008).

[15] See, for example, Gigerenzer (2002), Lofstedt and Cvetkovich (1999), or Ropeik and Gray (2002).

individual definitions of these terms vary significantly (see Peters 2008; Fischhoff 2013). The perceptions of these concepts are molded by personal experiences and societal framework of each stakeholder, regardless of the individual role within the communication process. This is true for talking to public and political audiences as well as for communicating within the scientific community itself. Therefore, it seems obvious to integrate a definition of key concepts such as uncertainty, probability, or variability into science communication.

One of the crucial requirements for the communication of risk and uncertainty is the strong coupling of both issues with resulting options of action. Risk and uncertainty cannot be divorced from choice. The context for communicating uncertainty should always contain the contrast of cost and benefit (both in economic and sociological terms) as well as the balance between the results of the communicated uncertainty and risk resulting from actions taken to minimize the uncertainty (Finkel 2008; Stirling 2008; Wang 2008).

Finally, the influence of disagreement among scholars and experts and its impact on uncertainty has to be analyzed in detail. Until today, the factor of disagreement on uncertainty perception is not fully understood, but it represents an important component in creating a neutral notion towards uncertainty. These tools can help to communicate uncertainty and risk more effectively without offering toeholds for abuse by propaganda communication. Nevertheless, the public's concepts and consequences of uncertainty, probability, and confidence in scientific results have to be addressed alongside fostering numerical and scientific literacy in order to increase the awareness, acceptance, and appreciation of uncertainty in science communication.

Currently, social science is analyzing and formulating new models and theories on the influence of different cases of uncertainty to precipitation and communication. Chow and Sarin (2002) introduced the *unknowable uncertainty*, complementing the long-existing assumption of *known* and *unknown uncertainty*. Known uncertainty is related to uncertainty with precisely known probabilities. Unknown uncertainty similarly is meant to describe uncertainty with unknown, but estimated probabilities. The unknowable uncertainty is defined as the uncertainty, "where the missing information is unavailable to all" (Chow and Sarin 2002, p. 127).[16] Science communication is rapidly evolving. The appreciation of professional science communication increases. But within this process, science communicators should not focus only on new demands from a changing media landscape. Professional science communicators should also have the communication of scientific uncertainty on their agenda.

[16]Additional cases of uncertainty include fuzzy uncertainty or hybrid uncertainty (Warren 2007) and other approaches to classify uncertainty.

References

Bostrom A, Anselin L, Farris J (2008) Visualizing seismic risk and uncertainty: a review of related research. In: Tucker WT, Ferson S, Finkel A, Long TF, Slavin D, Wright P (eds) Annals of the New York Academy of Science 1128, Strategies for risk communication evolution, evidence, experience. Wiley-Blackwell, Boston, pp 22–40

Bresciani S, Eppler MJ (2008) The risk of visualization: a classification of disadvantages associated with graphic representations of information. ICS Working paper #1/2008, http://www.knowledge-communication.org/pdf/bresciani-eppler-risks-visualization-wpaper-08.pdf. Accessed 23 Oct 2014

Cameron FR (2012) Climate change, agencies and the museum and science centre sector. Mus Manag Curatorship 27(4):317–339

Cho CH, Martens ML, Kim H, Rodrigue M (2011) Astroturfing global warming: it isn't always greener on the other side of the fence. J Bus Ethics 104(4):571–587. doi:10.1007/s10551-011-0950-6

Chow CC, Sarin RK (2002) Known, unknown, and unknowable uncertainties. Theor Decis 52:127–138

Cox JR (2010) Beyond frames: recovering the strategic in climate communication. Environ Comm 4(1):122–133. doi:10.1080/17524030903516555

Einhorn HJ, Hogarth RM (1985) Ambiguity and uncertainty in probabilistic inference. Psychol Rev 92:433–461

Elliot AJ, Friedmann R, Moller AC, Maier MA, Meinhartd J (2007) Color and psychological functioning: the effects of red on performance attainment. J Exp Psychol Gen 136(1):154–168

Finkel AM (2008) Perceiving other's perception of risk: Still a task for sisyphus. In: Tucker WT, Ferson S, Finkel A, Long TF, Slavin D, Wright P (eds) Annals of the New York Academy of Science 1128 Strategies for risk communication evolution, evidence, experience. Wiley-Blackwell, Boston, pp 121–137

Fischhoff B (2013) Communicating uncertainty: fulfilling the duty to inform, Social Science and Environmental Policy, Issues in Science and Technology, 2013. http://issues.org/28-4/fischhoff/. Accessed 16 Jul 2014

Frisch D, Baron J (1988) Ambiguity and rationality. J Behav Decis Mak 1:149–157

Funtowicz S, Ravetz J (1991) A new scientific methodology for global environmental issues. In: Constanza R (ed) The ecological economics. Columbia University Press, New York, pp 137–152

Gifford R, Comeau LA (2011) Message framing influences perceived climate change competence, engagement, and behavioral intentions. Glob Environ Chang 21(4):1301–1307. doi:10.1016/j.gloenvcha.2011.06.004

Gigerenzer G (2002) Calculated risk: how to know when numbers deceive you. Simon & Schuster, New York

Hochpöchler U, Schnotz W, Rasch T, Ullrich M, Horz H, McElvany N, Baumert J (2013) Dynamics of mental model construction from text and graphics. Eur J Psychol Educ 28(9):1105–1126. doi:10.1007/s10212-012-0156-z

Hodge B (2011) Climate chance and the museum sector: 10 reflections from the 'Hot Science, Global Citizens' symposium. Blog-post, retrieved from: http://www.uws.edu.au/ics/news_and_media/blog/180511. Accessed 16 Jul 2014

Kahneman D, Tversky A (1979) Prospect theory: an analysis of decision under risk. Econometrica 47(2):263–292

Kuhn KM (1997) Communicating uncertainty: framing effects on responses to vague probabilities. Organ Behav Hum Decis Process 71(1):555–583

Kurz-Milcke E, Gigerenzer G, Martignon L (2008) Transparency in risk communication: graphical and analog tools. In: Tucker WT, Ferson S, Finkel A, Long TF, Slavin D, Wright P (eds) Annals of the New York Academy of Science 1128 Strategies for risk communication evolution, evidence, experience. Wiley-Blackwell, Boston

Lofstedt R, Cvetkovich G (1999) Social trust and the management of risk. Earthscan, London
Manning MR (2003) The difficulty of communicating uncertainty. Clim Change 61:1–8
May R (2001) Risk and uncertainty: at the frontier of science, we don't always know what may happen. Nature 411:891
Nisbet MC (2009) Communicating climate change: why frames matter for public engagement. Environ Sci and Policy Sustain Dev 51(2):12–23. doi:10.3200/ENVT.51.2.12-23
Patt AG, Schrag DP (2003) Using specific language to describe risk and probability. Clim Change 61:17–30
Peters E (2008) Numeracy and the perception and communication of risk. In: Tucker WT, Ferson S, Finkel A, Long TF, Slavin D, Wright P (eds) Annals of the New York Academy of Science 1128 Strategies for risk communication evolution, evidence, experience. Wiley-Blackwell, Boston
Reber BH, Berger BK (2005) Framing analysis of activist rhetoric: how the Sierra Club succeeds or fails at creating salient messages. Public Relat Rev 31(2):185–195. doi:10.1016/j.pubrev.2005.02.020
Rödder S, Schäfer MS (2010) Repercussion and resistance: an empirical study on the interrelation between science and mass media. Communications 35:249–267. doi:10.1515/COMM.2010.014
Ropeik D, Gray G (2002) RISK: a practical guide for deciding what's really safe and what's really dangerous in the world around you. Houghton Mifflin, New York
Schmälzle R, Häcker F, Renner B, Honey CJ, Schupp HT (2013) Neural correlates of risk perception during real-life risk communication. J Neurosci 33(25):10340–10347
Slovic P (2011) The perception of risk. Science New Series AAAS 236(4799):280–285
Stirling A (2008) Science, precaution, and the politics of technological risk: converging implications in evolutionary and social science perspectives, Annals of the New York Academy of Science 1128 Strategies for risk communication evolution, evidence, experience. Wiley-Blackwell, Boston
Thompson KM (2002) Variability and uncertainty meet risk management and risk communication. Risk Anal 22(3):647–654
Tucker WT, Ferson S (2008) Strategies for risk communication: evolution, evidence experience. In: Tucker WT, Ferson S, Finkel A, Long TF, Slavin D, Wright P (eds) Annals of the New York Academy of Science 1128 Strategies for risk communication evolution, evidence, experience. Wiley-Blackwell, Boston
Merriam-Webster. www.merriam-webster.com. Accessed 1 Jul 2014
Walker WE, Harremoes P, Rotmans J, van der Sluijs JP, van Asselt MBA, Janssen P, Krayer von Kraus MP (2003) Defining uncertainty: a conceptual basis for uncertainty management in model-based decision support. Integr Assess 4(1):5–17
Wang X-T (2008) Risk communication and risky choice in context: ambiguity and ambivalence hypothesis. In: Tucker WT, Ferson S, Finkel A, Long TF, Slavin D, Wright P (eds) Annals of the New York Academy of Science 1128 Strategies for risk communication evolution, evidence, experience. Wiley-Blackwell, Boston
Warren L (2007) On modelling hybrid uncertainty in information. Command and Control Division, DSTO Defense Science and Technology Organisation, Department of Defence, Australia. http://www.dtic.mil/get-tr-doc/pdf?AD=ADA467752. Accessed 16 Jul 2014
Weingart P (2005) Die wissenschaft der öffentlichkeit. Velbrück, [The science of the public] Weilerswist, Germany
Webster M (2003) Communicating climate change uncertainty to policy-makers and the public. Clim Change 61:1–8

Chapter 18
Science Diplomacy in the Geosciences

A. Ester Sztein

Abstract Scientists who work with international colleagues in their own country or outside its boundaries are, ultimately, practicing science diplomats. This chapter will explore the definition of science diplomacy and its functions, describe the types of individuals that practice science diplomacy, suggest potential resources and list potential challenges, discuss the key role of diaspora in science diplomacy, and outline best practices. It will also describe the global organizations that specifically promote geoscience engagement, such as the international unions for the geosciences, and others working on Earth system science issues, including both academic programs and global bodies such as the Intergovernmental Panel on Climate Change and UNESCO. From this point of view, it will describe the roles and activities of the US National Committees for the international geoscience unions hosted by the US National Academy of Sciences and will present case studies of geoscience diplomacy excellence, such as the US Geological Survey-Office of Foreign Disaster Assistance's Volcano Disaster Assistance Program (VDAP). Though this discussion will be done largely from the US perspective, the underlying concepts should be applicable around the world. The issues of global climate change and natural hazard risk communication in many instances involve interactions across cultures and across boundaries. An understanding of the diplomacy element in those interactions, and the integration of those approaches in the planning and establishment of those programs will increase their likelihood of success.

Keywords Science diplomacy • Diplomat • Geosciences • International scientific collaboration • International Scientific Unions • US National Committees • Capacity building • Volcano Disaster Assistance Program

---

The opinions expressed in this chapter are those of the author and do not necessarily reflect the views of the Board on International Scientific Organizations or the National Academies.

A.E. Sztein, Ph.D. (✉)
Board on International Scientific Organizations,
The National Academy of Sciences, Washington, DC, USA
e-mail: esztein@nas.edu

J.L. Drake et al. (eds.), *Communicating Climate-Change and Natural Hazard Risk and Cultivating Resilience*, Advances in Natural and Technological Hazards Research 45, DOI 10.1007/978-3-319-20161-0_18

18.1 Introduction

Humanity now confronts a series of serious global challenges affecting populations and natural systems. Many of them, such as natural hazards, global environmental change, water and food availability, health, and energy, have their roots in the geosciences. These global challenges transcend political and scientific borders and therefore cannot be addressed by a single country or by a single scientific discipline. International scientific cooperation is necessary for the advancement of science and society in the United States and abroad. Throughout history, science policy and science diplomacy have served to build relationships among countries, raise the status of science across borders, and produce concrete scientific and societal results (National Research Council [NRC] 2012; Sztein and Casadevall 2013).

Science diplomacy can be defined as "the use of scientific collaborations among nations to address the common problems facing twenty-first century humanity and to build constructive international partnerships" (Fedoroff 2009, p. 9), yet the term can be applied far more broadly. Most recently, science diplomacy has become an umbrella term used to describe a number of formal or informal technical, research-based, academic, or engineering exchanges, which some also refer to as *global science cooperation* (NRC 2012, p. 1). Even in cases where political relationships between governments are strained, relationships between scientists, either formal or informal, tend to continue—thus maintaining open communication channels, a key diplomatic asset. In fact, informal meetings at international scientific conferences help scientists in certain countries keep in touch with their international peers and keep abreast of advances and developments during even the most difficult political periods.

The Royal Society held a meeting in 2009 in partnership with the American Association for the Advancement of Science (AAAS) to discuss issues pertaining to science diplomacy and its role in the twenty-first century and published a report (The Royal Society 2010). In that document they state that *science diplomacy* refers to three types of activities: *science in diplomacy*, where science can provide advice to inform and support foreign policy objectives; *diplomacy for science*, where diplomacy can facilitate international scientific cooperation; and *science for diplomacy*, where scientific cooperation can improve international relations (p. 4).

Science diplomacy *policy* is defined by governments and their agencies through science and technology (S&T) agreements and partnerships that could be bi- or multilateral, but the *practice* of science diplomacy is generally in the hands of the scientists themselves. They establish collaborations that range from one-on-one relationships through small research group interactions to bilateral agreements between universities to activities within large programs or international scientific organizations (NRC 2012). International collaboration benefits all countries involved, not just the more affluent countries in the partnership. We can and do learn from each other. Diversity in the composition of the groups that tackle complex scientific problems leads to better, more nuanced solutions, since variety in scientific and cultural training allows for richer discussions (Chua et al. 2011).

18.2 Scientists as Diplomats

A scientist acts as a diplomat, in the broadest sense of the word, when he/she works in close interaction with foreign colleagues outside or within his/her own country. In the United States, they can be working on government initiatives at the political level through bilateral/multilateral science and technology agreements; in governmental entities such as the US Agency for International Development (USAID), the US Department of Agriculture (USDA), the US Geological Survey (USGS), the National Science Foundation (NSF), the National Aeronautics and Space Administration (NASA), the US Forest Service (USFS), or the Department of State; at universities and other institutions establishing one-on-one collaborations or small/large research group collaborations; or in activities organized under the auspices of larger programs (e.g., scientific unions or international organizations/ initiatives).

Scientists can also provide valuable input to government delegations by briefing them as part of their preparation for their diplomatic endeavors. Dialogue between the scientific and the diplomatic communities increases awareness of their respective fields and creates potential opportunities for advancement of their respective missions. One such example is the first course on science diplomacy for scientists and diplomats, organized in 2014 by AAAS and The World Academy of Sciences (TWAS). This workshop, attended by professionals from 32 countries, was held at the International Centre for Theoretical Physics in Trieste, Italy (Lempinen 2014).

Scientists are also active in science diplomacy through their work in a variety of independent, nonprofit organizations ranging from grassroots, such as Geoscientists without Borders (http://www.seg.org/groups-communities/gwb) and the Climate Institute (http://www.climate.org/), to organizations such as AAAS and, in particular, its Center for Science Diplomacy (http://www.aaas.org/program/center-science-diplomacy) and *Science and Diplomacy* journal (http://www.sciencediplomacy.org/), CRDF Global (http://www.crdfglobal.org/), and the US National Academy of Sciences (NAS, http://www.nasonline.org/), to name several. In some cases, non-profit organizations can be more flexible and nimble in their operations than governmental agencies. Nongovernmental organizations are key to successful science diplomacy (Kramer 2010). They facilitate scientific exchanges and provide unbiased, independent scientific advice to the stakeholders that require it.

As an example of *diplomacy for science*, scientists sometimes solicit the help of diplomats to secure access to research sites under national jurisdiction, to obtain permission for collecting data at those sites, and for making mutually agreeable arrangements for sample collection, transport permits, and the archiving of such data and samples. Diplomats help S&T cooperation through the establishment and maintenance of long-term relationships in the country in which they serve. Scientists can also serve as science attachés in embassies, where they have a dual mission: advise the policy corps (science in diplomacy) and assist scientists in their work (diplomacy for science). The Department of State for decades assigned science attachés to many US embassies, where they assisted in the success of many geoscience initiatives (El-Baz 2010). The US program was discontinued in 1997, but some of these functions are covered by scientists in short-term assignments (Linkov et al. 2014). The Science, Technology, and Innovation Expert Partnership (STIEP) Speaker Series is a program established in 2012 by the US Department of State in partnership with professional scientific organizations to facilitate the engagement of US scientists with foreign audiences and stakeholders (American Geosciences Institute 2014). Several US geoscientists have already participated in the program.

When considering *science for diplomacy*, besides bilateral/multilateral S&T agreements, there are other programs that have demonstrated high effectiveness. USAID has funded programs in all development sectors so individuals from developing countries could earn graduate degrees in the United States (USAID 2010). Prime examples are USAID's African Graduate Fellowship Program (1963–1990) and its successor, Advanced Training for Leadership and Skills Program (1991–2003). Over these four decades, very qualified professionals from African countries earned graduate degrees in the sciences and engineering in the United States. In turn, many of those graduates achieved leadership positions in their respective countries (Aguirre International 2004). Their training in the United States and refined research skills made critical differences in the development of their countries and beyond, and their knowledge of the US educational system and society became instrumental in the establishment of positive relationships with the United States.

Climate change knows no national boundaries, and the efforts to address it effectively should be international. Issues of global governance such as those surrounding climate engineering require the collaborative actions of both scientists and diplomats so that the international policy actions are deeply informed by the best scientific knowledge available.

Some of the largest US professional scientific societies, such as the Geological Society of America (GSA) and the American Geophysical Union (AGU), are developing their international offices, and some are entering into agreements with their counterparts around the world. In addition, they are cultivating foreign membership: 39 % of AGU's 62,000 members are international. Many geological societies from around the world have joined GSA as associated societies. The American Geosciences Institute, an umbrella organization of 49 US geoscience societies, now also includes societies based outside of the United States as international associate members.

Ultimately, we are all parts of groups—we represent (wittingly or unwittingly) our institutions, our scientific discipline communities, our scientific national communities, and our countries. How to manage these affiliations and the perceptions of our partners while doing successful international science is the crux of the issue. This work is an international asset for diplomacy, but it also presents challenges.

18.3 Challenges in International Scientific Collaborations

International scientific work has its own challenges beyond those expected in the normal course of the actual research activities. These may range from the purely operational to the more political:

- Lack of institutional support and recognition for work done abroad or with international partners, which might prevent interested researchers from undertaking this type of research (NRC 2008, 2012).
- Difficulties securing visas, access to research sites, and transport of scientific samples and materials across borders (El-Baz 2010; NRC 2012).
- Insufficient funds to start or to complete work already underway.
- Cultural differences among the project participants and inaccurate assumptions at any stage of the project (NRC 2014b).
- Power asymmetries, which can occur when well-known scientists from more developed countries attempt to set up research projects with collaborators in less affluent countries. These asymmetries can cause problems at the research set-up stage, difficulties in communication, and potential resentment because of the differential in resources available to the various researchers (NRC 2008).
- Push and pull between assistance and cooperation (Neureiter 2004). Assistance activities are usually unidirectional and targeted towards the country/area that just suffered a disaster and, therefore, more short term and designed to address an urgent issue. Cooperation involves contributions from all parts towards a common goal and, ideally, involves the establishment of long-term relationships that may become the foundation for future work. The balance between assistance and cooperation is more a matter of policy rather than practice, but is, nonetheless, a challenge for the practitioner on the ground.
- Clarity and transparency are paramount in the establishment of effective relationships (NRC 2012). Every effort should be made to build trusting relationships among scientists. Official policy may occasionally have motives beyond those overtly stated. The individual scientist is not normally involved in those decisions and works without that knowledge, but those policy considerations could affect the way his/her work will ultimately be perceived by partners.

18.4 Resources for US Science Diplomacy: A Few Examples

International scientific work with the participation of US scientists can originate in several different sources. Government initiatives based on agencies such as the USGS, NSF, NASA, USDA, USAID, and Department of State provide different streams of funding and emphasize different aspects of the work, such as capacity building, disaster assistance, or international scientific cooperation. US universities are expanding their international S&T activities by establishing partnerships with universities abroad and by establishing campuses in other countries, especially in the Middle East (http://www.globalhighered.org/index.php), though recently there has been a shift to Asia (Morgan 2012).

The US Department of State sponsors international scholar exchanges such as the Fulbright Program, which provides a suite of programs for US and international scholars to do research and teaching in other countries and the United States, respectively (Institute of International Education 2014), and the US Jefferson Fellowship Program (NAS 2014a), whose participants can spend part of their yearlong appointment providing scientific advice at US embassies abroad. Jefferson Fellows increase scientific capacity in the policymaker/diplomatic corps, and several have made great contributions in the geosciences arena, especially in the areas of disaster relief and capacity building (A. Reynolds, personal communication, October 3, 2014). Norman Neureiter, the first Science and Technology Adviser to the Secretary of State, noted the absence of effective government-wide funding mechanisms for international S&T cooperation (Neureiter 2004), though bilateral cooperation projects are now being established.

A program of special interest because of its wide scientific reach is USAID's Partnerships for Enhanced Engagement in Research (PEER) program, administered by the NAS (2014b). PEER is a competitive grants program that allows scientists in developing countries to apply for funds for research and capacity-building activities to be conducted in partnership with US-based collaborators with current US government agency awards. These agencies currently include NASA, NSF, Smithsonian Institution, USDA, USGS, and the National Institutes of Health. Research supported by this program encompasses many areas, including geoscience topics such as climate change and disaster research. Another initiative is NSF's Partnerships for International Research and Education (PIRE), a broad program covering all areas of science and engineering, promoting excellence in research and education through international collaboration and looking to develop a diverse, globally engaged US science and engineering workforce. Some of these projects are co-funded by governmental agencies in Asia, Europe, and Latin America (NSF 2014).

Private grant-making organizations support different aspects of work in the geosciences. The United Nations Foundation (UNF) (http://www.unfoundation.org/what-we-do/issues/energy-and-climate/tackling-global-climate-challenge.html) centers its geoscience efforts on the issue of climate change. The Rockefeller Foundation supports work on resilience and adaptation in the face of climate change in Asia and Africa (http://www.rockefellerfoundation.org/our-work/current-work/climate-change-resilience). The Bill and Melinda Gates Foundation (http://www.gatesfoundation.org/What-We-Do/Global-Development/Emergency-Response#TheOpportunity) does not have specific programs in the geosciences, but it does support emergency response to natural disasters.

Corporations are also involved in international scientific cooperation and sometimes are involved in projects that can be very costly for a single country or a single institution and involve professionals from many countries (Neureiter 2004). Small amounts of money for research coordination activities are also available from organizations such as international scientific unions (International Union for Quaternary Research (INQUA): http://www.inqua.org/awards.html; International Union of Geodesy and Geophysics (IUGG): http://iugg.org/programmes/grants.php). Some opportunities are geared towards certain groups, such as students and early career scientists. US government agencies such as NSF enable US scientists to conduct research abroad (such as PIRE, described above) and can provide funding to discuss research needs and present research results at international meetings. This funding can be especially helpful to early career scientists to develop their own networks and receive input on their research from a wide range of international scientists. Some of those programs are organized and managed by the US National Committees for the geosciences based at the NAS (2014c), usually with additional funding from professional scientific societies and others.

And last, but not least, scientists sometimes use their personal funds to undertake work that they see as crucial but for which there is no other available financial support.

18.5 Diasporas and Science Diplomacy

The US scientific endeavor has been enriched for many decades by the contributions of foreign-born students and scientists (National Science Board 2014) and those who, though born in the United States, have strong links to the countries of their ancestral origins. These professionals, members of the science diaspora, apply their knowledge, training, and diverse approaches in the development of a robust US scientific enterprise.

Diasporas usually form important financial and diplomatic ties with their countries of origin and/or ancestry, though not all diasporas are equally active. The Pew Research Center (2014) reports that international migrants sent $529 billion in remittances back to their home countries in 2012. The United States is the top sending country, with over $123 billion in remittances, according to estimates provided by the World Bank (Pew Research Center 2014). Scientific diasporas have a role in the economic growth, research, innovation, and education of their adoptive country and might play a similar role in their native countries. The United States is building various fora for engagement with countries of origin or ancestry. The Network of Diasporas in Engineering and Science (NODES) is a partnership between the Department of State, the AAAS, the US National Academy of Sciences, and the US National Academy of Engineering (Burns 2013). Its goals are to discuss science diplomacy experiences and devise ways to support scientific diasporas. The International diaspora Engagement Alliance (IdEA) organizes activities such as the 2014 Global Diaspora Week (http://diasporaalliance.org/), which had its origins in joint efforts of the Department of State and USAID such as the Global Diaspora Forums held in 2012 and 2013.

18.6 Successful Science Diplomacy in the Geosciences

Common threads among many successful programs are strong multidisciplinary and interdisciplinary components and the incorporation of social science elements. These programs also emphasize the needs and input of the users and have specific outreach strategies, both to disseminate information and to receive input from the various stakeholder communities (Sztein and Casadevall 2013). Following are some examples of successful science diplomacy activities in the geosciences.

18.6.1 United Nations Programs and Panels

UNESCO This organization includes a number of major activities aimed at international cooperation in the geosciences. Among them is the Intergovernmental Oceanographic Commission, which includes the Tsunami Unit, which works with other UN agencies and nongovernmental organizations and its member states to build sustainable tsunami early warning systems that provide protection at the local, regional, and global scales (http://www.unesco.org/new/en/natural-sciences/ioc-oceans/sections-and-programmes/tsunami/). Another such initiative is the International Hydrological Program, devoted to water research, water resources management, and education and capacity building (http://www.unesco.org/new/en/natural-sciences/environment/water/ihp/). UNESCO's Earth Science program includes numerous activities, including the Global Geoparks Network, composed of unified areas ("geoparks") with geological heritage of international significance (http://www.unesco.org/new/en/natural-sciences/environment/earth-sciences/global-geoparks/), and the International Geoscience Programme (IGCP) in collaboration with the International Union of Geological Sciences, which provides funding for international research projects (http://www.unesco.org/new/en/natural-sciences/environment/earth-sciences/international-geoscience-programme/).

The Intergovernmental Panel on Climate Change (IPCC) IPCC is the leading international body constituted to provide periodic, rigorous climate change assessments based on climate science literature. It was established in 1988 by the United Nations Environment Programme (UNEP) and the World Meteorological Organization (WMO) to provide the world with a clear scientific view on the current state of knowledge in climate change and its potential environmental and socioeconomic impacts (http://www.ipcc.ch/). It is composed of 195 countries, whose governments are part of the review process and adopt and approve the reports. IPCC's process is comprehensive, objective, open, and transparent; its work is policy relevant but never policy prescriptive. The IPCC has helped institutionalize international coordination and cooperation with respect to policy and has also fostered unprecedented international collaboration among thousands of climate scientists from all nations by allowing them to contribute to these assessments as equals (Overpeck 2011). The IPCC scientific community also builds the capacity needed for national to local-scale climate change adaptation and mitigation, including the effective use of development funds (Overpeck 2011).

18.6.2 Major International Initiatives

One Geology This is an international initiative of 110 geological surveys from around the world, the Commission for the Geological Map of the World, the International Union of Geological Sciences, the United Nations Educational,

Scientific, and Cultural Organization (UNESCO), and the International Framework of the International Steering Committee for Global Mapping. One Geology's mission statement is to "Make web-accessible the best available geological map and other geoscience data worldwide at the best possible scales, starting with at least 1:1 million scale" (http://www.onegeology.org/what_is/home.html).

Africa Soil Information Service (AfSIS) AfSIS is developing continent-wide digital soil maps for sub-Saharan Africa by using new types of soil analysis and statistical methods including the compilation and rescue of legacy soil profile data, collection and analysis of new data, and large-scale soil mapping systems using remote sensing and crowd-sourced ground observations. The project area covers ~17.5 million km^2 of continental sub-Saharan Africa, which includes more than 90 % of Africa's human population living in 42 countries. Funded by the Bill and Melinda Gates Foundation, this work involves scientific, operational, and implementation partnerships with the Earth Institute of Columbia University, the World Agroforestry Centre (ICRAF), and the ISRIC – World Soil Information (http://www.africasoils.net/about/who-we-are).

Group on Earth Observation (GEO) and Its Global Earth Observation System of Systems (GEOSS) The goal of GEOSS is to connect the producers of data and decision-support tools with end users, such as scientists, public and private sector decision makers, resource managers, and emergency responders, to enhance the relevance of Earth observations to global issues. The resulting global public infrastructure generates comprehensive, near-real-time environmental data, information, and analyses for a wide range of users (http://www.geoportal.org/web/guest/geo_home_stp). The observation systems need to have common standards for architecture and data sharing so they can be interoperable, and the exchange of data, metadata, and products has to be full and open. GEO has 75 country members and 77 participating organizations, which maintain ownership of the data and task components they contribute to the project. Of the nine areas of emphasis, the ones more closely related to the geosciences are climate, weather, water, and disasters (https://www.earthobservations.org/index.php).

18.6.3 Bilateral and Multilateral Programs

Past Natural Climate Variability This work, undertaken by US, Russian, German, and Austrian scientists on Lake El'gygytgyn in Arctic Russia, is an excellent example of multinational and multidisciplinary collaboration in harsh environments where no single nation has all the capabilities and financial resources to undertake such an enterprise alone (Brigham-Grette et al. 2013). This type of successful collaborative work has nonetheless pointed out the difficulties of working among circumpolar countries and resulted in the creation of the Arctic Council's Scientific Cooperation Task Force (http://www.arctic-council.org/index.php/en/about-us/working-groups/task-forces).

Water and Science Diplomacy As a result of the remote sensing work done by Farouk El-Baz and his collaborators, the "1,000 Wells for Darfur" initiative has the goal of creating new groundwater resources to help establish peace and economic security in the region (El-Baz 2008). This plan has the support of the governments of Sudan and Egypt, which will support this initiative by drilling some of the wells. Approximately 100 wells have been drilled already, some with the support of non-governmental organizations. Science diplomacy was fundamental to ensure the participation of the governments in this endeavor (http://www.bu.edu/remotesensing/research/dafur/). However, current political instability in the area has temporarily stopped the progress of this program.

National Academies of Sciences These organizations around the world organize activities that can be bilateral, multilateral, or inscribed within the framework of the InterAcademy Partnership (http://www.interacademies.net/). Among the many examples of collaboration activities, the US NAS is working with the UNF to expand international awareness of the findings of the IPCC's Fifth Assessment Report's three Working Groups. This is being done through the organization of events around the world and by providing authoritative information on climate change to different stakeholders including policymakers, national academy counterparts, other scientists, and media. Among the many events held in Africa, Europe, Asia, and Latin America, a special event discussing the findings of WGs II and III was co-organized by the Mexican Academy of Sciences and the NAS at the Senate of the Republic in Mexico City in 2014. This workshop included the participation of noted scientists from both countries and resulted in extensive press coverage across Mexico (Academia Mexicana de Ciencias 2014).

18.7 Volcano Disaster Assistance Program

One important US disaster assistance program is the Volcano Disaster Assistance Program (VDAP). This team was established after the 1985 disaster in Colombia, where a lahar triggered by the Nevado del Ruiz volcano eruption swept through the town of Armero and killed 23,000 people (USGS 2011). VDAP, the world's only volcano crisis response team, was originally developed by USGS and USAID's Office of Foreign Disaster Assistance to respond to volcanic crises at short notice anywhere in the world. It has deployed teams of scientists and equipment in response to volcanic eruptions and unrest in Central and South America, the Caribbean, Southeast Asia, Africa, and the Middle East (Pallister 2011). Since 1986, VDAP has responded to 27 major crises and helped to build capacity in 12 countries (USAID 2013).

When the host country requests assistance, VDAP consults with in-country colleagues and performs a rapid hazard assessment of the volcanically threatened area and activates any available remote sensing resources. VDAP prepares forecasts of potential activity using the volcano's history, local or regional seismic network data collected during the current unrest, and its knowledge of eruptive activity worldwide. The team relays its assessment of the situation to the host country's volcano hazard agency, which leads the response (USGS 2011). VDAP scientists always work in support of its counterpart agencies and are a scientific and technical part of US humanitarian and foreign aid activities with the overarching goal of saving lives and reducing property losses (Pallister 2011).

In its early days, VDAP acted as a *mobile volcano observatory*, and its activities were geared exclusively towards assistance after crises; but in later years, it developed disaster prevention activities with a strong capacity building component to increase the knowledge base and the preparedness level of the local scientists and communities before a disaster. As the international partner observatories become better equipped to deal with emergencies, the traditional VDAP *mobile observatory* response is deployed only in special circumstances. VDAP's capacity building activities include construction of volcano-monitoring infrastructure and education programs in monitoring, hazard assessment, and eruption forecasting. VDAP scientists can also serve as external advisors to foreign governments that request help in improving their volcanic hazard risk-reduction programs (USGS 2011). All monitoring equipment provided is donated to the counterpart agencies.

VDAP organizes workshops and on-the-job training activities in host countries as well as at USGS volcano observatories. VDAP has helped build and sustain volcano observatories and monitoring programs in more than a dozen countries (Pallister 2011). VDAP's role has grown to include remote consultation, enhancement of monitoring infrastructure, and increased capabilities in eruption forecasting. Education and training programs are integral to the program and contribute to the enhancement of the local partners' capabilities. In addition, VDAP participates in research collaborations with its partners in the Philippines and Chile (USGS 2011).

In the course of these collaborations, bi-and multinational partnerships are established, and multicultural understanding and enduring friendships develop. Such international relationships constitute essential elements of science diplomacy (Pallister 2011).

18.8 International Geoscience Scientific Unions

Other important vehicles for international geosciences are the International Geoscience Scientific Unions (http://www.icsu-geounions.org/). These include the International Astronomical Union (IAU), the International Geographical Union (IGU), the International Union for Quaternary Research (INQUA), the International Society for Photogrammetry and Remote Sensing (ISPRS), the International Union of Geodesy and Geophysics (IUGG), the International Union of Geological Sciences (IUGS), the International Union of Soil Sciences (IUSS), and the International Union of Radio Science (URSI).

Among their main functions are to stipulate scientific standards and nomenclature, maintain important worldwide databases, promote international capacity building, provide professional development opportunities and encourage early career scientists, support outreach activities, disseminate scientific research by organizing international meetings and publishing journals, initiate/develop/promote international research programs, and provide a neutral meeting forum. The interactions that take place, in both formal and informal settings, foster international understanding and help establish collaborations that result in important initiatives that benefit humanity. For example, IUGG was one of the main organizers of the International Geophysical Year (1957–58), which provided a great opportunity for formal and informal discussions at a delicate time during the Cold War and led to, among many other initiatives, the establishment of the Antarctic Treaty (Nicolet 1982). IUGG has eight constituent semi-autonomous associations that organize their own activities in addition to those organized by IUGG's commissions and services—widening even more the scientific spectrum covered by the union (http://www.iugg.org/). All these unions have national members (countries) and affiliate members (such as professional scientific societies).

Many country members have constituted national committees for these unions. In particular, the US National Committees (USNCs) for the various geoscience unions engage the US scientific community in the important work of these unions and their associations and commissions. The USNCs fulfill membership responsibilities in the unions through representation to and involvement in governance meetings, promotion of Union programs and priority activities in the US, and organization of activities for the benefit of the US community, emphasizing participation by early career scientists (NAS 2014c). The USNCs also actively involve national professional societies, organize and promote education and outreach activities, and co-organize activities with professional societies, universities, and other stakeholders. The USNCs contribute to and monitor the international arena and promote the international engagement of the US geoscience community. The 2011 AAAS symposium "The Practice of Science Diplomacy in the Earth Sciences," organized by the USNC for Geological Sciences, highlighted the collaborative work of geoscientists from academia, government, and nonprofits with their counterparts around the world on natural hazards and natural resources issues (Casadevall et al. 2011).

18.9 Best Practices in International Scientific Collaborations

In addition to the many examples of international geoscience collaborations, geoscientists can also learn from a growing body of work on best practices in international collaboration and team science efforts from the biomedical and from the behavioral and social sciences (see National Cancer Institute n.d.; NRC 2008, 2011, 2014b).

- *Culture*. Ideally, scientists understand local history and customs (NRC 2014a). For example, geologic research in areas of cultural or religious significance needs to be approached with the advice of local scientists in order to proceed in the most sensitive way.

- *Language*. Speaking, reading, and writing are different, though related, skills. It is advantageous to speak another language, which is a window into the other culture (according to Barrett et al., as cited in NRC 2014b). Misunderstandings in communication, even within variants of the same language (such as the Caribbean, US, and Indian English), could result in collaboration problems (NRC 2014a).
- *Mutual Trust*. Human interactions are central to scientific collaboration relationships at all stages: negotiation, establishment, conduct of research, and publication and dissemination. Trust issues at the negotiation stage might make or break a potential project. Scientists need to trust that their partners will conduct research in an effective, ethical, and accurate manner. The development and maintenance of those relationships is greatly aided by face-to-face interactions (Chua et al. 2011; NRC 2008, 2014a, b)
- *Topic Relevance*. The subject of the study should be relevant to the local partner and not only to the partner from the more developed country (NRC 2008).
- *Implied/Explicit Agreements*. Scientists should define who, what, when, why, and how (NRC 2011) with special attention to research protocols and intellectual property issues (NRC 2014b). In some cases, formal consortium agreements are required, and coordination among funding agencies can make the research project possible (NRC 2014b).
- *Not Being a Parachutist*. Taking advantage of local scientists and resources without a true collaborative partnership sours the current relationship and prevents other potential scientific collaborations from taking place (NRC 2008, 2014b).
- *Capacity Building when Appropriate*. Scientists should provide training and educational support to the local scientific community when it is not sufficiently developed or knowledgeable in a given topic. Part of the collaboration involves training *in situ* or travel to the more developed partner country for specialized training.
- *Bidirectional Learning Opportunities*. Different schools of thought, traditions of training, and knowledge of different parts of the world enrich the team by informing the research and making scientific process and interpretation more nuanced and, more likely than not, better.
- *Engagement*. Engaging with scientists, nonscientists, and local populations is the key in the course of the work (Kunen 2011).
- *Involvement of Early Career Scientists*. Early career scientists who gain experience in international collaborations have the opportunity to help answer the research question at hand and to develop their own networks and careers (NRC 2012, 2014b).
- *Expectations of Authorship and Credit on Publications (Local Language and English)*. Publications are an important part of the scientific process and a way to recognize the participation of the full research team (NRC 2008, 2014b). While it is understood that publications in English are highly valued by the scientific establishment, not all scientists in a country are fluent in English or have access

to foreign journals. Quality outlets for scientific research exist in many languages, and publication in such journals is also valuable.

- *Consideration of Additional Outreach Pathways*. Research results might also be of great importance to the public at large (e.g., for natural hazard research). In those cases, a local campaign with the appropriate message in the local language is a highly effective way to disseminate results (NRC 2014b). Extension activities with active outreach and educational components have been undertaken by the US soil community for over 75 years (NRCS 2014).
- *Speak the Policy Language*. Speaking the policy language helps link scientific research to outcomes that are valued by policy experts, especially in projects established in response to national priorities (Kunen 2011).

18.10 Conclusion

All scientists who work outside of their countries' borders or work with international colleagues are, in effect, science diplomats. They will have an impact in their host institution, their colleagues, and communities. Knowingly or not, scientists act as ambassadors of their institutions, disciplines, cultures, and countries. While these are not easy endeavors, there are many successful examples of science diplomacy in the geoscience realm. These initiatives may result from the efforts of institutions and governments, but always succeed because of the hard work and enthusiasm of individual scientists who have the vision, the ideas, and the wisdom to carry them to completion. The advantages of international scientific work are manifold, but the most important are the development of shared global scientific knowledge and societal advancement. Positive international scientific collaborations improve relationships among nations and advance global science for the benefit of all.

Acknowledgment The author wishes to thank Melody Brown Burkins (Dartmouth College), Thomas Casadevall (USGS, Emeritus), and Farouk El-Baz (Boston University) for their very helpful comments and suggestions. Additionally, the author dedicates this work to the memory of her friend, colleague, and mentor Louis B. Brown.

References

Academia Mexicana de Ciencias (2014) Reflexiones, propuestas y muchas tareas al concluir foro sobre cambio climático en el senado. [Reflections, proposals, and many tasks at the conclusion of the forum on climate change in the Senate] Boletín AMC/140/14, México, DF, 22 Apr 2014. http://www.comunicacion.amc.edu.mx/comunicados/reflexiones-propuestas-y-muchas-tareas-al-concluir-foro-sobre-cambio-climatico-en-el-senado/. Accessed 20 Sept 2014

Aguirre International (2004) Generations of quiet progress: the development impact of US long-term university training on Africa from 1963 to 2003. http://pdf.usaid.gov/pdf_docs/PNADB130.pdf. Accessed 31 Oct 2014

American Geosciences Institute (2014) Science, technology, and innovation expert partnership. http://www.americangeosciences.org/policy/STIEP. Accessed 3 Oct 2014

Brigham-Grette J, Melles M, Minyuk P, Andreev A, Tarasov P, DeConto R, Koenig S, Nowaczyk N, Wennrich V, Rosen P, Haltia-Hovi E, Cook T, Gebhardt T, Meyer-Jacob C, Snyder J, Herzschuh U (2013) Pliocene warmth, polar amplification, and stepped Pleistocene cooling recorded in NE arctic Russia. Science 340(6139):1421–1427. doi:10.1126/science.1233137

Burns WJ (2013) The potential of science diasporas. Science and Diplomacy. http://www.science-diplomacy.org/perspective/2013/potential-science-diasporas. Accessed 17 Sept 2014

Casadevall T, Sztein E, Burkins MB (2011) The practice of science diplomacy in the Earth sciences. Symposium at the annual meeting of the American Association for the Advancement of Science, Washington, DC, 17–21 Feb 2011. (convener T Casadevall and E Sztein, MB Burkins co-conveners). https://aaas.confex.com/aaas/2011/webprogram/Session2855.html. Accessed 12 Oct 2014

Chua RYJ, Morris MW, Mor S (2011) Collaborating across cultures: cultural metacognition and affect-based trust in creative collaboration. Harvard Business School working paper 11–127. http://www.hbs.edu/faculty/Publication%20Files/11-127.pdf. Accessed 3 Oct 2014

El-Baz F (2008) Remote sensing of the Earth: implications for groundwater in Darfur. The Bridge 38(3):5–13

El-Baz F (2010) Science attachés in embassies. Science 329(5987):13. doi:10.1126/science.1189621

Fedoroff N (2009) Science diplomacy in the 21st century. Cell 136:9–11. doi:10.1016/j.cell.2008.12.030

Institute of International Education (2014) Fulbright programs. http://www.iie.org/fulbright/. Accessed 28 Sept 2014

Kramer D (2010) Science diplomacy enlisted to span US divide with developing world. Phys Today 63(12):28–30

Kunen J (2011) Science diplomacy for development at USAID. Paper presented at the annual meeting of the American Association for the Advancement of Science, Washington, DC, 17–21 Feb 2011. Abstract retrieved from https://aaas.confex.com/aaas/2011/webprogram/Paper3091.html. Accessed 14 Oct 2014

Lempinen EW (2014) Building a bridge between diplomacy and science. The World Academy of Sciences, 26 Jun 2014. http://www.twas.org/article/building-bridge-between-diplomacy-and-science. Accessed 2 Oct 2014

Linkov I, Trump B, Tatham E, Basu S, Roco MC (2014) Diplomacy for science two generations later. Science and Diplomacy, 13 Mar 2014. http://www.sciencediplomacy.org/perspective/2014/diplomacy-for-science-two-generations-later. Accessed 26 Oct 2014

Morgan J (2012) Study documents geographic shift in branch campuses. Inside Higher Ed, 12 Jan 2012. https://www.insidehighered.com/news/2012/01/12/study-documents-geographic-shift-branch-campuses. Accessed 9 Oct 2014

National Academy of Sciences (2014a) Jefferson Science Fellowship. http://sites.nationalacademies.org/PGA/Jefferson/. Accessed 29 Sept 2014

National Academy of Sciences (2014b) Partnerships for Enhanced Engagement in Research (PEER). http://sites.nationalacademies.org/PGA/dsc/peerscience/index.htm. Accessed 29 Sept 2014

National Academy of Sciences (2014c) Board on International Scientific Organizations. http://sites.nationalacademies.org/pga/biso/. Accessed 5 Sept 2014

National Cancer Institute (n.d.) Team Science Toolkit. https://www.teamsciencetoolkit.cancer.gov/Public/Home.aspx. Accessed 20 Sept 2014

National Research Council (2008) International collaborations in the behavioral and social sciences (report of a workshop). The National Academies Press, Washington, DC

National Research Council (2011) Examining core elements of international research collaboration (summary of a workshop). The National Academies Press, Washington, DC

National Research Council (2012) US and international perspectives on global science policy and science diplomacy (report of a workshop). The National Academies Press, Washington, DC
National Research Council (2014a) Culture matters: international research collaboration in a changing world (summary of a workshop). The National Academies Press, Washington, DC
National Research Council (2014b) Building infrastructure for international collaborative research in the social and behavioral sciences (summary of a workshop). The National Academies Press, Washington, DC
National Science Board (2014) Immigration and the S&E workforce. In: Science and engineering indicators 2014. National Science Foundation (NSB 14–01), Arlington. http://www.nsf.gov/statistics/seind14/index.cfm/digest/trends.htm#6. Accessed 27 Oct 2014
National Science Foundation (2014) Partnerships for international research and education. http://www.nsf.gov/funding/pgm_summ.jsp?pims_id=505038. Accessed 21 Sept 2014
Natural Resources Conservation Service (NRCS) (2014) http://www.nrcs.usda.gov/wps/portal/nrcs/main/national/about/history/. Accessed 13 Oct 2014
Neureiter NP (2004) Science and technology in the Department of State. Technol Soc 26:303–320
Nicolet M (1982) The international geophysical year 1957/58. World Meteorol Org Bull 31:222–231. https://www.wmo.int/pages/mediacentre/documents/Int.GeophysicalYear.pdf. Accessed 26 Oct 2014
Overpeck JT (2011) The IPCC, international science diplomacy, and the challenge of climate change. Paper presented at the annual meeting of the American Association for the Advancement of Science, Washington, DC, 17–21 Feb 2011. Abstract retrieved from https://aaas.confex.com/aaas/2011/webprogram/Paper3165.html. Accessed 13 Oct 2014
Pallister JS (2011) Volcano science diplomacy. Paper presented at the annual meeting of the American Association for the Advancement of Science, Washington, DC, 17–21 Feb 2011. Abstract retrieved from https://aaas.confex.com/aaas/2011/webprogram/Paper3132.html. Accessed 13 Oct 2014
Pew Research Center (2014) Remittance flows worldwide in 2012, 20 Feb 2014. http://www.pewsocialtrends.org/2014/02/20/remittance-map/. Accessed 6 Oct 2014
Sztein E, Casadevall T (2013) Science diplomacy in the geosciences. Invited paper presented at the American Geophysical Union Fall Meeting, San Francisco, CA, 9–13 Dec 2013. Abstract #PA42A-01 retrieved from http://adsabs.harvard.edu/abs/2013AGUFMPA42A..01S. Accessed 2 Oct 2014
The Royal Society and AAAS (2010) New frontiers in science diplomacy. The Royal Society, London. https://royalsociety.org/~/media/Royal_Society_Content/policy/publications/2010/4294969468.pdf. Accessed 6 Oct 2014
USAID (2010) Agriculture long-term training: assessment and design recommendations. http://pdf.usaid.gov/pdf_docs/PNADT511.pdf. Accessed 31 Oct 2014
USAID (2013) Natural and technological risks: geological hazards update. http://www.usaid.gov/sites/default/files/documents/1866/geohazards_update_october2013.pdf. Accessed 12 Oct 2014
USGS (2011) Volcano Disaster Assistance Program. http://volcanoes.usgs.gov/vdap/. Accessed 20 Sept 2014

Chapter 19
Stormy Seas, Rising Risks: Assessing Undisclosed Risk from Sea Level Rise and Storm Surge at Coastal US Oil Refineries

Christina Carlson, Gretchen Goldman, and Kristina Dahl

Abstract As the world works to build resiliency to the impacts of climate change, it is increasingly important that companies, in addition to communities, consider and manage risks from such impacts to their assets and infrastructure. Discussion of climate change as a financial risk presents opportunities to engage companies and their investors on climate-related issues. Coastal oil-refining operations, in particular, face tremendous risks from the impacts of climate change, which can damage or destroy coastal energy facilities, curtail or stop production, and inundate nearby communities. Here, five oil refineries' risk to sea level rise, and climate change-enhanced storm surge is assessed for the present day, 2030, 2050, and 2100. Risk assessments are compared against the degree to which the five companies have publicly disclosed climate-related risk to the US Securities and Exchange Commission, as recommended. Finally, suggestions are offered for better company consideration and disclosure of climate risks.

Keywords Climate change • Climate risk • Storm surge • Sea level rise • SLOSH modeling

19.1 Introduction

Oil and gas companies often have large refining operations at or near the coastline. Many of these facilities are in low-lying areas—on land less than 10 ft above the high tide line (Strauss and Ziemlinski 2012). Climate change impacts, including sea level rise and changes in storm intensity, add to the risks that these coastal facilities face now and in the future.

C. Carlson, A.B. (✉) • G. Goldman, Ph.D.
Center for Science and Democracy, Union of Concerned Scientists, Washington, DC, USA
e-mail: ccarlson@ucsusa.org; GGoldman@ucsusa.org

K. Dahl, Ph.D.
Dahl Scientific, San Francisco, CA, USA

J.L. Drake et al. (eds.), *Communicating Climate-Change and Natural Hazard Risk and Cultivating Resilience*, Advances in Natural and Technological Hazards Research 45, DOI 10.1007/978-3-319-20161-0_19

Past weather- and climate-related refining outages have had significant impacts on gasoline prices and affected, in turn, the US economy as a whole. In 2005, for example, Hurricanes Katrina and Rita devastated the Gulf Coast, shutting down 23 % of the US refining capacity causing a significant drop in gasoline production and resulting in a 50 % jump in the weekly average spot price of conventional gasoline (Kirgiz et al. 2009; Reuters 2005).

In the USA, the East and Gulf coasts have the fastest rates of local sea level rise, due in part to local subsidence and changes in ocean currents (Ezer et al. 2013; NOAA 2014; Sallenger et al. 2012). Recently published projections suggest that, under a mid-range scenario of future warming, the Gulf of Mexico may experience 3–4 ft of sea level rise by the end of this century, while many locations along the Mid-Atlantic and Northeast coasts are projected to experience 2–3 ft (Kopp et al. 2014).

In addition, climate change may affect the strength of the coastal storms themselves. Recent studies have shown a substantial increase in the proportion of more intense hurricanes (i.e., Category 4 and 5) in the North Atlantic basin since 1975 (Holland and Bruyere 2014; IPCC 2013a). And as the climate continues to warm, it is likely that the most intense categories of hurricanes will occur more often (IPCC 2012). It is also likely that global warming will cause hurricanes to have higher rainfall rates by the end of the twenty-first century, further increasing flood risks (GFDL 2013; Knutson and Tuleya 2004, 2008).

As a result of growing climate risks and companies' failure to disclose or prepare for climate impacts at vulnerable facilities, the SEC, the investors, and the public have increased demands on companies for more transparency around their climate preparedness.

The SEC (2010) issued guidance to companies for considering and discussing such risks in their annual Form 10-K filing, explicitly stating:

> significant physical effects of climate change … have the potential to have a material effect on … business and operations. These effects can impact … personnel, physical assets, supply chain, and distribution chain. They can include the impact of changes in weather patterns, such as increases in storm intensity, [and of] sea-level rise. (p. 6)

19.2 Methodology

This analysis focuses on the top five US energy companies with respect to their total crude-refining capacity: Valero, Phillips 66, Exxon Mobil, Marathon Petroleum, and Chevron (Brelsford et al. 2013). One coastal-refining facility for each of the five companies was chosen for analysis based on perceived risk, which was determined by vulnerability of location and historical storm damage (see Table 19.1).

Methods used in this report draw from the best practices established by the NOAA Coastal Services Center and laid out in the "Mapping Coastal Inundation Primer" (NOAA 2012). The extent to which each company disclosed was assessed using (a) the SEC's EDGAR database; (b) targeted keyword searches in the Ceres/CookESG SEC Climate Disclosure Search Tool; and (c) 2013 SEC Form 10-K fil-

Table 19.1 Facility statistics and company disclosure for refineries analyzed

US refining rank of company by crude capacity	Global refining rank of company by crude capacity	Company	Site location	Crude capacity at facility analyzed (barrels per calendar day)	Company disclosure of physical climate risk
1	6	Valero	Meraux, Louisiana	125,000 b/cd	None
2	10	Phillips 66	Linden, New Jersey	238,000 b/cd	Poor
3	1	Exxon Mobil	Baytown, Texas	560,500 b/cd	None
4	13	Marathon Petroleum	Texas City, Texas	84,000 b/cd	None
5	9	Chevron	Pascagoula, Mississippi	330,000 b/cd	None

ings, which note all of the companies' disclosed material risks (for the most recent year for which data were available) (Ceres/CookESG 2014; SEC 2014). Specifically, each company was assigned a disclosure score for its reporting on physical risk to the SEC (in its 2013 Form 10-K filings) based on methodology established in the Ceres 2012 report (see Table 19.1).

19.2.1 Data Sources

The capacity (listed in barrels per calendar day (b/cd)) and location information for each petroleum refinery is reported by the US Energy Information Administration (EIA) as of January 1, 2014 (EIA 2014) (see Table 19.1). Refinery property lines were determined using county-level parcel-ownership data, when available. Digital Elevation Model (DEM) data were obtained from the US Geological Survey National Map. The extent of future sea level rise was mapped for the years 2030, 2050, and 2100 using recently published, localized sea level rise projections (Kopp et al. 2014). The projection used for each site analysis was from the geographically closest tide gauge evaluated by Kopp et al. Maps featured in this report show the median (50th percentile) projection for the Representative Concentration Pathway (RCP) 4.5 scenario from the Intergovernmental Panel on Climate Change Fifth Assessment Report (IPCC 2013b).

19.2.2 Sea Level Rise and Storm Surge Models

Inundation from storm surge was estimated using the National Weather Service Sea, Lake, and Overland Surges from Hurricanes (SLOSH) model's maximum of maximums (MOMs) at Gulf and East coast sites. Storm surge maps produced from

SLOSH MOMs show worst-case-scenario flooding given all possible storm paths for a hurricane of a particular strength. MOM data for each relevant SLOSH basin used was relative to the vertical datum NAVD88.

To examine specific effects sea level rise might have on storm surge, a SLOSH MOM model of a hurricane category that already affects each facility was selected in order to compare how risk today changes with estimated sea level rise in 2030, 2050, and 2100. Projected sea level rise values were added to the flood layers created by the SLOSH model at each site for the relevant hurricane category.

Additional analysis not presented in this chapter was done to asses site vulnerability including: sea level rise projections alone in 2030, 2050, and 2100 at RCP4.5; extent of storm surge flooding for Category 1–5 hurricanes or up to the maximum category that completely inundated the facility; and depth of flooding from a Category 3 hurricane at present-day levels.

Available online at ucsusa.org/risingrisks is a detailed methodology for this work.

19.3 Results

19.3.1 Valero Energy Corporation

Valero Energy Corporation's Meraux, Louisiana, refinery (125,000 b/cd) sits below sea level 10 miles east of New Orleans and has risks both from sea level rise and storm surge. By 2050, sea level rise will make the facility vulnerable to Category 2 hurricanes (see Fig. 19.1), which are not currently projected to flood the facility at all. With sea level rise to date, a Category 3 storm could put parts of the facility under 10 ft of water.

After Hurricane Katrina in 2005, the then Murphy Oil-owned Meraux refinery spilled 25,000 barrels of oil and was shut down for several months (DOE 2009; EPA 2006). Meraux city canals and more than a square mile of neighborhood were oiled, resulting in a $330 million settlement for Murphy Oil (EPA 2006; MNS 2006). The Meraux facility again saw damages from the 2008 hurricane season and was shut down for many days (DOE 2009). Following these adverse events, Murphy Oil disclosed in its 2010 SEC Form 10-K that "the physical impacts of climate change present potential risks for severe weather (floods, hurricanes, tornadoes, etc.) at our Meraux … refinery in southern Louisiana and our offshore platforms in the Gulf of Mexico" (Murphy Oil 2011, p. 34).

Yet Valero has not disclosed such climate risks since it acquired the Meraux facility from Murphy Oil on October 1, 2011. Valero's 2013 SEC filing noted only that there could be "weather conditions that disrupt the supply of and demand for refined products," (p. 41) mentioning Hurricane Sandy as an example; however, direct discussion of climate change concerns was limited to the impacts of climate-related regulations (Valero 2014).

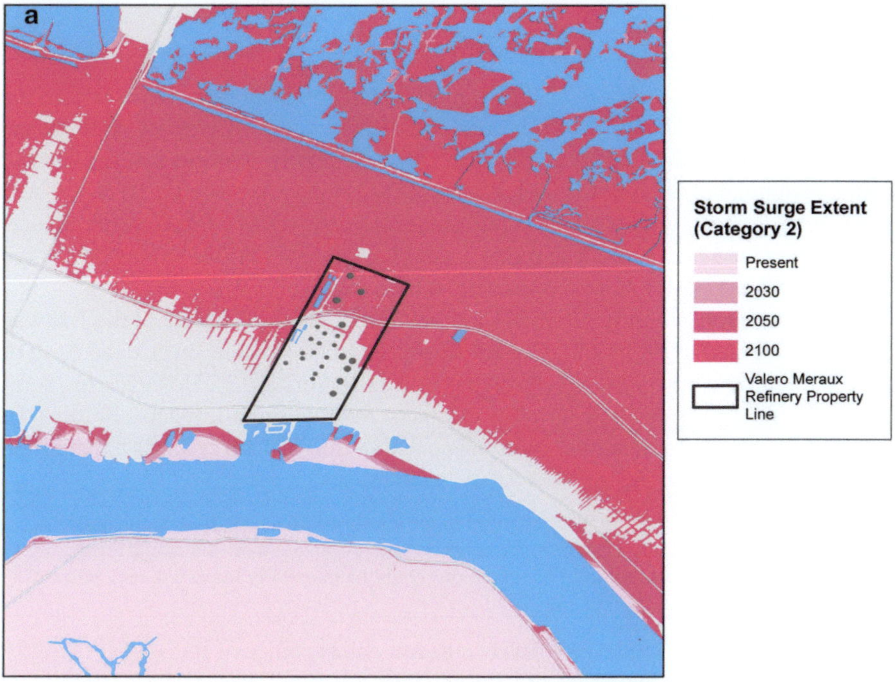

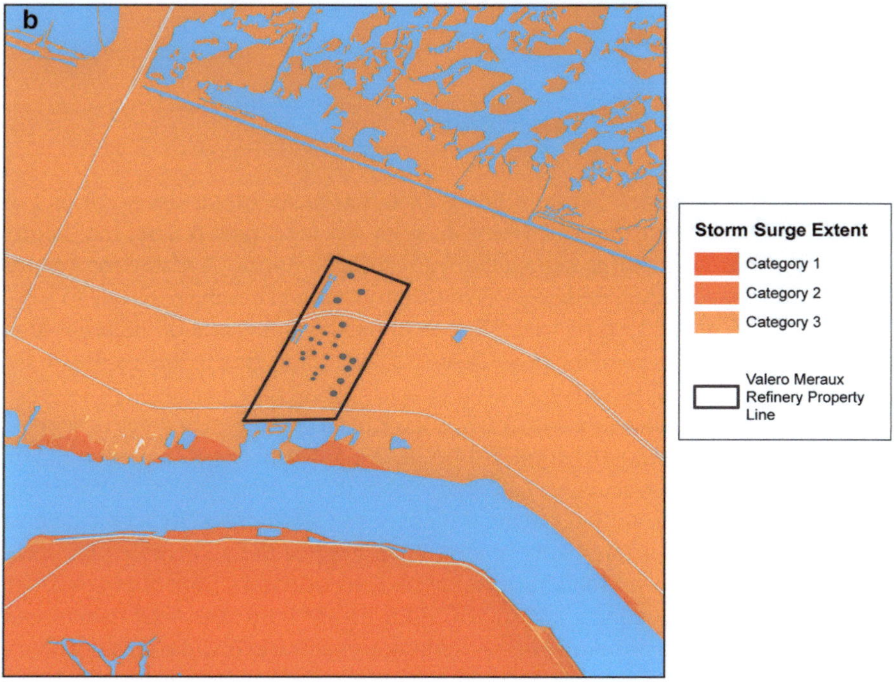

Fig. 19.1 (**a**) Projections for storm surge extent from a Category 2 hurricane today and with sea level rise by 2030, 2050, and 2100 at Valero Energy's Meraux, Louisiana, refinery. (**b**) Extent of storm surge flooding today from Category 1–3 hurricanes at Valero Energy's Meraux, Louisiana, refinery

19.3.2 Phillips 66

Phillips 66's Bayway refinery (238,000 b/cd) in Linden, New Jersey, sits within the New York metropolitan area. With sea level rise to date, a Category 1 hurricane could put parts of the facility under water. If a Category 3 storm hit, the facility could be inundated, with potential for parts of the refinery to be under 10 ft of water. Sea level rise, along with changes in hurricane intensity, could worsen storm impacts in the future (see Fig. 19.2). The refinery could also be vulnerable from storm surge as a result of nor'easter storms occurring at high tide.

After Superstorm Sandy hit in 2012, some 7,800 gal of oil spilled at the Bayway refinery (modeled in Fig. 19.2). The refinery was shut down for 3 weeks because of flood damage and power outages from the storm (Reuters 2012). However, the company has not fully reported its physical risks from climate impacts. Phillips 66 (2014) SEC filings contain only minimal consideration of such risks, as exemplified by the following:

> To the extent there are significant changes in the Earth's climate, such as more severe or frequent weather conditions in the markets we serve or the areas where our assets reside, we could incur increased expenses, our operations could be materially impacted, and demand for our products could fall. (p. 18)

The company has made no reference to the vulnerability of its coastal facilities to climate-related sea level rise and storm surge, nor has it provided any significant discussion on how, or if, it is preparing for those risks.

19.3.3 Exxon Mobil Corporation

Exxon Mobil's Baytown, Texas, refinery (560,500 b/cd) sits at the north end of Galveston Bay, 25 miles east of Houston. With sea level rise to date, the facility faces risks from storm surge associated with stronger storms. A Category 3 hurricane, for example, could inundate parts of the property and has the potential to leave some structures under 15 ft of water. Sea level rise and increases in the proportion of more intense storms could make such storms more damaging to this facility in the future (see Fig. 19.3).

The Baytown complex (see Fig. 19.3) is the largest petroleum and petrochemical complex in the USA (Exxon Mobil 2014a). In 2005, Hurricane Rita caused Exxon Mobil's Baytown and Beaumont facilities to shut down (Reuters 2005). With continued sea level rise as well as potential increases in storm intensity as the climate warms, more shutdowns are likely in the future. Despite the vulnerable placement of these facilities, Exxon Mobil has not reported physical risks from climate change impacts. Though the company's 2013 SEC filing noted "hurricanes may damage our offshore production facilities or coastal refining and petrochemical plants in vulnerable areas," (p. 4) the only direct reference to climate-related risks discussed how climate regulation could affect the company's finances (Exxon Mobil 2014b).

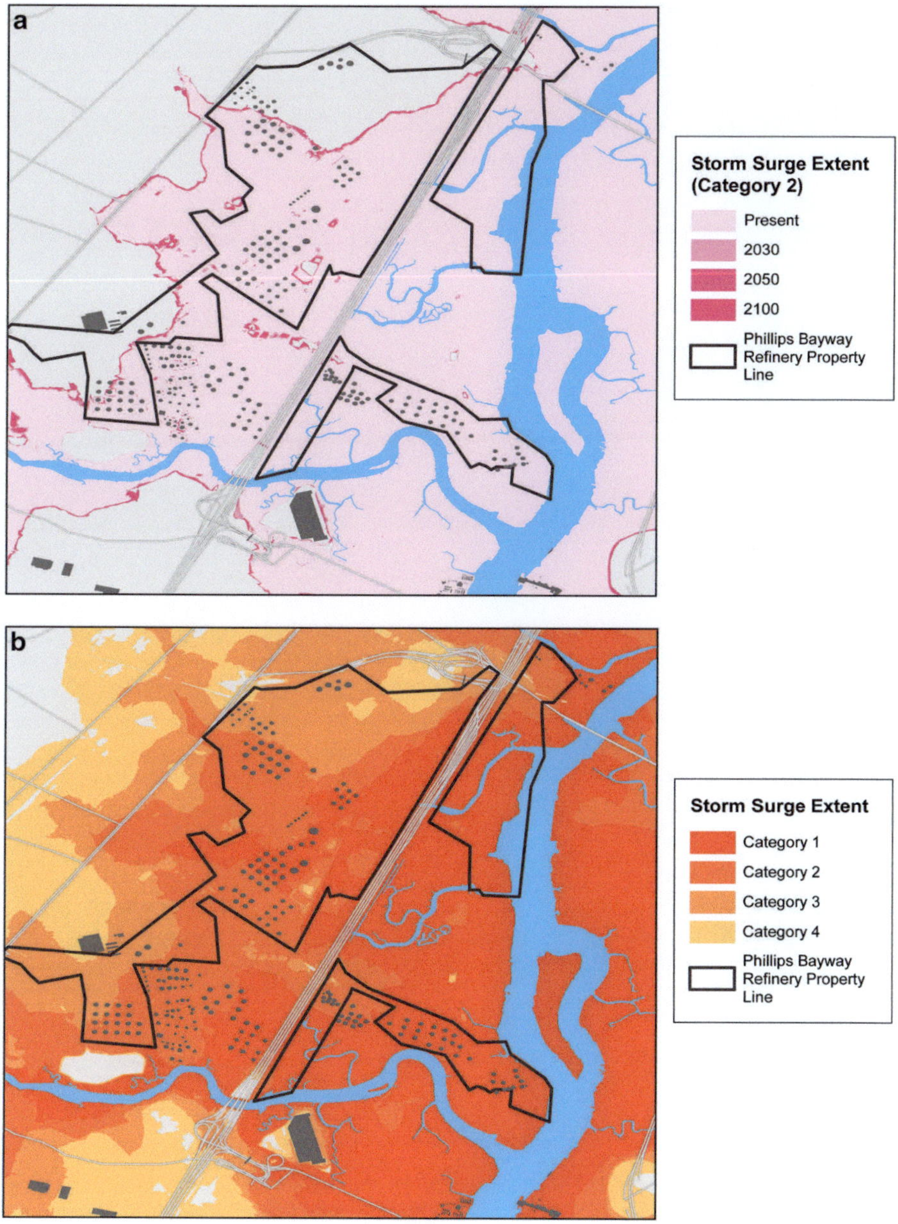

Fig. 19.2 (**a**) Projections for storm surge extent from a Category 2 hurricane today and with sea level rise by 2030, 2050, and 2100 at Phillips 66's Linden, New Jersey, refinery. (**b**) Extent of storm surge flooding today from Category 1–4 hurricanes at Phillips 66's Linden, New Jersey, refinery

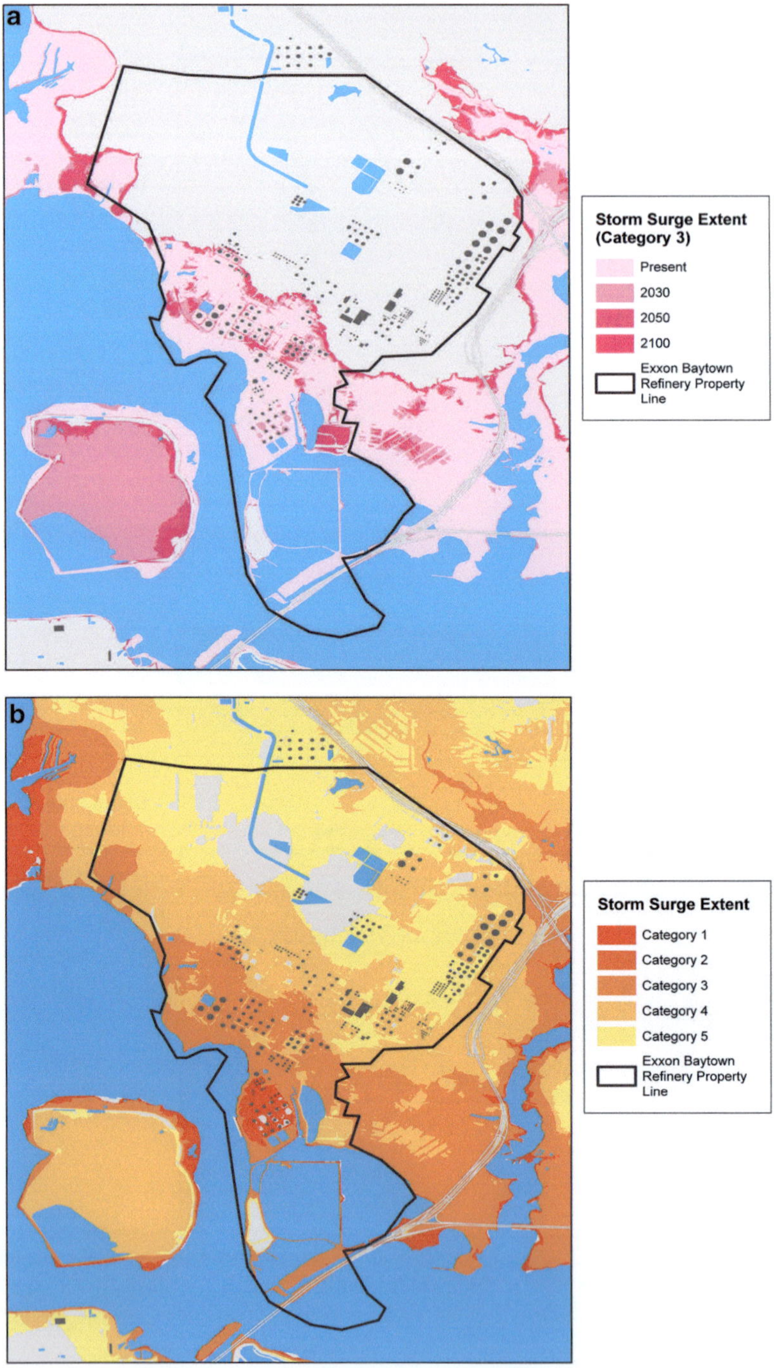

Fig. 19.3 (**a**) Projections for storm surge extent from a Category 3 hurricane today and with sea level rise by 2030, 2050, and 2100 at Exxon Mobil's Baytown, Texas, refinery. (**b**) Extent of storm surge flooding today from Category 1–5 hurricanes at Exxon Mobil's Baytown, Texas, refinery

19.3.4 Marathon Petroleum Corporation

Marathon Petroleum's Texas City refinery (84,000 b/cd) sits adjacent to its Galveston refinery to the west and Valero's Texas City refinery to the south. With sea level rise to date, the Marathon Texas City refinery and those around it could see storm surge impacts from a Category 3 storm or higher with operations closest to the coast being inundated. In the future, sea level rise and changes in storm intensity could put the facility at greater risk for storm damages (see Fig. 19.4).

In the wake of Hurricane Isaac in 2012, Marathon Petroleum's Garyville, Louisiana, refinery—one of the world's largest—experienced a significant reduction in operating capacity. The company took a loan of one million barrels of crude oil from the federal government's emergency reserves to support its refining operations after the storm (Gardner and Schneyer 2012). Despite such incidences,

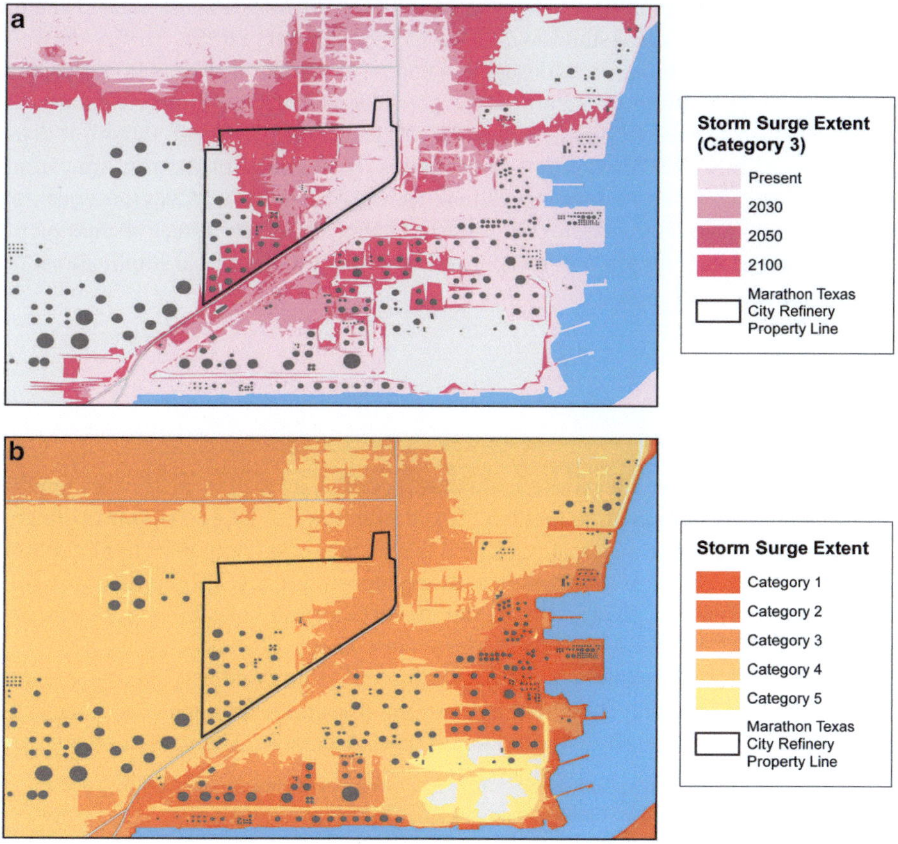

Fig. 19.4 (**a**) Projections for storm surge extent from a Category 3 hurricane today and with sea level rise by 2030, 2050, and 2100 for Marathon Petroleum's Texas City, Texas, refinery. (**b**) Extent of storm surge flooding today from Category 1–5 hurricanes at Marathon Petroleum's Texas City, Texas, refinery

Marathon Petroleum has not disclosed any risks at its facilities from climate change impacts, including sea level rise and storm surge. The company did note in its 2013 SEC filing the potential for severe "local weather conditions" and "natural disasters such as hurricanes and tornadoes" (p. 24); however, its only direct discussion of climate change impacts was from regulation (Marathon Petroleum 2014).

19.3.5 Chevron Corporation

Chevron's refinery (330,000 b/cd) in Pascagoula, Mississippi, is vulnerable to storm surge impacts. Given its location in the Gulf of Mexico, even a Category 1 hurricane could penetrate the facility, and a Category 3 storm could leave parts of the facility under 10 ft of water. Such storm impacts are likely to be worse in the future, as sea levels encroach upon the property (see Fig. 19.5), subsidence in the Gulf continues, and storms may become more intense. In 2005 Hurricane Katrina caused major problems and an extended shutdown of the facility (Reuters 2005).

Despite the vulnerable locations of this and other Chevron-owned facilities, the company has not publicly reported its physical risk from climate change. In Chevron's (2014) SEC filings, it makes no disclosures of this kind, though it does note the possibility of "disruptions at refineries or chemical plants resulting from unplanned outages due to severe weather" (p. FS-5). However, Chevron does not consider these risks in the context of climate change. The only direct discussion of climate change in Chevron's risk reporting surrounds the political implications of climate-related regulations.

19.4 Discussion

The storm surge and sea level rise estimates modeled, as well as historic examples of storm damage, demonstrate that US coastal oil refineries face risks from climate-related impacts, despite a lack of or limited disclosure from companies in SEC Form 10-Ks, annual reports, and voluntary climate reporting data. Yet, company shareholders have requested greater consideration and disclosure of climate-related risks in recent years.

For the 2015 shareholder season, Phillips 66 shareholders, led by Calvert Investment Management, have filed a resolution asking the company for better disclosure of risk from the physical impacts of climate change. The resolution states:

> Diminished refining utilization rates, potential downtime or closure of facilities due to direct damage to facilities, danger to employees, disruption in supply chains, and power supply [outages] due to storm surges or sea level rise could have a material impact on the company's production and related cash flows. This was made evident when the company's Bayway refinery lost power after Superstorm Sandy, was shut down for several weeks due to flood damage from the storm, and incurred significant maintenance and repair expenses. (Calvert Investment Management 2014, para. 5)

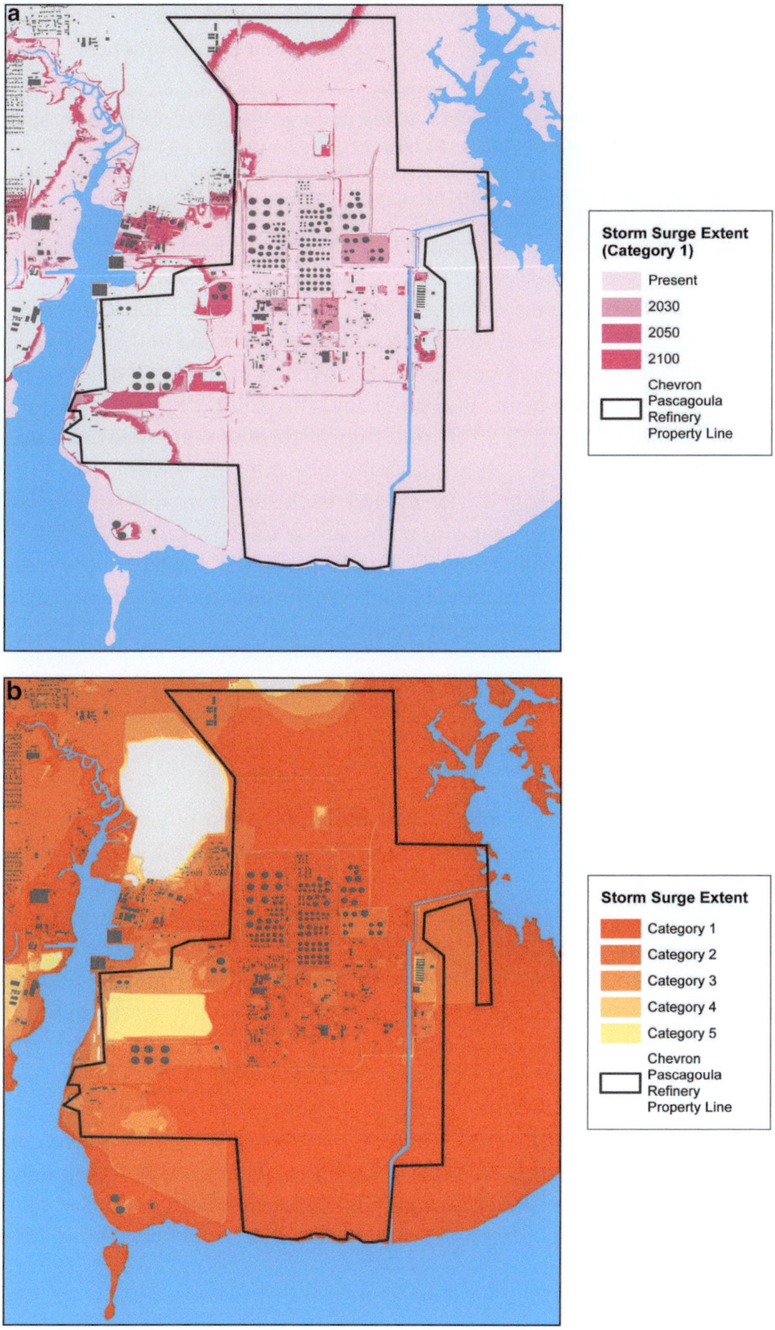

Fig. 19.5 (**a**) Projections for storm surge extent from a Category 1 hurricane today and with sea level rise by 2030, 2050, and 2100 for Chevron Corporation's Pascagoula, Mississippi, refinery. (**b**) Extent of storm surge flooding today from Category 1–5 hurricanes Chevron Corporation's Pascagoula, Mississippi, refinery

In 2013, Exxon Mobil and Chevron shareholders, both led by the Christopher Reynolds Foundation, filed resolutions, requesting that the companies:

> review the exposure and vulnerability of [their] facilities and operations to climate risk and issue a report that … estimates the costs of the disaster risk management and adaptation steps the company is taking, and plans to take, to reduce exposure and vulnerability to climate change and to increase resilience to the potential adverse impacts of climate extremes. (Christopher Reynolds Foundation 2013a, para. 13, 2013b, para. 8)

Such resolutions have led companies to a greater focus on climate-related issues in communications with their shareholders and the public. Due to the lack of publicly available information on refineries' level of preparedness for storm impacts, it is unknown the degree to which storm surge like that modeled in this analysis would cause material damage to facilities; however, given historical examples of physical damage to refineries from tropical systems and the extent of storm surge modeled here, it can be estimated that current and future storm surge, potentially enhanced by sea level rise, at the five locations modeled may pose significant risks to refining operations. More demands for climate-related disclosure from investors and citizens may lead companies to better consider and disclose risks from the physical impacts of climate change. In addition, future storm impacts and further improvements in climate-related impact assessment may make such risks and their associated costs, more evident to companies and their investors.

19.5 Conclusions

Five US oil refineries were found to face climate-related physical risks from sea level rise and enhanced storm surges. These five companies have disclosed little to no risk associated with climate change at these locations or any other, despite SEC guidance and shareholder demands. Nevertheless, the presentation of climate change as posing financial risks to business provides opportunities to engage companies, shareholders, decision-makers, and communities around climate change impacts. To encourage such engagement, the SEC could push companies to follow its guidelines for disclosing climate change risks, while also educating them about how to comply and on what full disclosure looks like. Further, the SEC could choose to go beyond guidance and issue a rule that requires companies to report physical climate risks. In turn, investors can pressure companies to better consider and disclose these risks by focusing on financial effects. With greater transparency and dialogue around financial risks associated with climate change, such as sea level rise and enhanced storm surge, companies and communities can better build resiliency to climate impacts.

References

Brelsford R, True WR, Koottungal L (2013) Western Europe leads global refining contraction. Oil Gas J, 2 Dec 2013. www.ogj.com/articles/print/volume-111/issue-12/special-report-worldwide-report/western-europe-leads-global-refining-contraction.html. Accessed 12 Dec 2014

Calvert Investment Management (2014) Phillips 66 physical climate impacts 2015 (Phillips 66 shareholder resolution). www.ceres.org/investor-network/resolutions/phillips-66-physical-climate-impacts-2015. Accessed 2 Jan 2015

Ceres (2012) Sustainable extraction? An analysis of SEC disclosure by major oil & gas companies on climate risk and deepwater drilling risk. www.ceres.org/resources/reports/sustainable-extraction-an-analysis-of-sec-disclosure-by-major-oil-gas-companies-on-climate-risk-and-deepwater-drilling-risk/view. Accessed 2 Jan 2015

Chevron Corporation (2014) 2013 SEC form 10-K filing. www.sec.gov/Archives/edgar/data/93410/000009341014000011/cvx-123113x10kdoc.htm. Accessed 2 Jan 2015

Christopher Reynolds Foundation (2013a) Chevron shareholder resolution: Chevron climate risk 2013. www.ceres.org/investor-network/resolutions/chevron-climate-risk-2013. Accessed 2 Jan 2015

Christopher Reynolds Foundation (2013b) Exxon Mobil shareholder resolution: ExxonMobil climate risk report 2013. www.ceres.org/investor-network/resolutions/exxonmobil-climate-risk-report-2013. Accessed 2 Jan 2015

Ceres & CookESG Research (2014) SEC climate disclosure search. www.ceres.org/resources/tools/sec-climate-disclosure/. Accessed 2 Jan 2015

Department of Energy (DOE) (2009) Comparing the impacts of the 2005 and 2008 hurricanes on US energy infrastructure. www.oe.netl.doe.gov/docs/HurricaneComp0508r2.pdf. Accessed 2 Jan 2015

Energy Information Administration (EIA) (2014) Table 3. Capacity of operable petroleum refineries by state as of January 1, 2014. www.eia.gov/petroleum/refinerycapacity/table3.pdf. Accessed 2 Jan 2015

Environmental Protection Agency (EPA) (2006) Murphy Oil USA refinery spill: Chalmette and Meraux, LA. Region 6 Oil Response Team US EPA. Archive document: presentation, from www.epa.gov/oem/docs/oil/fss/fss06/franklin_2.pdf. Accessed 12 Dec 2014

Exxon Mobil Corporation (2014a) About our facilities: Baytown area operations. corporate.exxonmobil.com/en/company/worldwide-operations/locations/united-states/baytown/about?parentId=da547204-0aaa-4776-850a-83fd38e3fc21. Accessed 12 Dec 2014

Exxon Mobil Corporation (2014b) 2013 SEC Form 10-K filing. www.sec.gov/Archives/edgar/data/34088/000003408814000012/xom10k2013.htm. Accessed 2 Jan 2015

Ezer T, Atkinson LP, Corlett WB, Blanco JL (2013) Gulf stream's induced sea level rise and variability along the US mid-Atlantic coast. J Geophys Res Oceans 118:1–13

Gardner T, Schneyer J (2012) U.S. loans emergency oil to refiner after Isaac. Reuters, 1 Sept 2012. in.reuters.com/article/2012/08/31/us-doe-reserve-marathon-idINBRE87U10Q20120831. Accessed 4 Jan 2015

Geophysical Fluid Dynamics Laboratory/NOAA (GFDL) (2013) Global warming and hurricanes: an overview of recent research results. www.gfdl.noaa.gov/global-warming-and-hurricanes. Accessed 2 Jan 2015

Holland G, Bruyere C (2014) Recent intense hurricane response to global climate change. Clim Dyn 42(3–4):617–627. link.springer.com/article/10.1007/s00382-013-1713-0/fulltext.html. Accessed 12 Dec 2014

Intergovernmental Panel on Climate Change (IPCC) (2012) Changes in climate extremes and their impacts on the natural physical environment. In: Field CB, Barros V, Stocker TF, Dahe Q (eds) Managing the risks of extreme events and disasters to advance climate change adaptation. Cambridge University Press, New York, pp 109–230

Intergovernmental Panel on Climate Change (IPCC) (2013a) Climate phenomena and their relevance for future regional climate change. In: Stocker TF, Qin D, Plattner G-K, Tignor M, Allen SK, Boschung J, Nauels A, Xia Y, Bex V, Midgley PM (eds) Climate change 2013: the physical science basis. Contribution of Working Group I to the fifth assessment report of the Intergovernmental Panel on Climate Change. Cambridge University Press, New York

Intergovernmental Panel on Climate Change (IPCC) (2013b) Long-term climate change: projections, commitments, and irreversibility. In: Stocker TF, Qin D, Plattner G-K, Tignor M, Allen SK, Boschung J, Nauels A, Xia Y, Bex V, Midgley PM (eds) Climate change 2013: the physical science basis. Contribution of Working Group I to the fifth assessment report of the Intergovernmental Panel on Climate Change. Cambridge University Press, New York

Kirgiz K, Burtis M, Lunin D (2009) Petroleum-refining industry business interruption losses due to Hurricane Katrina. J Bus Valuat Econ Loss Anal 4(2):3–3. https://www.cornerstone.com/GetAttachment/0e885cd0-888c-4b6b-ac0b-b38c5c92e6dc/Petroleum-Refining-Industry-Business-Interruption.pdf. Accessed 14 Jan 2015

Knutson TR, Tuleya RE (2004) Impact of CO_2-induced warming on simulated hurricane intensity and precipitation: sensitivity to the choice of climate model and convective parameterization. J Clim 17(18):3477–3495

Knutson TR, Tuleya RE (2008) Tropical cyclones and climate change: revisiting recent studies at GFDL. In: Diaz HF, Murnane RJ (eds) Climate extremes and society. Cambridge University Press, New York, pp 120–144

Kopp RE, Horton RM, Little CM, Mitrovica JX, Oppenheimer M, Rasmussen DJ, Strauss BH, Tebaldi C (2014) Probabilistic 21st and 22nd century sea-level projections at a global network of tide-gauge sites. Earth's Future 2:383–406

Marathon Petroleum (2014) 2013 SEC form 10-K filing. www.sec.gov/Archives/edgar/data/1510295/000151029514000004/mpc-20131231x10k.htm#sA5719E7262C104FE8345153A71110781. Accessed 2 Jan 2015

Murphy Oil (2011) 2010 SEC form 10-K filing. www.sec.gov/Archives/edgar/data/717423/000119312511049276/d10k.htm. Accessed 2 Jan 2015

Msnbc.com News Services (MNS) (2006) $330 million settlement deal in Katrina oil spill. Msnbc.com, Sept 25. www.nbcnews.com/id/15004868/ns/us_news-environment/t/ million-settlement-dealkatrina-oil-spill/. . Accessed 12 Dec 2014

National Oceanic and Atmospheric Administration (NOAA) Coastal Services Center (2012) Mapping Coastal Inundation Primer. http://coast.noaa.gov/digitalcoast/_/pdf/guidebook.pdf. Accessed 2 Feb 2015

National Oceanic and Atmospheric Administration (NOAA) (2014) NOAA's state of the coast: State of the coast, beach closures indicator. stateofthecoast.noaa.gov/vulnerability/localthreat.html. Accessed 12 Dec 2014

Phillips 66 (2014) 2013 SEC form 10-K filing. www.sec.gov/Archives/edgar/data/1534701/000153470114000042/psx-20131231x10k.htm. Accessed 2 Jan 2015

Reuters (2005) Rita and Katrina have shut 23 percent of US oil refining capacity. New York Times, 22 Sept 2005. www.nytimes.com/2005/09/22/business/RITA-FACTBOX.html. Accessed 4 Jan 2015

Reuters (2012) Bayway refinery to restart in 2 to 3 weeks—Phillips 66. www.reuters.com/article/2012/11/05/storm-sandy-energy-bayway-idUSL1E8M5FSM20121105. Accessed 2 Jan 2015

Sallenger AH Jr, Doran KS, Howd PA (2012) Hotspot of accelerated sea-level rise on the Atlantic coast of North America. Nat Clim Chang 2:884–888. doi:10.1038/nclimate1597

Securities and Exchange Commission (SEC) (2010) Commission guidance regarding disclosure related to climate change. Washington, DC. www.sec.gov/rules/interp/2010/33-9106.pdf. Accessed 2 Jan 2015

Securities and Exchange Commission (SEC) (2014) EDGAR company filings. www.sec.gov/edgar/searchedgar/companysearch.html. Accessed 2 Jan 2015

Strauss B, Ziemlinski R (2012) Sea level rise threats to energy infrastructure: a surging seas brief report by Climate Central. Climate Central, 19 Apr 2012. slr.s3.amazonaws.com/SLR-Threats-to-Energy-Infrastructure.pdf. Accessed 2 Jan 2015

Valero Energy (2014) 2013 SEC form 10-K filing. www.sec.gov/Archives/edgar/data/1035002/000103500214000008/vloform10-kx12312013.htm#sB8230C7869E4716607E733090035651C. Accessed 2 Jan 2015

Index

J.L. Drake et al. (eds.), *Communicating Climate-Change and Natural Hazard Risk and Cultivating Resilience*, Advances in Natural and Technological Hazards Research 45, DOI 10.1007/978-3-319-20161-0

MIX
Papier aus verantwortungsvollen Quellen
Paper from responsible sources
FSC® C105338

If you have any concerns about our products,
you can contact us on
ProductSafety@springernature.com

In case Publisher is established outside the EU,
the EU authorized representative is:
Springer Nature Customer Service Center GmbH
Europaplatz 3, 69115 Heidelberg, Germany

Printed by Libri Plureos GmbH
in Hamburg, Germany